The
MAP
CATALOG

Also from Vintage Books/Random House and Tilden Press:

The Air & Space Catalog
The Complete Sourcebook to Everything in the Universe

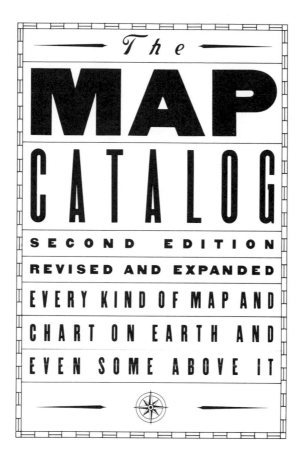

The MAP CATALOG

SECOND EDITION

REVISED AND EXPANDED

EVERY KIND OF MAP AND CHART ON EARTH AND EVEN SOME ABOVE IT

Joel Makower
Editor

Cathryn Poff
Laura Bergheim
Associate Editors

A Tilden Press Book

RANDOM HOUSE ■ NEW YORK

LIBRARY OF CONGRESS CATALOGING-IN-PUBLICATION DATA

The Map catalog: every kind of map and chart on earth and even some above it / Joel
        Makower, editor, Laura Bergheim and Cathryn Poff, associate editors. — 2nd ed., rev.
    and expanded.
        p.   cm.
        "A Tilden Press book."
        ISBN 0-394-58326-4: $27.50.
        1. Maps—Catalogs.   I. Makower, Joel, 1952– . II. Bergheim, Laura, 1962– . III. Poff,
    Cathryn, 1966– .
    Z6028.M23   1990
    [GA105.3]
    912'.0294—dc20   89-37566
                                                                                          CIP

Much of the information on pages 263-264 was excerpted from
"Globes: A Librarian's Guide to Selection and Purchase," by
James Coombs, from the March 1981 issue of the *Wilson Library
Bulletin*. Reprinted with permission.

Manufactured in the United States of America

10 9 8 7 6 5 4 3 2 1

Second Edition

About This Book

A book containing a list of every map published, or even every map presently available, should such a book be possible to compile, would be of little use to most people. In this book, we have attempted to provide information about the many types of maps available, the major sources of each map type, and descriptions or examples of the map products available from each source.

Much like a map itself, *The Map Catalog* is a reference tool, a portrait of the cartographic landscape. Like a map, it shows you possible destinations, but not always the exact directions by which to reach them. Ultimately, it is up to the reader to determine how the tool may best be used. Like a map, this book may be browsed and enjoyed for the pure pursuit of knowledge; or it may serve as a source for acquiring specific maps for specific needs.

We have divided the maps and charts in this book into several sections: "Travel Maps," "Maps of Specific Areas," "Boundary Maps," "Scientific Maps," "History Through Maps," "Utility and Service Maps," "Water Maps," "Sky Maps," "Images as Maps," "Atlases and Globes," and "Et Cetera," with each section itself divided into several chapters. The delineations are, admittedly, arbitrary at times. A map of U.S. forests, for example, could reasonably be listed under "Agriculture Maps," "Energy Maps," "Land-Use Maps," "Natural Resource Maps," "Tourism Maps," "United States Maps," or "Wildlife Maps"; we have chosen to list them under "Recreation Maps." To minimize confusion, there are cross-references within each subsection and a thorough index at the end of the book.

Most addresses of map sources appear within their respective listings. Addresses of most map-producing government agencies, however, are contained in the book's appendixes, to avoid repeating the same addresses many times throughout the book. The appendixes, moreover, contain addresses of map sources not necessarily referred to in the text—of selected map stores, for example, and major map libraries.

This second edition of *The Map Catalog* expands the original volume, with additional chapters and new product categories, including a sampling of "map stuff"—a wide range of products featuring cartographic images. This edition also reflects the comments and suggestions of hundreds of readers and reviewers who collectively contributed many valuable ideas and resources. We have been grateful for the response and continue to welcome comments and contributions, which we will incorporate in subsequent editions of this book. Please send them to The Map Catalog, c/o Tilden Press Inc., 1526 Connecticut Ave. NW, Washington, DC 20036.

Contents

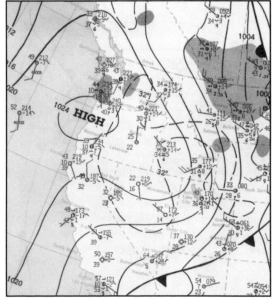

Acknowledgments

The editors would like to thank the following individuals and organizations who provided valuable support to this project:

Robert T. Aangeenbrug, Association of American Geographers
Dr. Paul S. Anderson, Department of Geography-Geology, Illinois State University
Frederick Argoff
Ron Beck, EROS Data Center
Ronald Bolton, National Oceanic and Atmospheric Administration
Michele Breslauer
André Caron, Canada Map Office
George Castillo, U.S. Forest Service
Alice Chen, Maps Alberta
Karen Craft
June Crowe, Urban Land Institute
Peter Cutts, Ankers, Anderson & Cutts
Jack Dodd, Tennessee Valley Authority
Melvin Dunaway, Defense Mapping Agency
Nancy Edwards, International Map Dealers Association
Ralph Ehrenberg, Geography and Map Division, Library of Congress
Mary Lee Elden, National Geographic Society
John Ferry, Buckminster Fuller Institute
C. Dallas Folsom
H. Kit Fuller, Geologic Inquiries Group, U.S. Geological Survey
Rose Gandy, American Public Transit Association
Phillip Guss, National Earth Science Information Center, U.S. Geological Survey
Matthew Hale, Map Link
William F. Hartwig, U.S. Fish and Wildlife Service
Maxine and Hans Hesse, Global Graphics
Joyce Hodel, Rand McNally
Mary Ingles, National Park Service
Barbara Jackson, The Map Center
David C. Jolly, David C. Jolly Publishers
Jerry Jones, Rand McNally

Jack Joyce, ITMB
Mike Keuss, U.S. Army Corps of Engineers
Elaine S. Larson, Travel Genie
Jesse Levine, Laguna Sales
Richard Lindeborg, U.S. Forest Service
Edward W. Lollis, National Map Gallery & Travel Center
Robert Marx, Bureau of the Census
Deborah McLaren
Sally Meyers, Association of American Geographers
Joanne Miller
M. Russell Miller
James Minton, U.S. Geological Survey Library
Tony Naden, Harvard Square Map Store
Paul Petruzzi, American Map Corporation
The Poff Family
Eric Riback, DeLorme Mapping Co.
Bob Richardson, National Archives
Dr. Arthur H. Robinson
Randy J. Rosenberg
Christopher J. Ryan, New England Cartographics
Rebecca Saletan, Random House
Paul Severson, EROS Data Center
Miranda Sherwin, Random House
Robert Sims, National Geographic Society
John P. Snyder, American Cartographic Association
Mary Kay Stoehr, Trails Illustrated
Norman Strasma, International Map Dealers Association
Milt Thomas, Modern School
Juan Jose Valdez, National Geographic Society
William E. Walling
Alva Wallis, Jr., National Climatic Data Center
Joan Wilson, National Ocean Service
Ian Woofenden, Pacific Puzzles

Maps ◆ Travel Maps ◆ Bicycle Route Maps ◆ Mass Transit Maps ◆ Railroad Maps
Maps ◆ Tourism Maps ◆ World Status Map ◆ County Maps ◆ Foreign Country Ma
and Maps ◆ State Maps ◆ United States Maps ◆ Urban Maps and City Plans ◆ Wo
orld Game Map ◆ Boundary Maps ◆ Congressional District Maps ◆ Land Owners
al Maps ◆ Scientific Maps ◆ Agriculture Maps ◆ Geologic Maps ◆ Land Use Map
rce Maps ◆ Topographic Maps ◆ Wildlife Maps ◆ Antique Maps ◆ Researching O
ic Site Maps ◆ History Maps ◆ Military Maps ◆ Treasure Maps ◆ Business Maps
Emergency Information Maps ◆ Energy Maps ◆ Utilities Maps ◆ Water Maps ◆ N
Ocean Maps ◆ River, Lake, and Waterway Maps ◆ Tide and Current Maps ◆ Sky
tical Charts ◆ Star Charts ◆ Star Magnitudes ◆ Weather Maps ◆ How to Read a W
mages as Maps ◆ Aerial Photographs ◆ Space Imagery ◆ How to Buy an Atlas ◆
Geography Educa   zations ◆ Map
Map Stu   he   ◆ How to Choose a Map ◆ Map Projections ◆
t Map Skills ◆ Copying Maps ◆ Travel Maps ◆ Bicycle Route Maps ◆ Mass Transi
Maps ◆ Recreation Maps ◆ Tourism Maps ◆ World Status Map ◆ County Maps ◆
Maps ◆ Indian Land Maps ◆ State Maps ◆ United States Maps ◆ Urban Maps an
World Maps ◆ The World Game Map ◆ Boundary Maps ◆ Congressional District M
nership Maps ◆ Political Maps ◆ Scientific Maps ◆ Agriculture Maps ◆ Geologic
e Maps ◆ Natural Resource Maps ◆ Topographic Maps ◆ Wildlife Maps ◆ Antique
ing Old Maps ◆ Historic Site Maps ◆ History Maps ◆ Military Maps ◆ Treasure Ma
Maps ◆ Census Maps ◆ Emergency Information Maps ◆ Energy Maps ◆ Utilities
aps ◆ Nautical Charts ◆ Ocean Maps ◆ River, Lake, and Waterway Maps ◆ Tide an
s ◆ Sky Maps ◆ Aeronautical Charts ◆ Star Charts ◆ Star Magnitudes ◆ Weather
ead a Weather Map ◆ Images as Maps ◆ Aerial Photographs ◆ Space Imagery ◆
tlas ◆ Atlases ◆ Globes ◆ Geography Education Materials ◆ Map Accessories ◆
tions ◆ Map Software ◆ Map Stuff ◆ The Turnabout Map ◆ How to Choose a Map
ns ◆ Learning About Map Skills ◆ Copying Maps ◆ Travel Maps ◆ Bicycle Route M
nsit Maps ◆ Railroad Maps ◆ Recreation Maps ◆ Tourism Maps ◆ World Status M
Maps ◆ Foreign Country Maps ◆ Indian Land Maps ◆ State Maps ◆ United States
aps and City Plans ◆ World Maps ◆ The World Game Map ◆ Boundary Maps ◆ C
strict Maps ◆ Land Ownership Maps ◆ Political Maps ◆ Scientific Maps ◆ Agricul
Geologic Maps ◆ Land Use Maps ◆ Natural Resource Maps ◆ Topographic Maps
◆ Antique Maps ◆ Researching Old Maps ◆ Historic Site Maps ◆ History Maps ◆
Treasure Maps ◆ Business Maps ◆ Census Maps ◆ Emergency Information Maps
Utilities Maps ◆ Water Maps ◆ Nautical Charts ◆ Ocean Maps ◆ River, Lake, and
Tide and Current Maps ◆ Sky Maps ◆ Aeronautical Charts ◆ Star Charts ◆ Star Ma
er Maps ◆ How to Read a Weather Map ◆ Images as Maps ◆ Aerial Photographs ◆
How to Buy an Atlas ◆ Atlases ◆ Globes ◆ Geography Education Materials ◆
ies ◆ Map Organizations ◆ Map Software ◆ Map Stuff ◆ The Turnabout Map ◆ H
Map ◆ Map Projections ◆ Learning About Map Skills ◆ Copying Maps ◆ Travel
oute Maps ◆ Mass Transit Maps ◆ Railroad Maps ◆ Recreation Maps ◆ Tourism
atus Map ◆ County Maps ◆ Foreign Country Maps ◆ Indian Land Maps ◆ State M

A WORLD OF MAPS

State of the Art

We have become a world awash in maps.

For more than five centuries, we have measured and documented virtually every square foot of our planet, not to mention the oceans and the heavens. And we have recorded our findings with astounding accuracy in graphic repre- sentations—in black and white and in glorious color—called "maps." We have maps of everything from airports to zip codes, from highways to hurricanes to hidden treasures.

Maps are so much a part of our everyday lives that we may think of them as being more real than the "real world" itself. Consider Huck Finn and Tom Sawyer, soaring high above the Midwest in a balloon in Mark Twain's *Tom Sawyer Abroad*. Estimating their present location, Huck claims they're still over Illinois; Tom thinks they've floated into Indiana. "I know by the color," says Huck. "And you can see for yourself that Indiana ain't in sight."

"What's color got to do with it?" asks Tom.

"It's got everything to do with it," explains Huck. "Illinois is green, Indiana is pink. You show me any pink down here, if you can."

"Indiana *pink*? Why, what a lie!"

"It ain't no lie; I've seen it on the map, and it's pink."

This is a sentiment that's been expressed by others even more worldly than Messrs. Sawyer and Finn. Astronaut John Glenn, approaching splashdown near the end of his historic 1962 Mercury space flight, informed Mission Control, "I can see the whole state of Florida, just laid out like on a map."

Maps have become a way of life.

How many maps are there? No one knows for sure, but some data from the federal govern- ment, the world's most skilled and prolific cartogra- pher, are reveal- ing. According to Uncle Sam, there are some 39 federal agen- cies involved in map- making. Together, they have produced nearly a quarter-million separate maps. In a typical year, the 12 largest map-making agencies alone distribute more than 161 million copies of their maps at a cost of just over a half-billion dollars. All 39 agencies expend about 13,000 worker-years of effort annually carrying out their map- making responsibilities. That's just the tip of the cartographic iceberg. Each year, Rand Mc- Nally, the world's largest nongovernment map-maker, sells about 400 million maps, through its 2,500 or so sheet maps, atlases, and globes. The 26-million-member American Automobile Association distributes about 35 million sheet maps a year, along with another 215 million "Trip Tiks." There are road maps galore from oil companies, state and local tourism offices, foreign embassies, and other sources. And there are countless other mapmakers around the world, producing anywhere from a handful to several hundred different maps each year for general or highly specific audiences. Some of these maps end up in collections. While literally hundreds of private and public libraries have map collections, the 70 largest collections contain nearly 20 million maps and about 22 million aerial photographs.

It's safe to say that the output of maps, globes, atlases, and related products is well over a half-billion copies a year in this country alone.

And such variety! The types of information being mapped these days seem nearly as endless as the world itself. If you doubt this, take a browse through *New Mexico in Maps* (2nd Edition), a massive 432-page reference book ($24.95 paperback) edited by Jerry L. Williams and published by the University of New Mexico Press (Albuquerque, NM 87131; 505-277-7564). The book, containing only government-produced maps of that state, features (unfortunately, in black and white) maps on some 131 *different topics*, from "Vacation Facilities" to "Vulnerable Aquifers," "Housing Characteristics" to "Horse Shows." Keep in mind that, except for the Rio Grande on its southern border, New Mexico is a land-locked state; given a substantial body of water, the number of map types available for that state might double.

There's good reason for this veritable map mania. In our information society, maps play an important role. A mere three-foot-square map can contain thousands of pieces of information on an endless number of topics, making maps highly efficient information-storage devices. But it is for more than just data that we turn to maps. For some, maps make fascinating reading. It's not unreasonable to curl up with a good map to try to put things into perspective, to determine where you are, where you're going (or would like to go), or where you've been. For many, maps are works of art, worthy of display on a wall, perhaps even framed. Indeed, some antique maps truly *are* works of art, with price tags to match, ascending into the thousands of dollars.

The pages that follow reveal the vast and varied world of maps, atlases, globes, and related products—who makes them, how they're compiled, how they're used, and where they can be found. All told, this information represents something more than a mere celebration of maps. Understanding the nature of, say, a topographic map can provide insight into our planet's structure and beauty. Even a colorless census map can speak volumes about the myriad forces that shape society. From the earliest sketches to the latest in digital imaging, maps tell the story of our world and our lives in ways words can't even begin to describe.

So, anybody need a map?

The Map Unfolds

The history of maps dates back to man's first realization that a picture really is worth a thousand words. Archaeologists and other social scientists have often marveled at early man's almost instinctive ability to produce rough but amazingly accurate sketches of his surroundings. Throughout the world's civilizations—from African tribesmen to Arctic Eskimos—there are examples of these early maps, drawn in the earth or on stones or animal skins, showing the relative positions and distances of landmarks and localities. The Babylonians, more than 2,000 years before Christ, surveyed land holdings on clay tablets. Known as cadastral maps, they represent one of the earliest forms of graphic expression. Those ancient surveys later became the basis for map-making in Europe during the Middle Ages and, four millenia later, for land plats produced by the U.S. government.

The modern-day craft of making maps can be traced to western Europe in the 13th century. The regional and local maps of the day represented radical changes from the drawings that preceded them: Rather than being derived from literary sources and mythology, they were based on observation and measurements, the first maps intended for practical use by travelers on land or sea. The second half of the 13th century produced the earliest surviving nautical charts and post-Roman road maps.

It was in the waters of the Mediterranean and the Black Sea that map-making made great strides. The development of the mariner's compass permitted angular measurement, enabling a level of accuracy in nautical charts that wouldn't be seen in land maps for several hundred years. Among those first efforts were the Italian portolan charts, which were sets of sailing instructions created on parchment around 1250 by a community of Italian draftsmen just becoming familiar with mathematics and measurement.

Many of the early European cartographers were recruited from the ranks of painters, miniaturists, and other artists, whose introduction to the profession consisted largely of copying and decorating existing maps. Later, they were able to compile their own. Italy, especially Florence, was a center of cartographic activity for several centuries. Here, a succession of explorers, artists, and mathematicians created new pictures that expressed an expanding world view.

The era of Christopher Columbus was another time of great map-making advances. The year 1492, in fact, saw the creation of the first modern terrestrial globe, the work of Germans Martin Behaim, a cosmographer, and Georg Holzschuher, a miniaturist. The 20-inch-wide globe showed the equator, the two tropics, and the Arctic and Antarctic circles. Another key innovation was the copperplate, which proved a far more effective medium for map reproduction than the woodcut and helped launch a booming map trade throughout Europe. By the early 1600s, the governments of Spain, Portugal, and England were among those recognizing the importance of maps, using them for property assessment, taxation, military planning, and to inventory national resources.

Mapping underwent radical changes in 17th-century France, due largely to an unquenchable thirst for maps and nautical charts. Such innovations as the telescope, the pendulum clock, and logarithm tables permitted accurate astronomical observations and the measurement of arcs on the Earth's surface. Both contributed to major advances in cartography. New standards of precision, in turn, led to other advancements, such as the creation of the bubble level, the aneroid barometer, and the theodolite, all of which resulted in great leaps forward in plotting topographic measurements and absolute altitudes. The 18th century brought several

The Turnabout Map

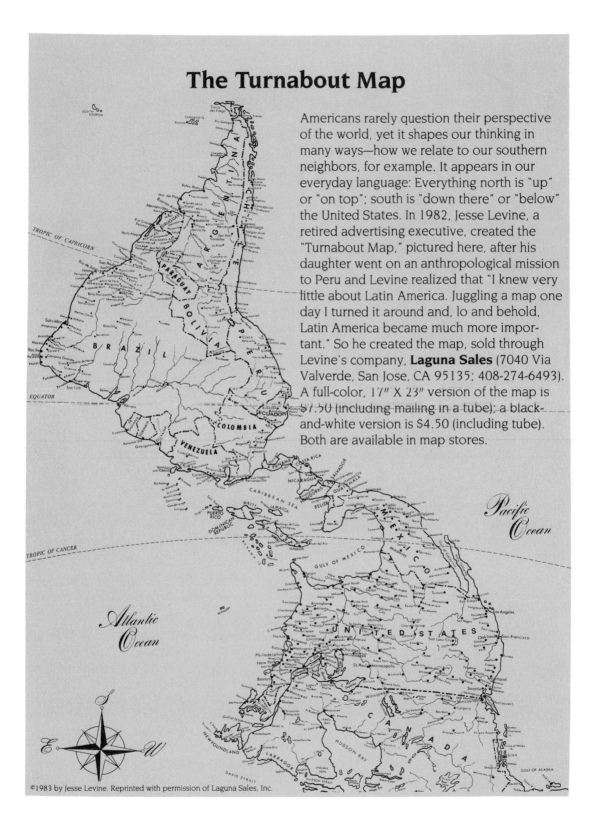

Americans rarely question their perspective of the world, yet it shapes our thinking in many ways—how we relate to our southern neighbors, for example. It appears in our everyday language: Everything north is "up" or "on top"; south is "down there" or "below" the United States. In 1982, Jesse Levine, a retired advertising executive, created the "Turnabout Map," pictured here, after his daughter went on an anthropological mission to Peru and Levine realized that "I knew very little about Latin America. Juggling a map one day I turned it around and, lo and behold, Latin America became much more important." So he created the map, sold through Levine's company, **Laguna Sales** (7040 Via Valverde, San Jose, CA 95135; 408-274-6493). A full-color, 17" X 23" version of the map is $7.50 (including mailing in a tube); a black-and-white version is $4.50 (including tube). Both are available in map stores.

Chinese woodcut world map first published in Korea sometime during the late 17th or early 18th century. Courtesy American Geographical Society.

advancements in printing, not the least of which was the introduction of chromolithography—the ability to print several colors at once—which enabled map-makers to enhance their works with color detail. All of these things aided the creation of such early cartographic masterpieces as Jacques Cassini's remarkable *Déscription géometrique de la France*. Published in 1783, it consisted of 182 engraved maps showing an entire nation in unprecedented detail—everything from canyons to channels to churches.

Meanwhile, in the newly formed United States of America, efforts were being made to take inventory of the burgeoning nation. As early as 1777, George Washington appointed a geographer and surveyor to the Continental Army to "take sketches of the country." This marked the first time that the U.S. government became involved in cartography. The first official large-scale surveying and mapping program was proposed by Thomas Jefferson and his congressional Committee on Public Land in 1784. This led to creation of the General Land Office, which produced a mountain of township plats and accompanying field notes. As president, Jefferson was concerned with the lack of information available about the newly acquired land west of the Mississippi River. During the War of 1812, his concerns led to creation by the War Department of an elite Bureau of Topographic Engineers, later the Army Corps of Engineers, which played a vital role in surveying and documenting the nation's lands and waters. Jefferson is further credited with creation of the Survey of the Coast, later the Coast and Geodetic Survey.

During the 19th century, while government surveyors measured and subdivided regions that were relatively well known, the War Department sent exploring parties into largely unmapped territory. Many of the documents that emerged were vital in building the roads, canals, and railroads needed to accommodate a prospering populace. The topographic surveying and mapping programs conducted by the U.S. Geological Survey (USGS) from its inception in 1879 were based on a complex system set up by the Coast and Geodetic Survey, the leading scientific agency in the federal government during that century.

Prior to the Civil War, government surveys were limited to the vast Midwest. The westward migration that followed the war created an urgent need for detailed information about the resources and natural features of the western United States. By the beginning of the 20th century, the USGS was undertaking a 20-year program to map nearly every inch of the nation in rigorous scientific detail.

Map-making excelled during World War I, when many USGS topographers were commissioned by the Army Corps of Engineers. Some played key roles in developing aerial-photography techniques used for military intelligence. Returning after the war, these topographers applied their new skills to cartography. Throughout the 1920s, experimenting with the new science of photogrammetry—the ability to take measurements from photographs—they succeeded in making maps from aerial photos. This development would change map-making forever.

A great surge in the application of photogrammetry came with the establishment in 1933 of the Tennessee Valley Authority. One of TVA's first needs was map coverage of the entire valley. Working with USGS, surveyors prepared planimetric maps of the area using state-of-the-art, five-lens aerial photographs and innovative radial-line plotting techniques. Their efforts began a revolutionary swing away from field methods as the basis of map-making, establishing aerial photos as the basis for all the maps that would follow.

After World War II, map-making innovations were rampant. Combining a variety of sophisticated measuring instruments with emerging computers, cartographers produced a treasure trove of new map types and products. The advent of space imagery and electronic imaging, along with the digitization of map data in computers, produced yet another revolution in map-making. And there would be more revolutions to come.

Mapping the Future

There are many who predict the demise of the map as we know it, the elimination of those familiar, hard-to-refold sheets of paper upon which we depend for so many things. Along with other technofuturistic predictions—that the computer will replace the printed book, for example, or that newspapers will someday appear only on video screens—this will never be completely true. Which is not to say that maps, and map-making, are not changing in dramatic ways. Or that the map of tomorrow won't appear in some rather innovative forms. But there will always be paper maps, although maybe not as many of them.

As it has with so many other things, the computer has revolutionized the world of mapping in myriad ways. Computerization has introduced an impressive list of new cartographic tools and techniques, such as electronic distance-measuring, inertial navigation, remote sensing, digital imaging, space science, and geographic-information systems. We are just beginning to learn how to use these tools. A new generation of cartographers is using computers' digital technology to make and modify maps. In the new high-tech cartography, map data are no longer entered by skilled draftsmen working on light tables but are created using satellite images by cartographers at keyboards. Map data are entered on computer tapes and floppy disks, among other things, which in turn generate visual displays for editing, or which can be printed out to use in producing conventional paper maps.

With map data in a computer, cartographers can easily modify maps, enlarge them, change their scale, isolate segments for use with other maps, and make them, or any part of them, instantly available to other cartographers and map users who can call up digital information on their computer screens around the world. Among the many benefits of computerization is speed: Combined with

other map-making techniques, computers can shorten considerably the four to five years it once took to produce a printed map from an aerial photograph or satellite image. Today, it can be done in a matter of days.

Two of the most ambitious digital-mapping programs come from the U.S. Geological Survey's Office of Geographic and Cartographic Research. The agency, which budgeted almost nothing on digitized mapping a decade ago, is now spending $13 million a year on it. By the year 2000, USGS expects to complete the National Digital Cartographic Data Base, which will include all information that now appears on the agency's maps. When completed, users will be able to illustrate almost instantly the relationships among such features as population density, water supplies, power lines, land-use patterns, and the presence of various natural resources. Moreover, USGS is working with the U.S. Bureau of the Census to produce a cartographic database called TIGER, for "Topologically Integrated Geographic Encoding and Referencing" system. Created for the 1990 census, TIGER will show every river, lake, highway, and railroad track in the United States as well as display a wide range of census data. Eventually, these databases will end up in the hands of businesses, which will use them to create still other maps and computer products.

Other digital-mapping technology comes from the Defense Mapping Agency, which uses map data to guide "smart" missile systems. The process involves loading data about routes to potential targets into a missile, then instructing that missile, before launching, on which route to take. DMA hopes eventually to reduce the entire world map to digits.

(There are many other computerized-mapping programs and geographic databases, a growing number available for use on personal computers; see "Map Software" elsewhere in this book.)

Ultimately, all this high-tech wizardry is expected to become available to individuals through what may well be standard equipment on the cars of the 1990s. With dashboard navigation systems—simplified, inexpensive versions of the systems that allow airline pilots and ship captains to know exactly where they are even in cloudy weather—drivers of the not-too-distant future may be able to choose the best route to a destination, even taking into account an accident or construction project, or more easily negotiate unfamiliar territory.

While automakers have talked for years about such sophisticated tracking systems, they are finally within view. In Germany, car-buyers already can purchase a computerized car-navigation system that sells for about $3,000. The device uses a compact disc to store its maps, which in the case of Germany include every road in every major city. The system figures out a vehicle's location by "dead reckoning": It senses the turning angle of the car's front wheels and the speed of the vehicle and combines those factors with its map-matching ability to fix location.

Such wizardry will soon be arriving on American shores. The German system, designed by Etak Inc., of Menlo Park, California, has been licensed to General Motors; it is already being used by ambulance drivers in some U.S. cities. Toyota Motor Corp. already offers a $2,300 factory-installed electronic map display on nearly a quarter of its Crown and Soarer models sold in Japan; the system is expected to be introduced in the United States soon. Another map system, being developed by Philips N.V. of the Netherlands, is expected to appear in the United States by 1992.

Eventually, say cartographic futurists, instead of stopping at a service station for directions, we'll be able to punch our destinations into the service station's computer, obtaining (for a fee, of course) a made-to-order map, complete with detailed a list of restaurants, motels, and attractions along the way.

The possibilities, thanks to computers, are endless. From car driving to city planning to coastal management, digital-mapping techniques are increasingly finding their way into our lives, making the world a bit more manageable and, perhaps, a little easier to understand.

But no less magnificent.

How to Choose a Map

Finding maps on just about anything is relatively easy. You simply consult your local map store or contact one of the hundreds of other resources listed in this book.

Finding a *good* map is another matter. There are maps that will suit your purpose, and maps that won't. Choosing from among the many available products requires a bit of insight and understanding of what makes a good map.

Selecting the right map can be somewhat like selecting a car: there are so many to choose from that the ultimate decision boils down to a combination of what's available and what suits your taste. But like cars, maps have a number of features (albeit far less technologically sophisticated) that you should consider when making a selection.

Among the things to consider are how the map will be used. A travel map, for example, must show sufficient detail to allow you to traverse unknown roads. If it shows only major highways, it will be of little use once you exit the main road in search of your destination. However, if all you plan to do is drive through an area on your way to some-place else, a highway map may do just fine.

Here are some additional considerations when choosing a map:

■ **Format.** Does the area covered require several sheets? This may make it more difficult to use while driving. On the other hand, a very large region—Los Angeles, for example—may be best suited for several less-comprehensive maps. If several maps are required, are they contained in one publication, or will you have to switch back and forth between several maps or books? If it is a folding map, can you easily refold it?

■ **Media.** Most people are used to maps printed on paper. But other media are sometimes appropriate. Some map collec-

tions or series are available on computer disk or in microfiche, for example, allowing for easy access to a great many map images with minimum storage.

■ **Materials.** A flimsy map can only be used under scrupulous conditions. Ideally, a map should be printed on good-quality paper, resistant to aging and tearing. If a map is intended for heavy use outside of a home, office, or library—in a car, on a trail, in a boat—it should be made of a water- and tear-resistant material (Tyvek is one such material), and perhaps be laminated to protect it from the elements. Lamination, however, makes it impossible to fold maps, which may be impractical. A good alternative is protective sprays that you can apply yourself; map covers and cases also may be helpful (see "Map Accessories").

■ **Design.** Are words and symbols large enough to read? Are they sufficiently distinct from their backgrounds and from one another to be easily seen and interpreted? Are colors, contrast, and patterns used to enhance the information, or do they make it cluttered and confusing?

■ **Currency.** A good map should be up to date. Keep in mind, however, that some maps—topographic maps, for example—don't go out of date very quickly. Most road maps, on the other hand, are of little use when not current. And aeronautical and nautical charts go out of date within a few weeks or months, making them dangerous (and often illegal) to use after they have become outmoded. Publication date alone is only one factor: a 1990 map will be of limited use if it is based on 1970 population data. Keep in mind that some maps are created on a one-time basis (due, usually, to a one-time appropriation of resources for this purpose).

■ **Scale.** Map scale is one of the most important aspects of any map, but one of the least considered by many map buyers. Scale, simply put, expresses the ratio of a distance on the map to the actual distance on the ground (or air or sea, for an aeronautical or nautical chart). So a map with a scale of 1:1 would be life-size, literally: every inch of the map would represent one inch of land. That's not feasible, of course, and defeats the whole purpose of maps—to represent a vast area in a relatively small space.

Map scales range widely, from 1:1,000 to 1:100,000,000 and smaller; the larger the second number, the smaller the scale. (Scale is sometimes expressed with a slash instead of a colon, as in 1/100,000.) Topographic maps range from about 1:20,000 to 1:1,000,000; the former are large-scale maps, frequently used by campers and hikers, the latter would likely be used for a topo map of the world. (See "Topographic Maps" for more on this.) City street maps fall in the 1:10,000 range. Some of the largest-scale maps, at 1:1,000 or even 1:500, are used by city governments for tax-collecting and property-assessing purposes.

To convert scale to actual size, use 1:60,000 as a rough base. At that scale, one inch on the map equals approximately one mile on land. (The exact ratio is 1:63,360.) So, on a 1:125,000 map, a mile would be covered in about a half-inch; a 1:500,000 map would cover about eight miles per inch.

MAP PROJECTIONS

For centuries, map-makers have grappled with the problem of how to depict a round world on a flat surface. Despite advancements in geometry and the creation of complex mathematical models, there remains controversy over the optimum method of producing maps of the world without distorting one or more sections of the Earth's surface.

For more than 400 years, the traditional view of the world was based on the models produced by Gerhardus Mercator, considered the leading cartographer of the 16th century.

His grid system of cartography—revealed in his 1569 map of the world—became the classic expression of cartography and has dictated our geographical world concept ever since. His map was revolutionary; among other things, the spherical nature of the globe, proved by Ferdinand Magellan's circumnavigation of the world, was clearly expressed in it.

But the Mercator projection has its problems—and its critics. For one thing, the polar regions appear grossly enlarged, as land masses and bodies of water are spread to fill the lines of latitude and longitude. Another problem is that the equator appears below the midpoint of the map, thereby enlarging the continents north of the equator.

Mercator's projection of the globe has been followed by many others. All told, more than 200 world-map projections have been produced, with about 100 receiving significant use. About a dozen projections have gained widespread use in the late 20th century. But until recently, the Mercator projection reigned supreme.

While there have been campaigns waged for decades to supplant the Mercator projection with other, more accurate versions, the movement reached a new peak in 1988 when the National Geographic Society, after reviewing nearly 20 proposals, unanimously chose the Robinson projection as its new official view of the world. Since 1922, the society had used maps that relied on a system developed by American engineer Alphons van der Grinten. But that projection, much like Mercator's, depicted Greenland as 554 percent larger than it is. The Soviet Union was depicted 223 percent larger and the United States 68 percent larger.

The Robinson projection, created by Arthur H. Robinson, professor emeritus of cartography and geography at the University of Wisconsin—Madison, was completed in 1963. Robinson had directed the U.S. Office of Strategic Services' map division during World War II and decided to devise a new projection system after encountering many problems in mapping a worldwide war.

The Robinson projection gained even greater official acceptance in 1989, when the American Cartographic Association and five major geographical organizations passed a resolution condemning the widespread use of rectangular maps of the world—such as Mercator's—and endorsed the Robinson version. In the resolution, which began "Whereas, the earth is round . . .," the cartographers urged book and map publishers, the news media, and government agencies to stop using rectangular maps.

Such official endorsements notwithstanding, there remain other projections that have garnered attention—and controversy—among cartographers. A map created in 1974 by German historian Arno Peters, for example, attempts to correct Mercator's distortions, which Peters says distort the shape of continents to overemphasize Caucasian countries in temperate regions. As a result, North America and Eurasia appear considerably larger than South America and Africa. Peters' notions, while not generally accepted by geographers and cartographers, caught the attention of the United Nations, which helped fund development of a detailed 51" X 35" color world map based on the Peters projection. Copies of the world map are available

for $7.95 (includes shipping) from **Friendship Press** (P.O. Box 37844, Cincinnati, OH 45237; 513-761-2100).

Another intriguing model is the Dymaxion projection, created by Buckminster Fuller, the self-described "engineer, inventor, mathematician, architect, cartographer, philosopher, poet, cosmologist, comprehensive designer, and choreographer." Fuller's projection uses triangular segments connected at the poles to show the world as a largely unified set of land masses. The map is available in a variety of forms, from flat maps to T-shirts, from the **Buckminster Fuller Institute** (1743 S. La Cienega Blvd., Los Angeles, CA 90035; 213-837-7710), which offers a free four-page catalog (see "World Maps" and "Map Stuff" for descriptions of Dymaxion maps and products).

The best sources for further reading on the subject of map projections are two publications by the Committee on Map Projections of the American Cartographic Association: *Which Map Is Best? Projections for World Maps* and *Choosing a World Map—Attributes, Distortions, Classes, Aspects*. They are available from the **American Congress on Surveying and Mapping** (210 Little Falls St., Falls Church, VA 22046; 703-241-2446). The price for each is $5.90 for ACSM members, $6.40 for others.

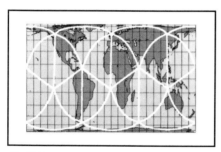

Gall-Peters

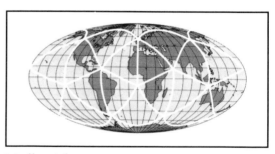

Mollweide

Mercator

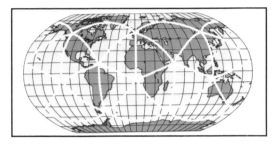

Robinson

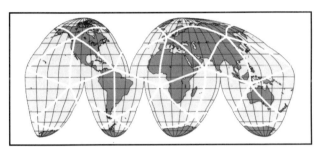

Goode Homolosine

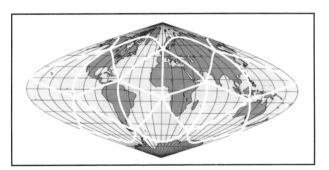

Sinusoidal

Six examples of map projection systems, courtesy American Congress on Surveying and Mapping.

Learning About Map Skills

by Dr. Paul S. Anderson

A "Peanuts" cartoon strip a few years ago features Lucy posing a question to Snoopy, who is reclining, as always, on the roof of his doghouse.

"Can you read a map?" she asks.

"Of course!" responds the dog.

"If you're going to visit your brother in Needles, you'll need a map," she says, offering the dog a folded road map.

Snoopy, sitting up and perusing Lucy's offering, reflects "I'm good at reading maps. I just don't know what all those squares and dots and lines and colors and numbers and names mean."

Like Snoopy, most people consider themselves to be good map readers. But when they hold a map in their hands, the map's message is often lost or only partially understood. There are three reasons for this:

■ *The subject matter.* Many topics, such as topography, geology, census data, and tidal currents, cannot be fully read from even a simple map if the basic concepts of the subject matter are not understood. Fortunately, maps themselves can help us learn some of the things we need to know.

■ *Lack of experience.* After grade school, few Americans have formal education with maps. At age 16, when one typically obtains a driver's license, there is a certain newfound interest in road maps, but even that has diminished over the years, thanks to well-marked interstate highways, with their big green-and-white signs.

■ *The "language" of maps.* Maps are *graphical* expressions of spatial relationships. They are different from books that are read from left to right, top to bottom, page by page, from beginning to end, with words made from combinations of 26 letters. Maps, in contrast, have no uniform starting point, no verbs, relatively few nouns, and hundreds of shapes

and colors. Even the sizes of symbols can have meaning on maps, and the spacings between objects are crucial bits of information. Without formal guidance or explanation of this language, each map user learns only a few basic expressions, something akin to speaking German with a vocabulary consisting of *Achtung, Aufwiedersehen, Bratwurst,* and *Volkswagen.*

Fortunately, the language of maps is easier to learn than German—or any other foreign language—and there are several sources for help. Concise, well-written explanations of map projections, scale, and basic symbols for small-scale (atlas-type) maps are commonly found in the introductory pages of most atlases. Map legends, often called "keys," serve as translators to explain what the diverse symbols represent. If the legend is not printed on a map (such as on topographic maps), one is usually available separately from the publisher.

A readily available source of information on the use of large-scale (a small area with much detail) topographic maps is the Boy Scouts of America. The main *Boy Scout Handbook* features about 10 pages on scales, symbols, directions, and general map use. In addition, the *Merit Badge Pamphlet for Orienteering* gives 32 pages that are useful for anyone planning to take a hike ($1.39; available from Boy Scouts of America, 1325 Walnut Hill Ln., Irving, TX 75038; 214-580-2280).

For those who desire to learn greater map skills, three books offer excellent how-to explanations:

■ *Map Reading and Land Navigation* (Government Printing Office, 008-020-00156-2; $6.50) is a U.S. Army field manual (FM 21-26) available for public use. Its illustrations are excellent, albeit with a military perspective.

Copying Maps? Beware!

Need to make a map? Creating one from scratch is easier these days than it used to be, thanks to "desktop mapping" programs for personal computers (see "Map Software"). But for many individuals, companies, and associations the temptation to copy an existing map is overwhelming.

Copying many maps is illegal. The same copyright laws that prohibit you from stealing other writers' words or artists' images protect map-makers as well. Simply put, according to the International Map Dealers Association, "It is illegal for any firm or individual to reproduce copyrighted works, in whole or in part, regardless of the final purpose of the reproduction, without permission of the copyright owner." Technically, even a single photocopy of a map is a copyright violation.

There are two exceptions. Maps produced by federal or state government agencies usually are in the public domain and may be reproduced without restriction. And copyright-free maps are available from several companies, notably Facts on File, which produces a set of photocopyable boundary maps called "Maps on File" (see "Boundary Maps" for details).

Some map publishers, seeking to thwart those who illegally reprint their maps, have taken to adding "sham" streets and other landmarks to their maps. Usually, these are innocuous, half-block-long alleys, sometimes bearing the names of the mapmakers (or their friends and relatives). When those streets show up on other maps, there is no question that the map was "stolen" from its creator.

The fact is, obtaining permission to reprint all or part of a copyrighted map usually is relatively simple and inexpensive. The alternative—the penalties allowed under federal copyright law for unauthorized use of copyrighted material—can be severe, and may include payment of triple royalties, damages, court costs, and attorney's fees.

For a free brochure, *Questions and Answers About Map Copyrights*, send a stamped, self-addressed legal-size envelope to the **International Map Dealers Association**, P.O. Box 1789, Kankakee, IL 60901.

■ *Map Use: Reading, Analysis, and Interpretation*, 2nd edition ($25; JP Publications, P.O. Box 4173, Madison, WI 53711; 608-231-2373) is the premier college-level text on the subject. Its author, Philip Muehrcke, has put together an appropriate mixture of technical explanations, quality illustrations, and comic wit.

■ *The Language of Maps*, by Phillip Gershmel, is currently available only as a photocopy edition from University of Minnesota Book Store ($17, Minneapolis, MN 55455; 612-625-6000). Its distinct advantage is a set of classroom exercises (using Minnesota examples) which are available from the Department of Independent Study (Westbrook Hall, University of Minnesota, Minneapolis, MN 55455; 612-624-0000).

Two other books focus on map appreciation (and less on skills) but may also be helpful:

■ *Map Appreciation*, by Mark Monmonier and George A. Schnell ($30; Prentice Hall, 200 Old Tappan Rd., Old Tappan, NJ 07675; 201-767-4970; 800-223-2348), is a well-illustrated college text organized by map themes such as landscape, population, politics, and municipal areas.

■ *Interpretation of Topographic Maps*, by Victor Miller and Mary Westerback ($25.95; Merrill Publishing, P.O. Box 508, Columbus, OH 43216; 614-890-1111; 800-848-6205), is devoted strictly to understanding physical landscapes using contour lines on topographic maps.

Learning how to use maps is a very worthy objective. You are encouraged to contact your local college or university about map-related courses, usually offered by the geography faculty. Schoolteachers and education-minded parents can also bring the benefits of map skills to students using some of these publications.

Dr. Anderson is an associate professor of geography in the Department of Geography-Geology at Illinois State University in Normal, Illinois.

TRAVEL MAPS

Bicycle Route Maps

Over the years, biking has grown from a Sunday pastime to a full-blown sport, complete with high-tech equipment, designed racing gear, and detailed route maps. Whether you're looking for a morning glide through a neighborhood park or a cross-country endurance test, there are maps to keep you on track.

The quality of bike route maps varies among publishers. Some maps are simply road maps with a line drawn to indicate a bike route. Other maps provide detailed information about weather conditions, repair or supply services available, and points of interest.

State highway departments or bureaus of tourism often produce bike route maps that are free or inexpensive. Write to the appropriate office (see Appendix A) or get in touch with a local bicycling group. Many groups publish their own maps and most know the best sources for maps in their area. If no bicycling group is listed in a local phone book, try one of the following national organizations, whose publications often cover biking activities and map products.

NATIONAL BICYCLE ORGANIZATIONS

Bicycle Federation of America (1818 R St. NW, Washington, DC 20009; 202-332-6986), a nonprofit organization that publishes a monthly newsletter, *ProBike News* ($18 a year), including reports on mapping and bicycle organizations.

Bicycle USA (a.k.a. The League of American Wheelmen, 6707 Whitestone Rd., Ste. 209, Baltimore, MD 21207; 301-944-3399), a membership organization that produces a monthly magazine, *Bicycle USA*, and an annual *Almanac* ($7.50; available only to members) containing a state-by-state listing of biking information, including organizations, mapping and travel services, and tourism departments. Membership is $22 for individuals, $27 for families.

BikeCentennial, the Bicycle Travel Association (P.O. Box 8308, Missoula, MT 59807; 406-721-1776) publishes a touring magazine, *BikeReport*, nine times a year and provides member discounts on maps and biking accessories. BikeCentennial produces a route-network series of state-of-the-art bike maps covering the U.S. The maps, most printed on special waterproof paper with waterproof ink, are sized to fit in handlebar map cases. Illustrated in shades of blue and green, the maps are drawn at a scale of about 1:250,000. They provide such information as local bike laws, weather, stopovers, and special attractions. The maps feature detailed riding information and matchlines that enable bikers to move effortlessly from one route map to the next. BikeCentennial maps are available in some map and sporting goods stores and directly from the organization (include $2 for shipping and handling). Bike routes published by BikeCentennial include:

■ "The TransAmerica Trail" ($6.95 each; $64.95 for the set of 12 maps), covering the 4,250-mile coast-to-coast bike route.
■ "The Washington to Minnesota Bicycle Route" ($6.95 each; $20.95 for the set of four maps), a 1,815-mile route through Washington, Idaho, Montana, and North Dakota.
■ "The Canada to California Bicycle Route" ($6.95 each; $10.95 for the set of two maps), a 780-mile route that skips through Canada and down the Pacific coast to Crescent City, California.
■ "The Great River North Bicycle Route" ($6.95 each; $10.95 for the set of two maps), an 875-mile river route from Fargo, North Dakota, to Davenport, Iowa.

BikeCentennial also distributes maps published by others, such as "Southern Overland Bicycle Route" ($8.95), a spiral-bound map of the 1,700 mile route from Austin, Texas, to Los Angeles.

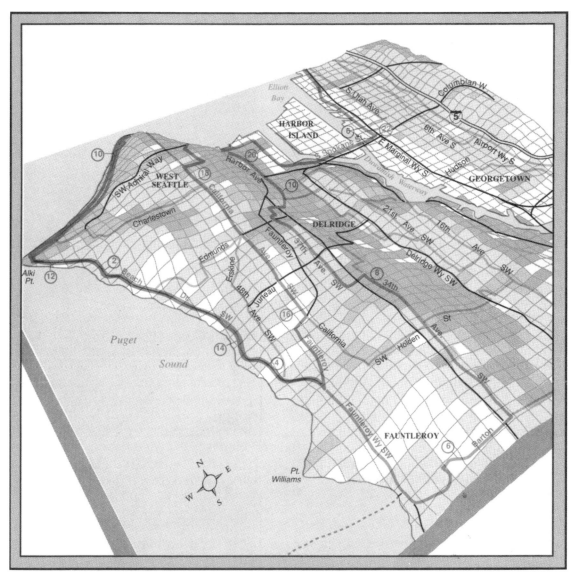

Bicycle route map of West Seattle and Alki Beach, from Terragraphics' Touring Seattle by Bicycle.

MAPS OF DOMESTIC BIKE ROUTES

There are thousands of maps created by hundreds of local bike groups and publishers. Following are producers of maps or books containing maps for a variety of routes throughout the U.S.:

ADC (6440 General Green Way, Alexandria, VA 22312; 703-750-0510; 800-232-6277)

publishes a "Washington Area Bike Map" showing bike routes in and around Washington, D.C. ($6.95).

American Youth Hostels (1332 I St. NW, Washington, DC 20005; 202-783-6161). Local AYH branches sometimes produce bike route maps for their areas. Write to AYH to obtain a list of branches.

The Butterworth Co. of Cape Cod Inc. (476 Main St., Harwichport, MA 02646; 508-432-8200) publishes a "Rail-Trail Map" ($2.50) detailing points of interest along the trail from Dennis to Eastham.

Computer Terrain Mapping Inc. (P.O. Box 4982, Boulder, CO 80306; 303-444-9391) produces "Mountain Bike Map of Boulder County," a three-dimensional map showing bike routes on one side and a topographic map of the same area on the other. Included are historical notes and practical tips for mountain bikers ($8.50 plus $4 shipping, rolled or folded).

DeLorme Mapping Co. (P.O. Box 298, Freeport, ME 04032; 207-865-4171) produces "Bicycling" ($2.95), a handlebar-bag-size booklet of full-color maps covering the Maine coast and inland trips.

Globe Pequot Press (10 Denlar Dr., Chester, CT 06412; 203-526-9571; 800-243-0495; 800-962-0973 in Conn.) sells the "Short Bike Rides" series ($7.95 to $14.95). These guides offer maps of different rides of varying length and difficulty. Guides available are "Cape Cod/Nantucket/The Vineyard," "Connecticut," "Greater Boston/Central Massachusetts," "New Jersey," and "Rhode Island."

Gulf Publishing (P.O. Box 2608, Houston, TX 77252; 713-520-4444) distributes "Bicycling in Texas" ($9.95), a 104-page map-laden guide to scenic routes between major cities as well as "spot tours" of special areas not centered in the cities.

Pantheon Books (201 E. 50th St., New York, NY 10022; 212-751-2600; 800-638-6460) publishes *Bicycle Touring in the Western United States* ($9.95). Authors Karen and Gary Hawkins provide maps for tours covering 7,500 miles of terrain through Arizona, California, Colorado, Idaho, Montana, Nevada, New Mexico, Oregon, Utah, and Wyoming. Also included is information about weather conditions, places to stay, repair and supply sources, and topographic details. The book is also available from **American Youth Hostels**, 1332 I St. NW, Washington, DC 20005; 202-783-6161.

The Rails-to-Trails Conservancy (1400 16th St. NW, Ste. 300, Washington, DC 20036; 202-797-5400) publishes guidebooks to the various rail-trails, abandoned railway routes converted into recreation paths, many of which are good bike paths. The "Guide to America's Rail-Trails" is a directory of existing rail-trails in the U.S. In addition to other useful information, the guide lists trail length, type of surfacing, and suitable uses ($5 members; $6.50 nonmembers); the "Sampler of America's Rail Trails" features maps and detailed descriptions of 12 of the nation's best rail-trails. Also included is information on types of trail use and places to eat and sleep along the trails ($2).

Section of BikeCentennial's Virginia-to-Florida Bicycle Route, *showing the Santee River area in South Carolina. Reprinted with permission of BikeCentennial, the Bicycle Travel Associaiton.*

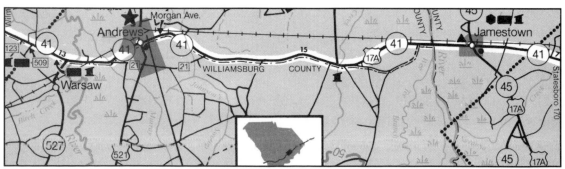

Southwest Trails Association (P.O. Box 191126, San Diego, CA 92119; 619-448-0884) publishes three books with mapped bike routes. *Southwest American Bicycle Route* ($9.95) is a 140-page book with several maps that trace a 1,748-mile route from Oceanside, California, to Larned, Kansas, including travel and topographic information. *Southern Overland Bicycle Route* ($7.95) is a 168-page booklet with 17 maps of varying scale covering a route of 1,607 miles from Austin, Texas, to Los Angeles, following the trails of western pioneers, along with historical, geographical, and travel information. *Bicycling Baja* ($12.95) has 260 pages with 17 maps that detail bike routes in Baja, Mexico. Sunbelt Publications also offers an extensive bike book list with many of the books containing detailed bike maps.

Terragraphics (P.O. Box 1025, Eugene, OR 97440; 503-343-7115) publishes unique bicycle touring guidebooks using three-dimensional maps, so bikers can easily see what terrain lies ahead. Two titles are available: "Touring the Islands: Bicycling in the San Juan, Gulf and Vancouver Islands," and "Touring Seattle by Bicycle" ($9.95 each). Titles on San Francisco, California's wine country, and the Northeast are planned.

MAPS OF FOREIGN BIKE ROUTES

Information about bike routes through foreign countries is often available from the tourist boards of those countries. Some additional organizations that publish foreign maps are:

American Youth Hostels (1332 I St. NW, Washington, DC 20005; 202-783-6161) AYH offers *Bicycle Touring in Europe* ($8.95) by Karen and Gary Hawkins, published by Pantheon Books. This 344-page book has 14 tour routes, ranging from "A Taste of Two Wines" to "Germany's Black Forest" to "The Grand Tour—London to Athens." A small map is provided

for each route, and sources for additional maps are listed. Extensive appendixes cover biking and travel organizations, mail-order map and equipment sources, and places to rent bikes abroad.

The Cycle Touring Co. (14101 Huckleberry Ln., Silver Spring, MD 20906; 301-871-8665) offers a self-guided tour through Switzerland and the surrounding countries ($7.50 postpaid). The 35 large-scale map cards and accompanying text trace a 750-mile route for all levels of cyclists through some of the most challenging bike trails in Europe. The map cards are sized to fit in a typical handlebar bag and are designed so they can be easily read while riding. The Cycle Touring Company also offers free touring advice by telephone to anyone interested in cycling in Switzerland.

Geoscience Resources (2990 Anthony Rd., P.O. Box 2096, Burlington, NC 27216; 919-227-8300; 800-742-2677) distributes Kümmerly & Frey's cycling maps of Switzerland. These maps provide information on such things as scenic routes, road gradients, and road surfaces. The set of 15 cycling maps covers western Switzerland from the West German to the French border ($14.95 for the set).

Gulf Publishing (300 Raritan Center Pkwy., Edison, NJ 08818; 201-225-1900) distributes the PAN/Ordnance Survey's "Cycling Britain" ($14.95), a comprehensive guide to cycling routes in England, Scotland, and Wales, which includes more than 200 two-color maps.

Travel Genie (3714 Lincolnway, Ames, IA 50010; 515-292-1070) distributes "Radtourenkarte," bicycle maps of Germany, published by Haupka & Co. of West Germany. The series includes 35 maps, on a scale of 1:100,000, showing roads with less traffic in bold lines and identifying long-distance bike routes throughout Germany.

Mass Transit Maps

Transit maps reflect where we are—and where we're going. Even a simple bus or subway map can reveal the changing patterns of a region: the newest communities, the growth of suburb-to-suburb commuting, the rebirth of downtown. But more likely, their purpose is considerably more utilitarian. By helping you to understand a city's bus or subway system at a glance, they can make the arrival into a major city far less forbidding. They may even make life in your own home town a bit easier.

Bus and subway maps are far more attractive than they used to be. Many of the graphic innovations in cartography have been pioneered at the local level, in attempts to overhaul and simplify maps of public-transit systems, many of which are far from simple. As energy and environmental matters have breathed new life into public transportation—even car-worshiping Los Angeles is getting a subway—map-makers have strived harder to reduce citizen resistance to learning these new or renewed systems by creating easy-to-understand maps.

While many transit systems have system maps readily available from bus drivers or subway attendants, you needn't arrive in town empty-handed. Here are addresses and ordering information for obtaining transit maps for major U.S. cities. They are free unless otherwise indicated.

Atlanta, GA
Metro Atlanta Rapid Transit Authority (MARTA). Numerous maps are available for Atlanta's rail and bus systems, including a systemwide map, individual rail route schedule/maps, individual bus route maps, the "85-Cent Enjoy Ride" book, and "Guide to MARTA." They can be picked up at the MARTA Ride Store (5 Points Rail Station) or obtained by writing or calling MARTA, Customer Service, 2424 Piedmont Rd. NE, Atlanta, GA 30324; 404-848-5077.

Baltimore, MD
Maryland DOT/Mass Transit Administration. "Ride Guides," featuring system maps of bus and subway routes, as well as timetables, are available in Metro stations and pass sales outlets or can be requested by writing the Mass Transit Administration (Customer Service Department, 300 W. Lexington St., Baltimore, MD 21201) or by calling Customer Service: 301-539-5000.

Boston, MA
Massachusetts Bay Transportation Authority (MBTA). Boston's bus and subway routes are shown on a system map, a wallet card, and individual bus schedule/maps. They can be picked up at the Park Street Station Information Booth (Park St. & Tremont St.) or requested by writing or calling MBTA Marketing Dept., Attn.: Map & Scheduling Information, 120 Boylston St., Boston, MA 02116; 617-722-3200; 617-722-5146 for the hearing impaired; 800-392-6100 in New England.

Buffalo, NY
Niagara Frontier Transportation Authority (NFTA). Various route maps are available for Buffalo, including a systemwide map, a rail system map, a downtown rail map, and individual bus route schedule/maps. Maps can be picked up at NFTA offices (Main Fl., 181 Elliott St.) or obtained by sending a self-addressed, stamped envelope to NFTA, Attn.: Metro Public Relations, 181 Elliott St., Buffalo, NY 14205. For more information call 716-855-7211.

Chicago, IL
Chicago Transit Authority. Various route maps are available from CTA: A "CTA Map," showing bus and subway routes in the city; an "RTA Map," illustrating bus and train routes in the city and suburbs; a "Systemwide" map; and maps for some individual bus routes. They can

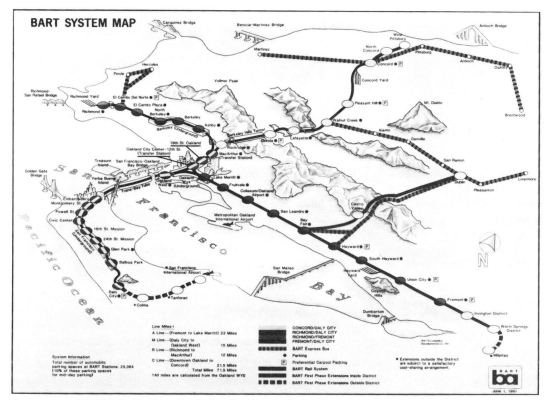

System map of BART, the Bay Area Rapid Transit system, in the San Francisco Bay area.

be picked up at CTA headquarters (Merchandise Mart Plaza, Chicago) or at the RTA (Regional Transit Authority) office (1 N. Dearborn, 11th Floor). Information and maps can also be obtained by calling 312-836-7000 or 800-972-7000 (in Illinois).

Cincinnati, OH
Southwest Ohio Regional Transit Authority. "RideGuide" bus schedules feature route maps. Information and RideGuides can be found at "METROCenter" (122 W. 5th St.), 7:30 a.m. to 5 p.m., Monday through Friday. Call 621-4455 for information.

Cleveland, OH
Greater Cleveland Regional Transportation Authority. System maps and individual route schedules/maps of buses in the Cleveland area

can be obtained from the Customer Service Center (2019 Ontario St.) or by writing or calling the Greater Cleveland Regional Transportation Authority, Attn.: Comm. Dept., 615 Superior Ave. NW, Cleveland, OH 44113; 216-621-9500.

Dallas, TX
Dallas Area Rapid Transit (DART). Maps showing bus and rail routes can be picked up at the DART Action Center (1701 N. Market St., Ste. 302) or obtained by mail or phone from DART Action Center, 601 Pacific Ave., Ste. 500, Dallas, TX 75202; 214-573-8500.

Denver, CO
Regional Transportation District. A system map of Denver's 149 bus routes, as well as individual timetables featuring route maps, can

be obtained at the Market Street Station (Market St. and 16th), the Civic Center Station (Colfax and Broadway), or at area K-Marts, Waldenbooks, and Vickers gas stations. They can also be obtained by mail by writing the Regional Transportation District (Customer Service, 1600 Blake St., Denver, CO 80202) or by calling the Customer Service Telephone Information Center, 303-778-6000.

Detroit, MI
Detroit Dept. of Transportation. Detroit's bus system is described on individual route maps and a system map. They can be found in the Cadillac Square Information Center (Cadillac St. & Bates) or requested by writing or calling the Dept. of Transportation, Bus Schedules, 1301 E. Warren, Detroit, MI 48207; 313-933-1300.

Ft. Lauderdale, FL
Broward County Transit. Ft. Lauderdale's bus system is revealed on a system map that includes points of interest; individual route maps are also available. They can be picked up at area libraries, chambers of commerce, or the main government center or ordered through the mail by writing or calling Broward Co. Transit, Attn.: Marketing, 3201 W. Copans Rd., Pompano Beach, FL 33069; 305-357-8400.

Houston, TX
Metro Transportation Authority of Harris County. Systemwide individual route maps are available at Metro offices (912 Dallas St. or 705 Fannin) or can be requested by writing or calling Marketing, P.O. Box 61429, Houston, TX 77208; 713-739-4000.

Indianapolis, IN
Indianapolis Public Transportation Corp. Single copies of system maps and individual route maps of Indianapolis' buses can be picked up at Indianapolis Public Transportation Corp. (Customer Service Center, 14 E. Washington St.) or requested by writing or calling Marketing Dept., P.O. Box 2382, Indianapolis, IN 46206; 317-635-3344.

Kansas City, MO
Kansas City Area Transportation Authority. System maps and schedules of buses in Kansas City can be picked up in downtown stores, libraries, and churches or can be requested by writing or calling the Kansas City Area Transportation Authority, Information Services, 1350 E. 17th St., Kansas City, MO 64108; 816-221-0660.

Los Angeles, CA
Orange County Transit District. A system map and individual route schedules and maps can be found in area stores, banks, post offices, libraries, and malls, or requested by writing or calling Public Information, 11222 Acacia Pkwy., Garden Grove, CA 92642; 714-636-7433.

Southern California Rapid Transit District (RTD). A system map, as well as nine sector maps, available for San Fernando Valley, San Gabriel Valley, South Bay, Western Region, Mid-Cities, South Central, Downtown L.A., East L.A., and Burbank/Glendale/Pasadena, can be picked up at any of 10 service centers (call for nearest location) or requested by writing RTD, Customer Relations, 425 S. Main St., Los Angeles, CA 90013.

Miami, FL
Metro-Dade Transit Authority. Miami's bus system is illustrated on a systemwide map; maps of individual bus routes are also available. Pick them up at the Metro Rail Station or the Government Center Station, call Maps by Mail (305-638-6137) or write Metro-Dade Transit Authority, Marketing Div., 111 NW 1st St., Ste. 910, Miami, FL 33128.

Milwaukee, WI
Milwaukee County Transit System. Systemwide maps and route maps of Milwaukee's transit system can be found in area libraries, banks, and stores, or requested by writing or calling Milwaukee County Transit System, 1942 N. 17th St., Milwaukee, WI 53205; 414-344-6711.

Minneapolis/St. Paul, MN
Metropolitan Transportation Commission.
Maps of transit routes in the Minneapolis/St. Paul metropolitan area are available at the MTC Transit Store (719 Marquette Ave.—on Baker Block, in Minneapolis) or American National Bank (Skyway Level, St. Paul) or can be requested by writing or calling MTC, Attn.: Schedule Info, 560 6th Ave., Minneapolis, MN 55411; 612-349-7400.

New Orleans, LA
Regional Transportation Authority.
A systemwide map is available for streetcar and bus routes in New Orleans, as well as individual schedules, which include map sketches. They can be picked up at the RTA office (101 Dauphine St., New Orleans) or obtained by calling (504-569-2700) or writing RTA RideLine, 101 Dauphine St., New Orleans, LA 70112.

New York, NY and Vicinity
Metro Suburban Bus Authority. A system route map and individual route schedule/maps are available for suburban New York areas by writing or calling Metro Suburban Bus Authority, Public Affairs, 700 Commercial Ave., Garden City, NY 11530; 516-222-1000. They can also be found in local libraries or on the buses.

New York City Transit Authority. A bus-system map is available for each borough; a map showing the subway system is also available. The maps can be found in libraries, hotels, token booths, tourist information centers, and the information booths at Penn Station and Grand Central Station. They can be mail-ordered by writing or calling the NYC Transit Authority, Consumer Information, 370 Jay St., Rm. 875, Brooklyn, NY 11201; 718-330-8757.

Port Authority Trans-Hudson. A "Guide to Trans Hudson Public Transit" and a "Guide to PATH (Rail) System" are available by calling 212-466-7649.

Philadelphia, PA
Southeastern Pennsylvania Transportation Authority (SEPTA). A city map and a regional map showing transit system routes can be picked up at the SEPTA office (Concourse, 15th & Market Sts.) or obtained by writing or calling SEPTA, Customer Service Dept., 841 Chestnut St., Philadelphia, PA 19107; 215-574-7800. If you include your Philadelphia itinerary, SEPTA will send all relevant route maps.

Phoenix, AZ
Phoenix Transit System. A "Busbook," including a tear-out systemwide map and packed with information on the Phoenix transit system, can be obtained by writing Phoenix Transit System, Attn.: Marketing, P.O. Box 4275, Phoenix, AZ 85030.

Pittsburgh, PA
Port Authority Transit. A countywide rail and bus route map with an inset map of the downtown area can be obtained at the Port Authority Transit Service Center (534 Smithfield St., Mellon Square) or by writing or calling Port Authority Transit, Marketing Dept., Meaver and Island Aves., Pittsburgh, PA 15233; 412-237-7139.

Portland, OR
Tri-County Metropolitan District of Oregon (Tri-Met). Systemwide maps of bus and light-rail routes can be found at the Tri-Met Customer Assistance Office (#1 Pioneer Courthouse Square, Portland; or call 503-238-4982); or write to Tri-Met Consumer Programs, 4012 SE 17th Ave., Portland, OR 97202. A Tri-Met guide with map is $2 (add 50 cents postage if requesting by mail).

St. Louis, MO
Bi-State Development Agency. Bus schedules with route maps are available for all bus lines in St. Louis at the St. Louis Union Station or in area stores. They can also be obtained by writing or calling the Bi-State Development Agency, Customer Service, 707 N. 1st St., Lacleds Landing, MO 63102; 314-231-2345; 800-223-3287 in Illinois.

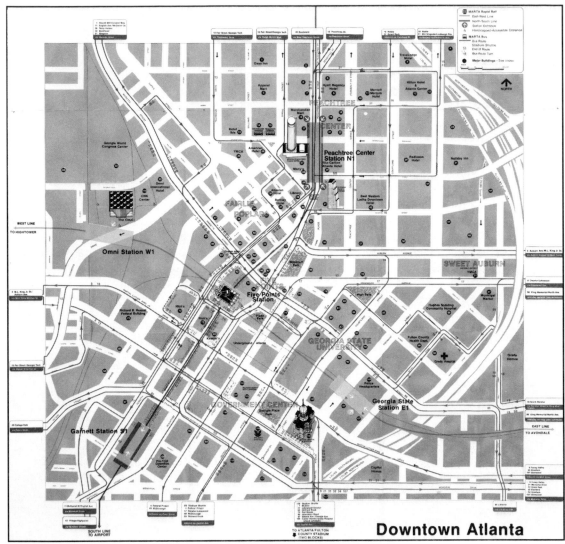

Portion of Metro Atlanta Rapid Transit Authority (MARTA) *map showing downtown Atlanta.*

San Antonio, TX
VIA Metropolitan Transit System. A system map of San Antonio's bus routes and individual schedule/maps can be picked up at VIA Information Center (112 N. Soledad) or obtained by writing or calling VIA Metropolitan Transit System, Customer Service, 800 W. Myrtle, San Antonio, TX 78212; 512-227-2020.

San Diego, CA
Metropolitan Transit Development Board. A regional transit map of San Diego, as well as

maps of individual bus and trolley routes, can be picked up at The Transit Store (449 Broadway, downtown San Diego) or requested by contacting MTDB, 1255 Imperial Ave., Ste. 1000, San Diego, CA 92101; 619-233-3004.

San Francisco, CA
Bay Area Rapid Transit District (BART). A system map of BART and bus routes in San Francisco is available by writing or calling BART, Attn.: Public Information, 800 Madison, Oakland, CA 94607; 415-464-7115.

Alameda-Contra Costa Transit District.
Maps of the East Bay bus system, as well as individual route schedules with route maps, are available by calling, writing, or stopping by AC Transit, Customer Relations, 1600 Franklin St., Oakland, CA 94612; 415-839-2882.

San Francisco Municipal Railway. A "Muni Map," showing BART stations, Golden Gate Transit routes, and train stations, is available for $2, payable in check or money order, from Muni Map, 949 Presidio Ave., Rm. 222, San Francisco, CA 94115. The maps can also be obtained in area bookstores and newsstands for $1.50.

San Jose, CA
Santa Clara County Transportation Agency.
A system map of San Jose bus routes and individual-route timetables with schematic maps are available at the agency's downtown office (4 N. 2nd St.) or can be obtained by writing S.C.C. Transit (Customer Service, P.O. Box 4009, Milpitas, CA 95035) or calling the Telephone Information Center, 408-287-4210.

Seattle, WA
Municipality of Metro Seattle. Seattle's bus system is illustrated on a systemwide map, which includes areas surrounding Seattle;

individual bus route maps are also available. They can be picked up at Metro Customer Assistance Offices (821 2nd Ave. or 1201 4th Ave.) or obtained by writing or calling Metro Customer Assistance Office, 821 2nd Ave., Mail Stop 42, Seattle, WA 98104; 206-447-4824.

Washington, DC
Montgomery County Transit Ride-On. A system map and individual bus route maps for the bus system in the Maryland/Virginia/Washington, D.C. area are available in local libraries, government service centers, and the Transit Information Center (101 Monroe St., 11th Fl., Rockville, MD 20850). They can be requested by writing to the above address or calling 301-217-7433, or 301-217-2222 for the hearing impaired.

Washington Metro Area Transit Association (WMATA). A poster-like full-color "All About Metro" system map of bus and subway routes in the Washington, D.C. area ($10) can be obtained by stopping at or writing WMATA (Metro Headquarters, 600 5th St. NW, Washington, DC 20001) or by calling Public Affairs, 202-962-1047. A system map of bus routes is available for $1.50 from WMATA headquarters or at the Metro Center subway station.

Railroad Maps

The railroads of America crisscross the country, connecting coast with coast and small town with metropolis. For 150 years they have carried settlers to the West and presidents to the White House. Maps of railroad lines are mirrors of America's past and, in some cases, projections for its future. Railroad maps are also useful tools for travelers, engineers, planners, military strategists, transportation buffs, and historians.

The Baltimore & Ohio Railroad was under construction—and surveyance—by 1830, opening 14 miles of track before the end of that year. Dozens of other railroads soon followed, as rail travel for passengers and freight became popular in the expanding nation. The first American railroad map was probably an 1809 survey of the Leiper Railroad in Pennsylvania. The original didn't survive, but a long-winded reproduction, "Draft Exhibiting...the Railway Contemplated by John Leiper Esq. from His Stone Sawmill and Quarries...to His Landing on Ridley Creek," can be found in an 1866 book, A *Short Account of the First Permanent Tramway in America*, by Robert P. Robins. The book itself is part of the **Library of Congress** collection.

Railroad maps are available from a variety of sources, including the federal government, commercial map publishers, the railroad lines themselves, and as reproductions from libraries and map collections. Some are made specifically for or about a railroad, while others are general-use maps that include railroad lines.

GOVERNMENT SOURCES, UNITED STATES

The **U.S. Geological Survey** publishes a number of maps that depict the U.S. rail system. Topographic maps of the states include past and present railroads (see "Topographic Maps"). The rails are depicted according to the condition and use of the tracks rather than by type of train travel: Mainline tracks are drawn as solid line-and-crossties, those under construction are shown as dashed lines, abandoned but still-intact tracks appear as double crossties, and dismantled tracks are indicated by a dashed trail symbol and the legend "Old Railroad Grade." The maps also distinguish between standard- and narrow-gauge tracks and single- and multi-rail routes. One advantage of topographic maps is that you can see at a glance the past and present railroads along a specific section of land.

Railroads are also depicted on other USGS maps. For example, "U.S. Base Map" ($3.10; 10-AO; 24" X 36") includes markings for railroads along with other standard features such as roads, parks, and cities; "U.S. General Reference" ($3.10; 19" X 28") is a colorful, single-sheet map from the *National Atlas* depicting major features, including railroad lines.

The **Tennessee Valley Authority (TVA)** has a topographic map of that region which includes railroad routes. The map ($10, plus $3 for mailing tube and handling; 48" X 62"), a 1979-80 full-color edition of TVA's principal base map, also is available with an overprint of an index to the maps available in the region. Also available through TVA is a blueline print of a map published in 1864, "Railroad and County Map of Tennessee" ($1).

The **Government Printing Office** distributes two books on railroad maps produced by the Library of Congress. *Railroad Maps of North America: The First 100 Years* ($28; 1984; S/N 303-004-00021-3) is a 208-page history of American railroad maps, beautifully illustrated and containing full-color as well as black-and-white map reproductions, as engrossing to scan as it is to study. *Railroad Maps of the United States: A Selected Annotated Bibliography of Original Maps in the Geography and Map Division of the Library of Congress* ($5.50; 1975; S/N 030-004-0014-1) is a 117-page paperbound volume presenting a concise history of railroad maps held in the Library of Congress, with commentaries on

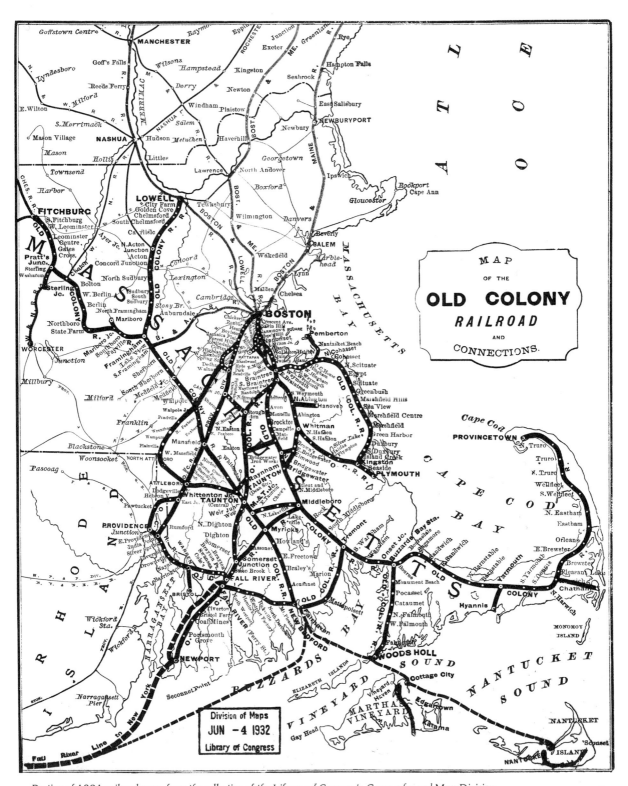

Portion of 1891 railroad map, from the collection of the Library of Congress's Geography and Map Division.

many. The illustrations are in black and white.

Like the Library of Congress, the **National Archives** holds thousands of railroad maps in its collection of maps and charts tracing the history of transportation in America, as surveyed by the government. The first railroad to request and receive federal assistance for mapping its line was the B & O in the early 1830s, quickly followed by numerous eastern railroads vying for government surveyors. The federal government also became involved in mapping railroads to show right-of-way privileges and land grants in the booming West. These and other railroad-related maps (including maps of postal routes that used the railroads to carry mail, and railroad routes through Indian reservations) are stored in the Archives' collections.

Central Intelligence Agency. The CIA produces maps of foreign countries, some depicting transportation systems. Examples are "USSR Railroads" ($11.50; PB 82-927925), "Senegal/Gambia, Israel, and West Berlin, Transportation Systems" ($11.50; PB 82-928044), "China: Cities, Transportation and Admin. Divisions" ($10; 503823), and "China Railroads" (PB 82-928203).

GOVERNMENT SOURCES, CANADA

Canada Map Office (615 Booth St., Ottawa, Ontario K1A 0E9; 613-952-7000) distributes "Canada in Motion" ($10 Canadian), a package of four maps from *The National Atlas of Canada* that supply fascinating information on Canada's network of rail, road, air, and water routes, along with text on transportation geography ($10 Canadian set; $5.50 Canadian each).

Maps Alberta (Land Information Services Div., Main Fl., Brittania Bldg., 703 6th Ave. SW, Calgary, Alberta T2P 0T9; 403-297-7389) publishes the *Alberta Resource and Economic Atlas*, a series of generalized thematic maps, including (as part of the "Service and Recreational Facilities" series) "Airlines, 1984" (D21) and "Transportation" (D24). The maps are 75 cents

Canadian each. Also available is a series of historical road maps of Alberta, spanning the years 1923 to 1961. The nine maps, varying in size and scale, cost $2.75 Canadian each. Maps Alberta also produces "Railway Networks," a 37$\frac{1}{2}$" X 71" map showing railway lines, operators, and stations in Alberta ($8 Canadian).

COMMERCIAL SOURCES

The following companies publish railroad maps in various forms. Some are contained in trade publications; others are history-teaching tools; still others are designed as business or travel maps. All are available directly through the publishers, although a few may also be found in book, travel, or map stores.

George F. Cram Co. (P.O. Box 426, Indianapolis, IN 46206; 317-635-5564) publishes a series of American history maps for classroom use, including two colorful depictions of the growth of the American railway network, "Transportation, Early Railroad Period, 1840-1880" and "Transportation, Principal Railroads Since 1880" ($63 to $72, depending on mounting; 52" X 40"), available singly or in sets.

Forsyth Travel Library (9154 W. 57th St., P.O. Box 2975, Shawnee Mission, KS 66201; 913-384-3440; 800-367-7984) distributes "Thomas Cook Rail Map of Europe" ($7.95; 27" X 36"), which shows all passenger rail lines throughout Europe, east to Moscow, and including North Africa to the south. The map depicts even small routes, with scenic lines in red, electrified, narrow gauge, and mountain railways, as well as hard-to-find route maps for Eastern Europe; and "Thomas Cook Rail Map of Great Britain & Ireland" ($8.95; 27" X 36"), showing all passenger lines and every station in the British Isles, with large-scale detail for each city on the reverse.

Hunter Publishing (300 Raritan Center Pkwy., Edison, NJ 08818; 201-225-1900) distributes several rail maps, including "Thomas Cook Rail Map of Europe" ($7.95; 27" X 36") and "Thomas Cook Rail Map of Great Britain & Ireland" ($7.95; 27″ X 36″); Kümmerly & Frey's "Rail Map of Europe" ($8.95) at a scale of 1:2,500,000; and *Trans-Siberian Rail Guide* ($12.95), a 224-page book with city and route maps as well as information on this romantic rail journey between Europe and Asia.

Interurban Press/Trans Anglo-Books (P.O. Box 6444, Glendale, CA 91205; 818-240-9130) publishes a variety of books on railroad history; many include maps. Especially note-worthy for their maps are: *Rails Through the Orange Groves* ($32; 144 pages), tracing the history of railroad development in Orange County, California; *Forty Feet Below* ($12; 84 pages), investigating the story of Chicago's underground freight tunnels; *Cajon-Rail Passage to the Pacific* ($45.95; 256 pages), tracing the 100-year history of California's Cajon Pass; and *Railroads of Arizona* ($44.95; 344 pages), focusing on copper-mining rail operations and Arizona shortlines. A catalog is available.

National Railway Publication Co. (424 W. 33rd St., New York, NY 10117; 212-714-3148; 800-221-5488) publishes, as part of the International Thompson Transport Press, a series of books for the travel and freight industries. The books may be purchased through subscription or on a single-copy basis. The maps in the books are utilitarian, black-and-white line drawings of specific routes. Three books include railroad maps: *The Official Railway Guide* (Freight Service Edition: six issues, $85; single copy, $35)—each of the six issues concentrates on a single freight line, with locator and route maps for all areas of the country serviced by freight railroads; *The Official Railway Guide* (Travel Edition, five issues, $58; single copy, $35), updated six times a year to provide the latest pricing and departure information from major passenger carriers including Amtrak, National Railways of Mexico, and VIA Rail Canada, as well as partial listings for a number of interna-tional carriers; *Railway Line Clearances* (1986-87

issues, $34), covering rail clearance and weight limitations across America. The maps identify routes and correspond to information in the accompanying charts and tables.

Nystrom (3333 Elston Ave., Chicago, IL 60618; 312-463-1144; 800-621-8086) publishes an American history map series, which includes "Transportation Unites the Nation," a series of four maps illustrating railroads, waterways, major highways, major air routes, and major oil and gas pipelines. Prices range from $41 to $63, depending on mounting.

The Rail Line (P.O. Box 4671, Dept. M, Chicago, IL 60680; no phone) distributes diagrams and maps of various U.S. railroads, such as "Amtrak's 14th Street Coach Yard ($11.50), "Illinois Central's Central Station Operating Diagram" ($11.95), and "Milwaukee Road Detour Map" ($26.95).

Rand McNally (P.O. Box 7600, Chicago, IL 60680; 312-673-9100) publishes *Handy Railroad Atlas of the United States* ($9.95), a 64-page atlas including distance tables and individual maps, in one and two colors, of major metropolitan areas in the U.S. and Canada. Also available from Rand McNally is a classroom teaching map, "Development of Transportation, 1829-1860" ($59). This full-color, 44″ X 38″ map, part of an educational American history series called "Our America," is markable and wash-able; special markers are included.

John Szwajkart (P.O. Box 163, Brookfield, IL 60513; 312-485-1222) produces "Train Watcher's Guides" to Chicago and St. Louis, which include maps and photos of rail lines. The *Train Watcher's Guide and Map to Chicago* is $15; the *Train Watcher's Guide and Map to St. Louis* is $11. Both guides are available for $24.50; extra maps are $2.50.

■ **See also: "Mass Transit Maps."**

Recreation Maps

Getting there, as they say, is half the fun. But without a good map, it can be half the headache. Whether it's a Sunday in the park or a month in the mountains, recreation maps can help find the best route to a nearby or faraway hiking trail, swimming hole, campsite, or amusement park—or they may reveal a new and exciting "discovery" you weren't expecting. Water recreation maps can point the way to the best fishing or the fastest boating.

Federal, state, and provincial agencies are a good source for recreation maps, as are tourist boards and organizations devoted to the preservation and enjoyment of the wilderness, such as the Sierra Club. Finding a recreation map can be as easy as walking into a local sporting goods store. Many retailers, especially chains such as Herman's World of Sports, Eddie Bauer, or REI, sell maps along with their canoes, skis, and parkas. Other sources include the many outdoors, conservation, and sports organizations that provide maps to members and the public, sometimes to promote their activities. Name an outdoor sport, hobby, or pastime and chances are there's a map to match.

GOVERNMENT SOURCES, UNITED STATES

Army Corps of Engineers, created in 1838 and responsible for mapping a lot of the Wild West, now spends much of its time preserving and mapping the water recreation lands under its jurisdiction. There are district offices all over the country (see Appendix B) and each has free or inexpensive maps and guides to the lakes, rivers, and beaches it maintains for public recreation and enjoyment.

Bureau of Land Management (BLM) administers nearly 450 million acres of public land known as the National Land Reserve. Most of BLM's holdings are in the West, although there are small sections in the South and North that fall under its jurisdiction. There are more than 3,600 miles of marked trails in BLM territory, as well as more than 3,000 federal recreation areas open to the public. BLM maps for these trails and recreation areas can be obtained from BLM offices (see Appendix B).

Fish and Wildlife Service administers the more than 90 million acres of wildlife refuge in this country that provide habitat for wild birds, mammals, and many other creatures. More than 300 refuges offer recreational opportunities including boating, fishing, hiking, and nature study. Free or inexpensive maps are usually available at the refuge, and the U.S. Geological Survey series map shows the boundaries of all the National Wildlife Refuges. "A Visitor's Guide to National Wildlife Refuges" is a fold-out map showing the locations of more than 300 refuges that offer recreational opportunities to the public. This brochure is available from the Government Printing Office for $1, or free from the Fish and Wildlife Service at its Washington, D.C. headquarters or regional offices. Maps of restricted or protected fishing areas are also available from the Fish and Wildlife Service.

National Ocean Service (NOS), part of the Commerce Department's **National Oceanic and Atmospheric Administration**, has sailing and boating maps covering most of the nation's water recreation areas. The NOS sailing and boating maps and charts are drawn at various scales and show sailing routes, flows of tides and currents, danger zones, and other important sailing and boating information for all craft sizes from yachts to canoes. Free nautical catalogs diagram the sailing and boating maps and charts available from NOS (see "Nautical Charts and Maps"). Also available from NOS are bathymetric fishing maps

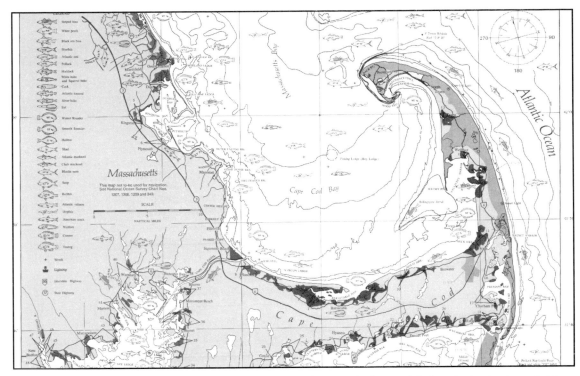

Two views of Cape Cod. Top: *fishing map showing species and known locations of indigenous fish, from GPO's* Angler's Guide to the U. S. Atlantic Coast: Fish, Fishing Grounds and Facilities: Nantucket Shoals to Long Island Sound *(out of print). Below: portion of* The Central New England Coastal Diving Map, *published by Gold Cartographics.*

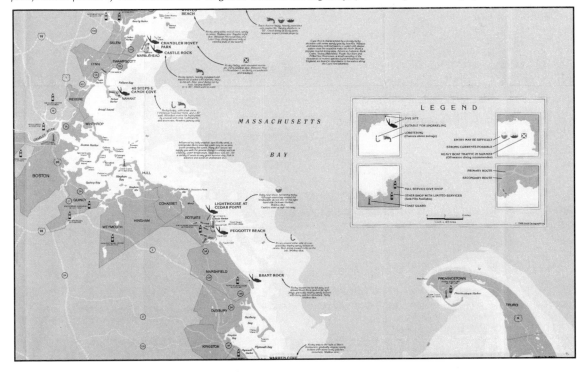

for most known fishing regions along U.S. coastlines. These are topographic depictions of the sea floor designed for use by recreational or commercial fishermen. The maps, which cost $4, illustrate and identify the distribution of bottom sediment and obstructions that may give clues to the location of fishing grounds. The NOS series is drawn at a scale of 1:100,000 and is described in a free catalog available from NOAA.

Names and addresses of retailers of water-recreation maps that sell government maps are printed in NOAA catalogs. There are thousands of boating and tackle shops around the country that distribute maps for their regions, and these are the best overall sources for finding local maps.

National Park Service administers millions of acres of park land, as well as nine national seashores and four national lakeshores. There are schematic and topographic maps available for most of these parks, as well as recreational and historic site maps. Nearly all national parks have free brochures containing maps and information, available either at the individual park headquarters or through the Public Inquiries Office of the National Park Service (see Appendix B). GPO publishes and sells several Park Service guides, including "Map and Guide: National Parks of the United States" ($1.25) and "The National Parks: Lesser-Known Areas" ($1.50). The National Park Service also has maps for the National Wild and Scenic River System.

Tennessee Valley Authority (TVA) produces color pocket-size recreation maps of TVA lakes, showing highways, roads, mileages, cities and rural communities, public access areas, commercial recreation areas, boat docks, private clubs, group camps, public parks, wildlife-management areas, boat-launching sites, and lands open to public use. Maps (75 cents each) are available for Cherokee/Douglas/Nollichucky Lakes, Chickamauga Lake, Fontana Lake, Fort Loudon Lake, Guntersville Lake, Kentucky Lake, Melton Hill Lake, Nickajack

Lake, Norris Lake, Pickwick Lake, Tims Ford Lake, Upper Hiwassee Lakes (Blue Ridge, Chatuge, Hiwassee and Nottely), Wheeler and Wilson Lakes, Watts Bar Lake, and Upper Holston Lakes (South Holston, Watauga, Boone, and Fort Patrick Henry).

U.S. Forest Service (USFS) maintains 156 national forests and has maps for all of them. The Forest Service produces various types of maps for its lands, such as forest visitors' guides and maps, wilderness area maps, and Special Designated Area maps. Maps can be ordered through the USFS regional and national offices, and selected maps are available through GPO. Requests for maps should, if possible, include a note about which parts of a particular forest interest you, because many of the larger forests are set out on several maps, divided by ranger station. USFS can be most helpful in selecting appropriate maps if you indicate what activities are planned for a given area; there are different maps for hikers, campers, and geologists.

U.S. Geological Survey (USGS) is the government's largest mapping agency, as well as its most complex. Maps depicting the topography, geology, and recreational use of national forests, parks, and refuges are available from USGS for a small fee, usually less than $4. Depending on the size and importance of the public-recreation area, details of these lands may be included in larger-scale maps that cover the area around the park. Detailed boundaries of these areas are usually shown if they are of substantial size. Smaller patches of public land may not be shown, but wildlife refuges and game preserves always appear on USGS maps.

USGS produces a "National Wilderness Preservation System" map ($3.10; 1987), showing locations and boundaries of wilderness areas administered by the U.S. Forest Service, the Bureau of Land Management, the Fish and Wildlife Service, and the National Park Service. An "Index to USGS Topographic Map Coverage of the National Park System" is

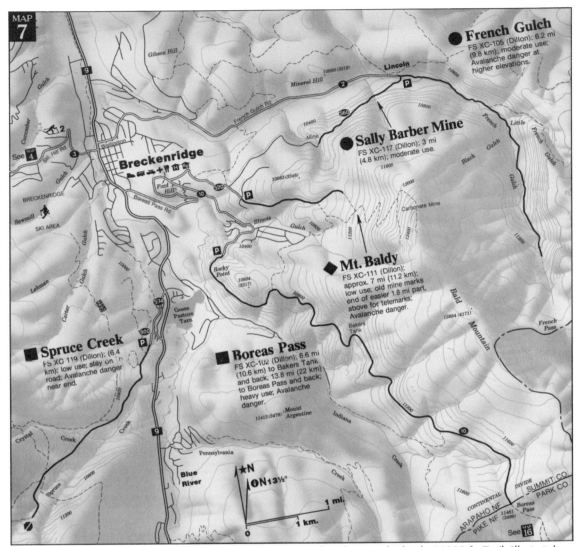

Portion of a cross-country skier's topographic trail map of the Frisco-Breckenridge area of Colorado. ©1988 by Trails Illustrated.

available free from USGS Earth Science Information Centers and Map Distribution Centers (see Appendix B).

GOVERNMENT SOURCES, CANADA

Canada Map Office (615 Booth St., Ottawa, Ontario K1A 0E9; 613-952-7000) distributes the Canadian National Park Maps series. The multicolor maps range in price from $4 to $5 Canadian, and are available for the following parks: Banff/Kootenay/Yoho (MCR 0220), Fundy (MCR 0215), Jasper (MCR 0221), Mount

Revelstoke & Glacier (MCR 0219), Prince Albert (MCR 0210SR), Riding Mountain (MCR 0207), Terra Nova (MCR 0214), and Waterton Lakes (MCR 0222). Maps of numerous areas in other Canadian national parks are also available; a map index/price list is available upon request.

The Canada Map Office also distributes National Topographic Series maps, which contain information vital to canoeists, hikers, and others. The maps illustrate in detail geographic and hydrographic features of the land, as well as population centers, road and rail networks, trails, and other notable information. Complete coverage of Canada is available at a scale of 1:250,000; these maps are recommended for wilderness canoeists. The Canada Map Office also distributes an informative pamphlet, "Maps and Wilderness Canoeing," produced by the Ministry of Energy, Mines and Resources, which provides useful information for canoeists, as well as guidelines on obtaining the appropriate maps for a canoe trip.

Canadian Hydrographic Service (Hydrographic Chart Distribution Office, 1675 Russell Rd., P.O. Box 8080, Ottawa, Ontario K1G 3H6; 613-998-4931) produces hydrographic charts of the coast and major navigable waterways and lakes of Canada, which are often useful for canoeists and water recreationists. Chart catalogs and price lists are available upon request.

Manitoba Natural Resources (Surveys and Mapping Branch, 1007 Century St., Winnipeg,

Manitoba R3H 0W4; 204-945-6666) produces lake-depth charts of Manitoba lakes designed for use by anglers. The maps are not intended for use in navigation as they do not show navigational aids or underwater structures; they do, however, illustrate lake depths and contours to offer some clues as to where fish can be found. The charts are available in "Angling Packages," which include the Manitoba Angling Map, showing a listing of the major sport-fishing lakes in Manitoba and the species of fish found in each lake. Angling Packages are available for the following regions: Duck Mountain, Flin Flon/The Pas, Grass River, Nopiming, Northern Whiteshell, Southern Whiteshell, Thompson/Split Lake, and Western Manitoba. The Manitoba Angling Map is also available separately.

The Surveys and Mapping Branch also publishes a series of Canoe Route Maps, created by Manitoba artist Real Berard. The illustrated, information-laden maps are not intended for navigational purposes, but they

Portion of Map of Yosemite Valley, *published by Tom Harrison Cartography.* ©1988 *by Tom Harrison.*

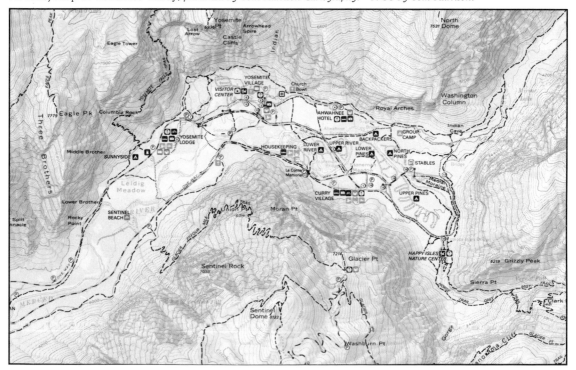

provide an excellent aid in planning a trip or learning about the natural and cultural history of an area. Maps are available for the following waterways: Assiniboine, Grass River, Kautunigan, Land of the Little Sticks, Little Grand Rapids, Middle Track/Hayes River, Mistik Creed, Oiseau, Riviere aux Rats, Sasaginnigak Canoe Country, Waterhen Country, Whitemouth River, and Winnipeg River.

Maps Alberta (Land Information Services Div., Main Fl., Brittania Bldg., 703 6th Ave. SW, Calgary, Alberta T2P 0T9; 403-297-7389) distributes many of the Canadian National Park series maps. It also distributes two maps of unique areas within the national parks, Columbia Icefields ($4.50 Canadian; IWD 1011) and Peyto Glacier ($4.50 Canadian; IWD 1010). The maps illustrate a variety of glacial features and include additional information on the physical environment and hiking trails in the areas.

Ontario Ministry of Natural Resources
(Public Information Centre, 99 Wellesley St. W., Rm. 1640, Toronto, Ontario M7A 1W3; 416-324-4841) publishes small- and medium-scale maps of many recreational areas in the province. Map types vary but may include hiking trails, canoe routes, portages, and campsites. Available maps are: Algonquin Provincial Park Canoe Routes ($2.31 Canadian); Algonquin Provincial Park Backpacking Trails/Topographic ($1.39 Canadian); Killarney Provincial Park/Topographic ($3.01 Canadian); Lake Superior Provincial Park ($2.60 Canadian); Leslie M. Frost Natural Resources Centre ($3.47 Canadian); Sleeping Giant Provincial Park/Topographic ($4.63 Canadian); and Quetico Provincial Park ($9 Canadian, printed on waterproof paper). A catalog of maps available through the ministry is furnished upon request.

The ministry also produces maps of more than 500 Ontario lakes. The maps depict lake contours, lake characteristics, fish species present, access, and facilities. Most maps are available as photostat copies, although some are printed on waterproof paper. Prices range from $1.30 to $2.40 Canadian.

COMMERCIAL SOURCES

There are a vast number of companies producing maps for recreational use; a sampling is below. For maps of bicycle routes, see "Bicycle Route Maps"; topographic maps can be found in "Topographic Maps."

Other sources include publishers of maps for business promotional use. These are often free or inexpensive and are created to illustrate the recreational or tourist facilities of a certain region. One such map-producing company is **The National Survey** (Chester, VT 05143; 802-875-2121), which produces a variety of recreation maps for state and local governments and retailers. One example is National Survey's free, ad-filled "Eastern Ski Map," showing ski and winter recreation resorts around the eastern United States.

Other sources include:

ADC (6440 General Green Way, Alexandria, VA 22312; 703-750-0510; 800-232-6277) produces numerous maps for recreation areas in Delaware, Maryland, North Carolina, Pennsylvania, South Carolina, and Virginia. Selections range from "The Chesapeake Bay Regional Road Map" ($2.95) to "Chincoteague-Assateague Fishing & Recreation Map" ($2.95) to *Saltwater Sportfishing & Boating in North Carolina Atlas* ($12.95). A free index is available.

Adirondack Mountain Club Books (RR 3, Box 3055, Lake George, NY 12854; 518-793-7737) has two books of maps for canoeing in the waters of the Adirondack's rivers: *Adirondack Canoe Waters—North Flow* ($14.95), with 34 maps; and *Adirondack Canoe Waters—South Flow* ($12.95), with 160 map-filled pages.

A.I.D. Associates Inc. (4378 Spring Valley Rd., Dallas, TX 75224; 214-386-6277; 800-442-6277) publishes regularly revised contoured-depth aerial maps depicting underwater features of ponds, lakes, and rivers, navigation markers, road systems, access points, and public areas. Maps are available for areas in Arkansas, Louisiana, Mexico, Mississippi, Missouri, Oklahoma, and Texas ($3.45 each).

Appalachian Mountain Club Books (5 Joy St., Boston, MA 02108; 617-523-0636) produces maps and guides to recreation areas throughout New England, including AMC *Massachusetts and Rhode Island Guide* ($14.95, 624 pages), an up-to-date guide with two full-color maps in a back pocket; "Map of Mount Washington and the Heart of the Presidential Range" ($8.95; 27 1/2" X 40"), a seven-color map drawn on a scale of 1:20,000; and the "AMC Trail Guide" series, recognized as a reliable and comprehensive source of trail information. The guides, which include maps and trail directions and distances, are available for "Maine Mountains," "Mount Desert Island and Acadia National Park," "Mount Washington and the Presidential Range," and "White Mountain." Maps contained in the guidebooks are also available separately, in paper ($2.50 each) or on waterproof, tearproof Tyvek ($3.95 each).

The Butterworth Co. of Cape Cod Inc. (476 Main St., Harwichport, MA 02646; 508-432-8200) publishes a two-color "Fresh Water Fishing Map" ($2.50; 24" X 36") showing fish types, water depth, contours, access roads, boat ramps, fishing tips, and pond locations for 110 Cape Cod and Plymouth lakes.

Clarkson Map Co. (1225 Delanglade St., P.O. Box 218, Kaukauna, WI 54130; 414-766-3000) distributes Canadian Boundary Waters maps ($13 each), including maps for Kabetogama Lake, Rainy Lake, and Lake of the Woods. It also sells lake maps for most lakes in Minnesota, Upper Michigan, and Wisconsin (prices vary); canoe charts for Wisconsin lakes ($4 each); *Lake Michigan Marine Maps* ($10.95), a complete set of maps showing shorelines, harbors, and offshore depths; *Wisconsin Muskellunge Waters* ($10.95), with county maps locating "muskie" (a type of fish) waters; and *Wisconsin Trout Waters* ($10.95), a collection of Wisconsin county maps showing the best trout streams.

Compass Maps Inc. (P.O. Box 4369, Modesto, CA 95352; 209-529-5017) publishes several recreation maps for California and Nevada,

including "Redwood Coast," "Ski Map of California and Nevada," "Tahoe Recreation Map," and "Wine Tour Map." A brochure is available upon request.

DeLorme Mapping Co. (P.O. Box 298, Freeport, ME 04032; 207-865-4171) publishes a variety of recreation maps for Maine, including topographic maps of many state and national parks. Examples are "Moosehead Lake Map & Guide," "Trail Map & Guide to the White Mountain National Forest," and "Baxter State Park and Katahdin Map & Guide"; *Fishing Depth Maps* ($3.95 each), 17 volumes of 1,700 Maine lakes and ponds, illustrating depths and fish species in each water; *Maine Fishing Maps* ($10.95 each), containing large-scale maps and descriptions of Maine's best fishing spots (Volume I focuses on lakes and ponds, while Volume II features rivers and streams); *New Hampshire Fishing Maps* ($10.95), a 112-page guide to the Granite State's rivers, streams, lakes, and ponds; and *Canoeing* ($3.95 each), a set of three 48-page guides to canoeing in Maine.

Geoscience Resources (2990 Anthony Rd., P.O. Box 2096, Burlington, NC 27216; 919-227-8300; 800-742-2677) distributes Kümmerly & Frey's hiking maps of Switzerland. These provide information on the national trail system, way stations, hikers' huts, and scenic routes. The set of 64 hiking maps ($12.95) covers most of Switzerland. Also available are government-produced maps of U.S. and Canadian national parks.

Globe Pequot Press (10 Denlar Dr., Chester, CT 06412; 203-526-9571; 800-243-0495; 800-962-0973 in Conn.) sells *Sixty Selected Short Walks in Connecticut* ($7.95; 180 pages), a guide to walks in and around the state's woods, mountains, lakes, waterfalls, and flower sanctuaries.

Gold Cartographics (P.O. Box 1813, Cambridge, MA 02238; 617-547-6538) publishes the award-winning "Central New England Coastal Diving Map" ($6.96), a 25" X 37" full-color guide/map of recreational diving in Massachusetts and eastern Rhode Island. The map locates 27 scuba and snorkeling sites, as well as access roads, dive clubs and shops, Coast Guard stations, decompression chambers, and sets out diving laws and weather information.

Gulf Publishing Co. (P.O. Box 2608, Houston, TX 77252; 713-520-4444) publishes a number of recreation guidebooks that include maps. A sampling is: *Camper's Guide to Texas Parks, Lakes, and Forests* ($9.95; 148 pages), *Hiking and Backpacking Trails of Texas* ($9.95; 148 pages), *A Guide to Texas Rivers and Streams* ($9.95; 120 pages), *Rock Hunting in Texas* ($9.95; 90 pages), and *Fishing in Texas* ($9.95; 128 pages).

Tom Harrison Cartography (333 Bellam Rd., San Rafael, CA 94901; 415-456-7940) produces shaded-relief topographic maps of state and national parks, forests, and recreation areas. Maps available include: "San Diego Backcountry" ($5.95); "Santa Monica Mountains NRA—Western Section" ($4.95); "Sequoia-Kings Canyon National Park" ($5.95); "Yosemite National Park" ($5.95); and "Yosemite Valley" ($4.95).

High Tech Caribbean Ltd. (P.O. Box 325, Road Town, Tortola, British Virgin Islands; 809-494-3811) publishes a colorful "Dive Map of the Virgin Islands," a 16" X 20" map illustrating choice diving spots ($2).

Hubbard Scientific (P.O. Box 104, Northbrook, IL 60062; 312-272-7810) produces more than 200 raised-relief maps of the country's mountainous areas—everything west of the Rocky Mountains and portions of the eastern U.S., from Alabama to Maine. Hubbard's maps are not as colorful as those of other companies, but they include high-quality detail. Maps vary in size, but are generally 21" square and represent approximately 110 X 70 miles. Prices are $15.95 unframed; $37.95 framed.

International Specialized Book Services Inc. (5602 NE Hassalo St., Portland, OR 97213; 503-

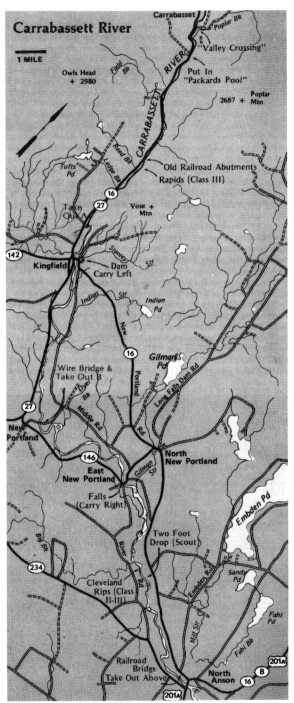

Canoeing map of the Carrabassett River in western Maine, showing rapids and other landmarks, from Maine Geographic Canoeing, Volume 2: Western Rivers. Copyright DeLorme Mapping Company. Reproduced with permission.

287-3093; 800-547-7734) distributes a variety of recreation guides—which include maps—for Australia, Canada, England, and New Zealand. A sampling includes: A *Guide to the Surf Beaches of Victoria, Australia* ($14.95; 127 pages), *Treks in New South Wales, Australia* ($10.95), and *Parkways of the Canadian Rockies: An Interpretive Guide to Banff, Jasper, Kootenay and Yoho Parks* ($14.95; 154 pages).

Kingfisher Maps Inc. (P.O. Box 1604, Seneca, SC 29676; 803-882-5840) publishes topographic-vicinity and lake-contour maps for anglers and boaters, printed on paper ($2.95 each) or waterproof paper ($4.95 each). They are available for a variety of lakes in Alabama, Florida, Georgia, North Carolina, and South Carolina.

Mountaineers Books (306 2nd Ave. W., Seattle, WA 98119; 206-285-2665) sells outdoor recreation guides, many containing useful maps. Examples are: *Best Hikes with Children In Western Washington & the Cascades* ($12.95; 240 pages); *Mt. St. Helens National Volcanic Monument* ($4.95; 120 pages), a guide for hikers, viewers and skiers; *103 Hikes in Southwestern B.C.* ($10.95; 224 pages); and *101 Best Routes on Scottish Mountains* ($25; 224 pages).

New England Cartographics (P.O. Box 369, Amherst, MA 01004; 413-253-7415) produces topographic maps of recreational areas in New England; titles available include "Mt. Greylock Trail Map" ($2.95), "Holyoke Range East" ($2.95), "Quabbin Reservation Guide" ($2.95). Guides include *Connecticut River Guide* ($5) and *Guide to the Taconic Trail System* ($5). Maps are available laminated or on waterproof paper.

Northern Cartographics (P.O. Box 133, Burlington, VT 05402; 802-655-4321) publishes *Access America* ($89.95), an atlas and guide to the 37 national parks for disabled visitors. Detailed, full-color maps, regional maps illustrating selected support services within a 100-mile radius of the park, and information on paths of travel, accessible parking, restroom evalu-

ations, road elevations, medical services, lodging, dining, and campgrounds are provided. It also distributes *The Atlas of Vermont Trout Ponds* ($11.95; 176 pages) and *Vermont Trout Streams* ($19.95; 128 pages), with six folded maps inside.

Pacific Historical Maps (P.O. Box 201, CHRB, Saipan, Mariana Islands 96950) publishes a "Dive Map of Ghost Fleet of Truk Lagoon" ($3.50), a combination map and pictorial depicting the 1944 U.S. carrier attack on Japanese fleets. Shipwrecks from this battle are located and described. The map is available from the **Department of Resources and Development**, P.O. Box 280, Moen, Truk, F.S.M. 96942.

Pelican Publishing Co. (1101 Monroe St., Gretna, LA 70053; 504-368-1175) distributes *The Frank Davis Fishing Guide to Lake Pontchartrain and Lake Borgne* ($7.95), a 160-page guide to these southern Louisiana lakes, supplemented with detailed maps.

Perigee Books (200 Madison Ave., New York, NY 10016; 201-933-9292; 800-631-8571) distributes Van Dam's "National Parks Unfold" map series of U.S. national parks. These colorful maps ($9.95 each) are pocket-size, unfold to 12 times the original size, and refold automatically; they are the only maps to hold a U.S. patent and a copyright as kinetic sculpture. The maps feature day trails, camping spots, sporting facilities, visitor information, and historical highlights, providing handy guides to exploring the nation's wilderness. Titles available are "The Everglades Unfolds," "The Grand Canyon Unfolds," "The Great Smokies National Park Unfolds," "Olympic National Park Unfolds," "Yellowstone Unfolds," and "Yosemite Unfolds."

Pierson Graphics Corp. (899 Broadway, Denver, CO 80203; 303-623-4299) publishes "Colorado Deluxe Road and Recreation Map" ($4.95), a 35" X 45" shaded-relief map with insets of major cities and national parks.

Pittmon Map Co. (732 SE Hawthorne Blvd., Portland, OR 97214; 503-232-1161; 800-452-3228 in western states) produces "Pittmon Recreation County Maps" ($2.99 each), indexed maps created for use by hunters, fishermen, four-wheelers, prospectors, and other recreationists. The maps are available for most counties in Oregon and Washington. Pittmon also distributes "Wilderness Maps" for areas in the Pacific Northwest, as well as USGS topographic and National Forest maps for areas in Oregon.

The Rails-to-Trails Conservancy (1400 16th St. NW, Washington, DC 20036; 202-797-5400) publishes guidebooks to rail-trails, abandoned railway routes converted into recreation paths. The *Guide to America's Rail-Trails* is a directory of existing rail-trails in the U.S. Among other useful data, the guide lists trail length, type of surfacing, and suitable uses ($5 members; $6.50 nonmembers). The "Sampler of America's Rail-Trails" ($2) features maps and detailed descriptions of 12 of the nation's best rail-trails. Also included is information on types of trail use and places to eat and sleep along the trails.

Rainbow Gold (8585 Top of the World Cir., P.O. Box 21681, Salt Lake City, UT 84121; 801-943-0298) produces "PhotoMaps" ($3.95 each), sheets of photos taken from the land and the air, along with a general map sketch of the area. PhotoMaps are available for Lake Powell, the San Juan River, and the Grand Canyon. These maps are not useful for finding your way around; they are more helpful for identifying the areas you are seeing.

Sierra Club Books (distributed by Random House, 201 E. 50th St., New York, NY 10022; 301-848-1900; 800-726-0600) publishes a series of paperback guides to parks and natural recreation areas, which include between 10 and 18 full-color maps per title. There are two series: *Sierra Club Guides to Natural Areas of the U.S.* (with separate titles for California; Colorado and Utah; Idaho, Montana, and Wyoming; New Mexico, Arizona, and Nevada; and Oregon and Washington) and *Sierra Club Guides to the National Parks* (with separate titles for parks of the desert Southwest; the East and Midwest; the Pacific Northwest and Alaska; the Pacific Southwest and Hawaii; and the Rocky Mountains). Another series, *Sierra Club Totebooks*, features hikers' guides that contain maps for a variety of wilderness areas, including the Great Smoky Mountains, the Grand Canyon, the North Cascades, the Wind River Mountains, the high Sierras, the Swiss Alps, the John Muir Trail, the Great Basin, Bigfoot Country, and Yellowstone backcountry.

Square One Map Co. (P.O. Box 1312, Woodinville, WA 98072; 206-485-1511) publishes various maps, most including park/campground charts, of Washington state's recreation areas. Titles include "San Juan Islands Map and Recreation Guide" ($4.95), "Mount Rainier—Central Cascades" ($3.98), and "The Okanogan—Central Washington" ($3.98).

Trails Illustrated (P.O. Box 3610, Evergreen, CO 80439; 303-670-3457) publishes more than 50 topographic maps of recreational areas in Colorado, as well as selected national parks. The waterproof, tearproof maps are updated annually and are available in retail stores or can be ordered directly. A catalog and price list is available. Trails Illustrated produces the following series:

■ "Colorado Series" ($7 each), including maps for Flat Tops, Indian Peaks, Kebler Pass West, Maroon Bells, Poudre River, Steamboat Springs, Tarryall Mountains, and Weminuche/Chicago Basin.
■ "National Park and Recreation Areas Series" ($7 each), including Dinosaur National Monument, Glacier/Waterton National Park, Grand Canyon National Park, Rocky Mountain National Park, and Sequoia/Kings Canyon National Park.
■ "Cross Country Series" ($5 each), maps for Aspen-Carbondale, Frisco-Breckenridge, Nederland-Georgetown, and Vail-Leadville.

Individual titles from Trails Illustrated include "Vail Ski Trail Map" and "Moab Area Mountain Bike Routes," a topographic map showing bicycle trails rated by difficulty.

University of Hawaii Press (2840 Kolowalu St., Honolulu, HI 96822; 808-948-8255) publishes On the Na Pali Coast: A Guide for Hikers and Boaters ($10.95; 112 pages), designed as a guide for both land and sea, containing detailed maps and narrative text.

Wilderness Press (2440 Bancroft Way, Berkeley, CA 94704; 415-843-8080) publishes many map-filled guides to recreational areas in California. A sampling is: Arizona Trails: 100 Hikes in Canyon and Sierra ($13.95; 320 pages); The Pacific Crest Trail, Volume I: California ($24.95; 480 pages); The Tahoe Sierra ($17.95; 320 pages); and Hiking the Big Sur Country ($14.95; 176 pages). It also distributes "Point Reyes National Seashore Map" ($5.95), a 24″ X 36″ "pictorial landform" map giving an aerial perspective of the Point Reyes Peninsula, and numerous topographic maps of California wilderness areas (see "Topographic Maps").

Wilderness Clubs
Following is a listing of outdoor groups that offer maps for hikers, bikers, campers, and nature lovers. There may be some restrictions on the availability of maps, especially to non-members of some membership organizations.

■ **Adirondack Mountain Club** (RR 3, Box 3055, Lake George, NY 12854; 518-793-7737) has maps of New York's Adirondack Mountains.
■ **Appalachian Trail Club** (5 Joy St., Boston, MA 02108; 617-523-0636) has maps of many parts of the Appalachian Trail.
■ **Appalachian Trail Conference** (P.O. Box 807, Harpers Ferry, WV 25425; 304-535-6331)

has maps of the Appalachian Trail's 14-state route.
■ **Buckeye Trail Association Inc.** (P.O. Box 254, Worthington, OH 43085) sells maps and trail guides of Ohio's 1,200-mile Buckeye Trail.
■ **Canadian Orienteering Federation** (1600 James Naismith Dr., Gloucester, Ontario K1B 5N4; 613-748-5649) sells orienteering maps for areas in Canada, published and distributed by individual clubs; a list of clubs is available.
■ **Colorado Mountain Club** (2530 W. Alameda St., Denver, CO 80219) has several trail maps of Colorado mountains, as well as hikers' guides to various wilderness areas in Colorado.
■ **Finger Lakes Trail Conference Inc.** (P.O. Box 18048, Rochester, NY 14618) sells maps of the Finger Lakes Trail.
■ **Florida Trail Association** (P.O. Box 13708, Gainesville, FL 32604) has trail and water recreation maps of the Florida Trail and the Florida National Scenic Trail.
■ **Green Mountain Club** (P.O. Box 889, 43 State St., Montpelier, VT 05601; 802-223-3463) offers maps of Vermont's 265-mile "Long Trail" as well as maps of day hiking trails throughout the state.
■ **Potomac Appalachian Trail Club** (1718 N St., NW, Washington, DC 20036; 202-638-5306) has maps of trails in Washington, D.C., Maryland, Pennsylvania, and Virginia.
■ **United States Orienteering Federation** (P.O. Box 1444, Forest Park, GA 30051) has orienteering maps, sold through individual clubs; a list of clubs and associate groups is available.

■ **See also: "Agriculture Maps," "Bicycle Route Maps," "Land-Use Maps," "Natural Resource Maps," "Nautical Charts and Maps," "River, Lake, and Waterway Maps," "Topographic Maps," "Tourism Maps and Guides," and "Wildlife Maps."**

Tourism Maps and Guides

The boom in travel maps has mirrored the boom in travel. It wasn't that long ago that choosing a touring map meant selecting between one or two potential products; often only one of them was readily available, making the "choice" somewhat easier. No more. Today there are dozens of maps available for some popular parts of the globe, many of which may be readily available at a well-stocked map or travel store.

Such competition has resulted in the improvement of tourism maps overall. Striving to distinguish themselves, the current crop of travel maps includes some impressive entries, featuring valuable travel information that goes beyond mere street names and landmarks. Color and detail on some maps are nothing short of spectacular. A few are printed in their country's native language, allowing you to more easily make sense of road signs along the way. There are laminated, pocket-size maps, pop-up maps, and maps that automatically fold themselves.

Equally impressive is the specialized nature of some of today's maps. The themes of today's travel maps range from the ridiculous to the sublime. There are a half-dozen or so maps of wine-making regions around the United States, as well as shopping maps, literary maps, even a jazz map. There is a map dedicated to the Loch Ness monster, a whiskey map of Scotland, and a map of Shakespeare's England. There's even an "Ernest Hemingway Adventure Map of the World."

The maps below are organized by continent. At the beginning of each section are listings of map publishers whose products cover the entire continent or several countries. Following that are descriptions of maps related to specific countries of each region.

AFRICA

Aaron Blake Publishers (1800 S. Robertson Blvd., Ste. 130, Los Angeles, CA 90035; 213-553-4535) publishes colorful, unique literary maps illustrating the homes and hangouts of famous or important authors and the locations of characters' exploits. These maps are great for planning a trip that includes literary sightseeing. The "Ernest Hemingway Adventure Map of the World," showing more than 200 locations from his novels and short stories, includes an inset map of Africa ($9.70 rolled; $5.75 folded).

American Map Corp. (46-35 54th Rd., Maspeth, NY 11378; 718-784-0055; 800-432-6277) distributes Hallwag maps, including:

■ "Country Road Maps" ($7.95 each), for Morocco/Canary Islands and Tunisia/Algeria.
■ "Regional Road Maps" ($7.95 each), for Kenya/Tanzania.
■ "Physical Reference Map" of Africa ($7.95), showing the physical contours of the terrain and physical boundaries.

Central Intelligence Agency. The CIA produces a surprising number of city- and tourist-related maps, including "Senegal/Gambia, Israel Transportation Systems" ($11.50; PB 82-928044) and the ever-useful "Standard Time Zones of the World" ($8.95; PB 88-928358). CIA maps can be ordered from the National Technical Information Service (see Appendix B).

Hammond Inc. (515 Valley St., Maplewood, NJ 07040; 201-763-6000; 800-526-4953) distributes Bartholomew's World Travel Map series ($9.50 each), featuring full-color reference maps for most countries of the world. The maps, in a variety of sizes and scales, are of frameable quality. Map titles include: "Africa," "Central and Southern Africa," "East Africa," "Northeast Africa," "Northwest Africa," "West Africa," "Egypt," "Kenya," and "Lebanon."

Hunter Publishing (300 Raritan Center Pkwy., Edison, NJ 08818; 201-225-1900) distributes Kümmerly & Frey "World Travel Maps" ($6.95 to $9.95), fully indexed maps that are updated annually. Maps are available for Africa and North and West Africa. Hunter also distributes "Hildebrand Travel Maps" ($6.96 to $7.95), containing travel information on climate, customs, currency, shopping, and more, in the margins, as well as detailed city maps. They are available for East Africa, Egypt, Morocco, South Africa, the Seychelles, Réunion, and Tunisia.

Michelin Travel Publications (P.O. Box 3305, Spartanburg, SC 29304; 803-599-3305; 800-423-0485) publishes "Main Road Maps" of Africa, including "Africa, North & West," "Africa North East-Arabia," "Central & South Madagascar," "Algérie-Tunisie," "Ivory Coast," and "Maroc."

Rand McNally (P.O. Box 7600, Chicago, IL 60680; 312-673-9100) publishes the Cosmopolitan map series ($1.95 to $2.95) for Africa. It also publishes the International map series, 52" X 34" maps of the United States and maps of the world ($2.95).

The Talman Co. Inc. (150 Fifth Ave., New York, NY 10011; 212-620-3182) distributes Ravenstein Verlag's "International Road Map" series of detailed, large-scale maps ($7.95 each) depicting topographic features, roads, towns, picturesque locales, cultural sights, and camping grounds, with legends in four to six languages. Available African titles are: "Morocco," "East Africa," "Africa," "South Africa," and "Egypt."

THE AMERICAS

Aaron Blake Publishers (1800 S. Robertson Blvd., Ste. 130, Los Angeles, CA 90035; 213-553-4535) publishes colorful, unique literary maps illustrating the homes and hangouts of famous or important authors, and the locations for characters' exploits. These maps are great for planning a day of literary sightseeing. American titles are: "Literary Map of Latin America," "Literary Map of New York," "Literary Map of the American South," "Literary Map of Los Angeles," "The Raymond Chandler Mystery Map of Los Angeles," "The John Steinbeck Map of America," and "The Beat Generation Map of America." The "Ernest Hemingway Adventure Map of the World" features inset maps of Cuba and Michigan, while the "Ian Fleming Thriller Map" identifies the locales where James Bond adventured, including many in North America. The maps measure approximately 20" X 27" and are available rolled ($9.70 each) or folded ($5.75 each).

Allmaps Canada (390 Steelcase Rd. E., Markham, Ontario, Canada L3R 1G2; 416-477-8480) distributes three road atlases: *North American Road Atlas* ($9.95; 128 pages); *Canada Road Atlas and Travel Guide/Atlas Routier et Guide de Voyage* ($5.95; 32 pages); and *Canada Pocket Road Atlas/Atlas Routier de Poche* ($3.95; 32 pages). Also available are "Travel Pacs," one- to five-pocket vinyl wallets containing the maps needed.

American Map Corp. (46-35 54th Rd., Maspeth, NY 11378; 718-784-0055; 800-432-6277) publishes annual road atlases that cover the U.S., Mexico, and Canada, including *United States Highway Atlas*, with enlarged insets for 37 major cities ($5.95) and *Premier Resorts Edition United States Road Atlas* ($5.95), locating points of interest such as "Regal Resorts," "Scenic Drives," "Hunting & Fishing Spots," and including a 48-page section of photos and text describing more than 180 vacation spots.

Hallwag maps of the Americas available

through American Map Corp. include:

■ "Country Road Map" ($7.95), for Mexico.
■ "Physical Reference Maps" ($7.95 each), showing the physical contours of the terrain and physical boundaries, available for Central America and South America.

Coastal Cruise Tour Guides (158 Thomas St., Ste. 11, Seattle, WA 98109; 206-448-4488) publishes "Cruise Tour Guides" ($9.95 each), designed for cruise-ship passengers. The guides feature fold-out, five-color shaded relief maps showing points of interest, resort locations, roads, airports, and other geographical features. The guides also contain illustrations, photos, and descriptive text and information on the culture, politics, religions, and arts of each area. Guides are available for the Eastern Caribbean, Alaska/Canada's Inside Passage, and Western Mexico.

Hammond Inc. (515 Valley St., Maplewood, NJ 07040; 201-763-6000; 800-526-4953) distributes Bartholomew's World Travel Map series ($9.50 each), featuring full-color reference maps for most countries of the world. The well-documented maps, in a variety of sizes and scales, are of frameable quality. Map titles include "North America," "South America," "Argentina/Chile/Paraguay/Uruguay," "Brazil/Bolivia," "Canada," "Mexico," "Peru/Colombia/Venezuela/Ecuador," "Western U.S.A.," and "Eastern U.S.A."

The maps in the Tourist Route Series ($9.50 each) have keys in English, German, and French, showing scenic areas, town plans, toll bridges, golf courses, beaches, and other tourist attractions. Titles include "North England," "South England," "Scotland," "Britain," and "Europe."

Portion of The Ernest Hemingway Adventure Map of the World, *one of a series of literary maps published by Aaron Blake Publishers.*

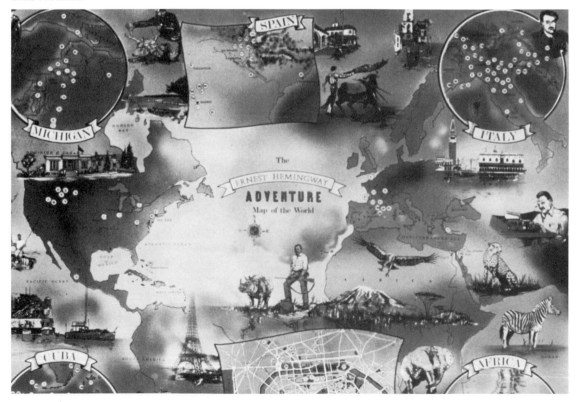

Hippocrene Books (171 Madison Ave., New York, NY 10016; 212-685-4371) is the U.S. distributor for several European publishers with foreign map series, including Hildebrand's maps of the Caribbean, Cuba, Haiti/Dominican Republic, Jamaica, and the U.S. ($5.95 each).

Hunter Publishing (300 Raritan Center Pkwy., Edison, NJ 08818; 201-225-1900) distributes Kümmerly & Frey "World Travel Maps" ($6.95 to $9.95), fully indexed maps that are updated annually. Maps are available for Canada and South America, as are world political and world physical maps. Hunter also distributes "Hildebrand Travel Maps" ($6.95 to $7.95), containing information on climate, customs, currency, shopping, and more, in the margins, as well as detailed city maps. Titles include "California," "The Caribbean," "Cuba," "Hispaniola," "Jamaica," "Mexico," "Puerto Rico," "The Southern Rockies," "Eastern U.S.," "Western U.S.," and "The U.S."

Lane Publishing Co. (80 Willow Rd., Menlo Park, CA 94025; 800-227-7346; 800-321-0372 in California) publishes Sunset Books atlases, including:

■ *Sunset Road Atlas* ($4.95), containing full-page maps of major metropolitan areas, color maps of the 50 states, and an interstate highway map. The easy-to-carry atlas measures 8 1/2" X 11 1/4".
■ *The Frequent Traveler's State & City Atlas* ($14.95), a spiral-bound atlas packed with full-color state maps; metro area and downtown city maps; an interstate highway map; and hotel, rental car agency, and airline listings.
■ *The Frequent Traveler's City Atlas* ($9.95), featuring 240 pages of detailed road and street maps of U.S. cities, along with an interstate highway map, and a listing of frequently used "800" telephone numbers.

Lane also distributes Sunset's Highway Services Directories:

■ "California Freeway Exit Guide" ($9.95),

featuring full-color maps along with safety tips from the California Highway Patrol.
■ "Highway Services Directory Exit-by-Exit (Southwest)" ($12.95), featuring full-color maps of Texas, Arizona, and New Mexico, and safety tips from state highway patrols.

MapArt/Peter Heiler Ltd. (72 Bloor St. E., Oshawa, Ontario, L1H 3M2; 416-668-6677) publishes a "Caribbean Treasure Map" ($5.95 Canadian), a 24" X 30" full-color map illustrating dates and possible locations of shipwrecks around the islands.

National Geographic Society (17th & M Sts. NW, Washington, DC 20036; 202-921-1200) produces a "Close-Up" series of plastic-coated maps showing places and events—festivals, historic sites, wilderness areas, resort areas, museums, etc.—including a mileage/kilometer guide for estimating driving time. Each Close-Up map comes with a 208-page index/guide. The maps make for good supplements to road maps. Close-Up maps ($4 each) are available for all regions in the U.S. and all Canadian provinces.

National Geographic offers various other maps of the Americas, such as:

■ "Visitor's Guide to the Aztec World" ($4; 02722), a 25" X 20" map of the Valley of Mexico and Mexico City, with a map of Mexico on the reverse side.
■ "Canada Political/Vacationlands" ($4; 20007), a 22 3/4" X 34" map printed on both sides.
■ "Heart of the Grand Canyon" ($4; 02803), a 35" X 36" topographic map depicting trails, roads, campgrounds, cliffs, and contours.
■ "West Indies/Tourist Islands" ($4; 02841), a 34" X 23" map printed on both sides.
■ "South America" ($4; 02014; 23" X 30 1/2").

Prentice Hall Press (200 Old Tappan Rd., Old Tappan, NJ 07675; 201-767-4970; 800-223-2348) sells Baedeker Road Maps ($6.95 each), the cream of the crop of international driving maps. Roadside attractions, motels, scenic

Overview map from San Francisco and Bay Area Illustrated Pocket Map, *published by* Travel Graphics International.

routes, and speed limits are all detailed. Titles include "Caribbean" and "Mexico."

Rand McNally (P.O. Box 7600, Chicago, IL 60680; 312-673-9100) publishes the Cosmo-politan map series ($1.95 to $2.95) for Canada, Mexico, South America, the West Indies/ Caribbean, and Alaska. It also publishes the International map series, 52" X 34" maps of the U.S. and the world ($2.95). Rand McNally also creates many travel and road atlases that are updated regularly and filled with maps and information to enhance any trip. Samplings include: *Road Atlas* ($6.95), a road and highway guide good for planning and taking vacations or business trips; *Road Atlas & Vacation Guide* ($14.95), an annual containing the full-color road and city maps of the *Road Atlas*, plus descriptions of 1,900 places of interest, ideal for family vacation planning; *Business Traveler's Road Atlas* ($14.95), an annual publication with full-color road and city maps, as well as major city information for business people; *Motor Carriers' Road Atlas* ($14.95), a road and highway atlas for truckers. Also available is "The Map Collection" ($22.50), a set of 39 Rand McNally

road maps covering all parts of North America.

Random House (201 E. 50th St., New York, NY 10022; 301-848-1900; 800-733-3000) publishes Fodor's "Flashmaps Instant Guides," a unique series of guides containing single-subject maps that are color-coded and cross-indexed to listings of addresses and phone numbers. The Boston Flashmap, for example, contains maps of hotels, restaurants, public transportation, museums and galleries, architecture, historic sites, colleges and universities, theaters, concert halls, nightspots, neighborhoods and suburbs, the Boston Marathon route, shopping, hospitals, and zip codes. Flashmaps are available for Boston, Chicago, Los Angeles, New York, Philadelphia, San Francisco, and Washington, D.C. ($5.95 each).

Reader's Digest Books (distributed by Random House, 201 E. 50th St., New York, NY 10022; 301-848-1900; 800-726-0600) offers *America from the Road* ($25.95), a book containing 125 tours designed to help road travelers see the most of America. 150 color-coded maps guide the way.

Schiffer Publishing Ltd. (Box E, Exton, PA 19341; 215-696-1001) distributes Laura McKenzie's "Travel Notes," 17″ X 22″ maps that fold to pocket-size. One side is a detailed, color-coded map with an index, while the other side includes useful information on traffic, weather, restaurants, shopping, the media, and some tips from McKenzie ($1.95 each). Travel Notes are available for a variety of North American cities and Caribbean islands, including Acapulco, Boston, Calgary, Las Vegas, New Orleans, Puerta Vallarta, Puerto Rico, and the Virgin Islands.

The Talman Co. Inc. (150 Fifth Ave., New York, NY 10011; 212-620-3182) distributes Ravenstein Verlag's "International Road Map" series of detailed, large-scale maps ($7.95 each) depicting topographic features, roads, towns, picturesque locales, cultural sights, and camping grounds, with legends in four to six languages. Maps of South America and North America are among the 55 titles available.

Travel Graphics International (1118 S. Cedar Lake Rd., Minneapolis, MN 55405; 612-377-1080) produces colorfully illustrated map/posters of popular vacation areas. The 17″ X 23″ Pocket Maps ($2.75 each), simplified maps showing tourist points of interest, are available for "The Islands of Hawaii," "Oahu," "Maui," "Acapulco," "Mexico City," "Cancun/Cozumel," "Puerto Vallarta," "San Francisco," "Las Vegas," and "Minnesota." Functional Pocket Maps (17″ X 23″), showing major roads and buildings in metropolitan areas, are available for Rochester, Minneapolis/St. Paul, St. Louis, San Francisco, and Indianapolis ($2.75 each). Colorful posters, although not useful as road or street guides, illustrate activities and things to see. Titles available are "The Islands of Hawaii," "Oahu," "Hawaii," "Maui," "Kauai," "Mexico," "The Bahamas," "Las Vegas," "Florida," and "Minnesota" ($6.50 each; 22″ X 28″ to 25″ X 38″). Prices are discounted for orders of more than one map or poster.

Portion of Wineries of the Napa Valley, *published by* Napa Valley Vintners Association.

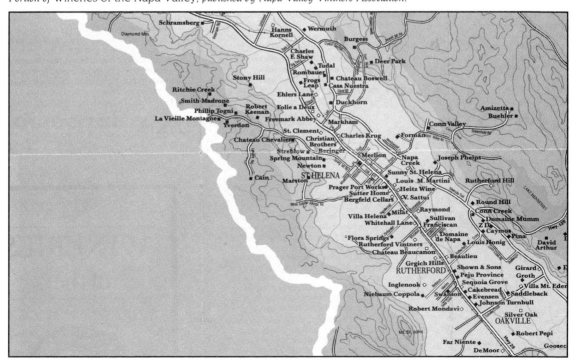

INDIVIDUAL COUNTRY MAPS

Canada

MapArt/Peter Heiler Ltd. (72 Bloor St. E., Oshawa, Ontario, L1H 3M2; 416-668-6677) publishes travel/road maps for Ontario, Toronto, Quebec, and the Atlantic Provinces and distributes Clearview Maps of Western Canada ($1.95 Canadian to $2.95 Canadian).

Signature Associates (314 Morey Ave., Bellingham, WA 98225; 206-671-4692) publishes tourist maps of British Columbia's Gulf Islands. Maps are available for Saltspring Island, North and South Pender Islands, Galiano Island, and Mayne and Saturna Islands. They can be purchased individually ($1.50) or as a kit ($2.70).

Mexico

American Map Corp. (46-35 54th Rd., Maspeth, NY 11378; 718-784-0055; 800-432-6277) distributes Guia Roji's maps of Mexico, including:

■ Detailed 20″ X 30″ "Road Maps" ($3.95 each), available for each of Mexico's 31 states.
■ *City Guides*, featuring special sections focusing on points of interest and tourist information, as well as up to 70 full-color map pages showing freeways, primary roads, and local streets. They are available for Mexico City, Monterrey ($7.95 each; perfect-bound) and San Diego/Tijuana ($22.95; spiral-bound).
■ Full-color "City Pocket Maps" ($3.95 each) showing local streets, state highways, and selected secondary routes are available for eight of Mexico's most visited cities.
■ "Mexico Highway Pocket Map" ($5.95) is a full-color detailed map of all major highways and through routes in the Republic of Mexico.
■ *Mexico Road Atlas* ($7.95), a 56-page bound atlas with full-color maps featuring freeways, interstates, and primary roads, along with listings of tourist and hotel information.

Marty Hiester Photo Design (4653 Winona Ave., San Diego, CA 92115; 619-283-6314; 619-583-9235) produces *Gringo's Guide to Baja*

California, Mexico, offering complete coverage of the peninsula on a topographic map, along with information on the culture, land, people, restaurants, hotels, and camping areas. The guide is available as a laminated wall mural ($17.95; 24″ X 36″) or in a "travel version," two 4 1/2″ X 11 1/2″ volumes ($8.95).

Warren Communications (P.O. Box 8635, San Diego, CA 92102; 619-531-0765) distributes topographic maps of Mexico published by the Mexican government at scales of 1:50,000 and 1:250,000., and "Tourist Maps/Cartas Turisticas," drawn at a scale of 1:1,000,000. Also available is *Atlas de Carreteras y Ciudades Turisticas* (*Road Atlas and Tourist Cities*), published in 1988 by Pemex, the Mexican government's petroleum company. Included are detailed road maps of the entire country, drawn at scales of 1:250,000 and 1:1,000,000, and a distance chart, emergency services information, and listings and descriptions of archaeological zones, colonial architecture, beaches, bathing resorts, natural landscape, arts and crafts, and events. The titles available vary; it's best to inquire. Individual copies of maps that Warren Communications imports can be ordered from **The Map Centre**, 2611 University Ave., San Diego, CA 92104; 619-291-3830.

United States

Basin Street Press (1627 S. Van Buren, Covington, LA 70433; 504-893-1130) produces a "Jazz Map of New Orleans" ($8.06, including postage) by Dr. Karl Koenig, which pinpoints the city's existing jazz clubs; the accompanying 30-page booklet outlines a walking tour of Storyville, the French Quarter, and St. Charles Street, tracing the rich history of jazz in New Orleans. Also available is *Just A Closer Walk* ($7.50), a guidebook to a walking jazz tour of the French Quarter, with street maps throughout.

Butterworth Co. of Cape Cod Inc. (476 Main St., Harwichport, MA 02646; 508-432-8200) publishes helpful street maps of the Cape Cod area. "Cape Cod Map & Guide" ($2.50) shows

Section from "Jazz Map of New Orleans." The circles indicate places of historical interest. Map courtesy of Dr. Karl Koenig, reproduced with permission of Basin Street Press.

highways and roads along with beaches, golf courses, historic sites, and other points of interest; "Hyannis Map & Guide" ($2.50) is a 10-page map showing highways and points of interest; *Cape Cod & Islands Atlas and Guide Book* ($12.95) includes detailed street maps for every town, as well as information on points of interest, Cape Cod living, outdoor recreation activities, and fishing maps of 80 Cape Cod lakes and ponds; and *Southeastern Massachusetts Atlas* ($12.95), contains maps of all communities in the southeastern part of the state. Butterworth also produces pocket street maps for Cape Cod, the Islands, and southeastern Massachusetts areas ($2.50 each).

Cypress Book (U.S.) Co. Inc. (Paramus Place, Ste. 225, 205 Robin Rd., Paramus, NJ 07652; 201-967-7820) distributes a "Map of the United States," published in the Republic of China, with text in both Chinese and Roman characters ($2.95).

DeLorme Mapping Co. (P.O. Box 298, Freeport, ME 04032; 207-865-4171) produces an "Atlas & Gazetteer Series" of full-color, large-scale, well-detailed and accurate atlases. Back roads, as well as major roads and highways, are shown, along with geographic features such as forests, mountains, lakes, ponds, rivers, and trails. Each 11" X 15" atlas includes a detailed place-name index. The gazetteer sections of the atlases list hundreds of places to go and things to do, from hiking, canoeing, and fishing spots to museums, historic sites, and golf courses. DeLorme Atlas & Gazetteers are available for Florida, Maine, Michigan, New Hampshire, New York State, Northern California, Ohio, Pennsylvania, Southern California, Vermont, Virginia, Washington, and Wisconsin. DeLorme also publishes the "Maine Map and Guide" ($1.50), a state highway map including a mileage chart and listings of travel and recreational services, as well as inset maps of major cities and towns.

Ferrytale Productions (P.O. Box 1004, Friday Harbor, WA 98250; 206-378-2648) produces a ferry guide and map of Washington state's San Juan Islands ($2.50), showing ferry routes and giving descriptions of each island. The map is available on the ferries, in local stores or by mail (add $1 postage).

Funmap Inc. (705 W. 6th Ave., Ste. 209, Anchorage, AK 99501; 907-272-6773) publishes colorful two-sided tourist maps of areas in Alaska. One side is a general overview map of the area; on the reverse is a pictorial street map of the metro region. The maps ($1.99 each, available through local retail outlets or directly from Funmap) have the added bonus of easily folding to fit in their pocketbook-size cover. Maps are available for Anchorage, Seward, Fairbanks, Juneau, Ketchikan, Sitka, Homer, Kodiak, Nome, and the entire state.

General Drafting Co. (P.O. Box 161, Convent Station, NJ 07961; 201-538-7600; 800-733-6277) produces the "Travel Vision" line of road maps and atlases, with maps and sets for the entire United States as well as for individual states and cities. Products include:

■ "Road Maps USA" ($21.95), a set of 25 road maps covering the entire United States.
■ "State Maps" ($1.50 each), available for all states.
■ "City Maps" ($1.50 each), available for many major cities in the United States.
■ *Road Atlas* ($3.50), covering the United States, Canada, and Mexico.

Gold Cartographics (P.O. Box 1813, Cambridge, MA 02238; 617-547-6538) publishes "The Illustrated Guide to New England's Bed & Breakfasts" series, road maps showing B&B options and providing descriptive information on each location. Special inset maps are included for Boston and surrounding areas. Volume I, *Northern New England* ($6.95) features a Boston-area inset map, while the award-winning Volume II, *Southern New England* ($7.95) includes useful color-coded symbols on the maps corresponding to B&B price ranges.

Government Printing Office. GPO distributes "United States Road Symbol Signs" ($2.25; S/N 050-000-00152-1), a useful guide to supplement highway maps and atlases. This folder illustrates different kinds of symbol signs, colors, shapes, and highway route markers for each state. GPO also has a small sampling of maps and guides for tourism in the United States, including:

■ "Welcome to Washington" (S/N 024-005-00823-3), maps of downtown Washington, D.C., and the metropolitan Washington area, with information on possible day trips, available only in quantities of 100 for $42.
■ "Devils Tower" ($3.50; S/N 024-005-00899-3), a history and travel guide that includes maps and tourist information for the famous volcanic rock tower in the Black Hills of Wyoming.

Graphic Concepts (1148 State Farm Dr., Santa Rosa, CA 95403; 707-545-5751) publishes a full-color "Wine Country Tour Map" ($2) of 138 wineries in the Napa, Sonoma, Mendocino, and Lake counties of California. The reverse side of the map lists addresses, phone numbers, hours, and activities available at each winery.

MapArt/Peter Heiler Ltd. (72 Bloor St. E., Oshawa, Ontario, L1H 3M2; 416-668-6677) distributes many Gousha and Rand McNally state and city maps of the U.S. in Canada. Call or write for an index.

Napa Valley Vintners Association (P.O. Box 141, St. Helena, CA 94574; 707-963-0148) distributes a "Wineries of the Napa Valley" map, available as a rolled poster ($7.35) or as a folded map/brochure (free).

Pelican Publishing Co. (1101 Monroe St., Gretna, LA 70053; 504-368-1175) publishes "Plantations on the Mississippi River: From Natchez to New Orleans," a full-color engraved map showing ownership of plantations along the Mississippi in 1858 ($25; 32" X 54").

Phillip A. Schneider, Cartographer (2109 Plymouth Dr., Champaign, IL 61821) produces "U.S. Virgin Islands," a full-color shaded relief tourist map of St. Thomas, St. John, and St. Croix ($4.95).

Phoenix Mapping Service (2626 W. Indian School Rd., Phoenix, AZ 85017; 602-279-2323) publishes the "Metro Phoenix Recreation & Shopping Map" ($1.95; 25″ X 18″), locating more than 225 recreation and shopping opportunities, including golf courses, parks, riding stables, popular shopping centers, and other points of interest. Phoenix Mapping is a division of Wide World of Maps, which stocks maps and travel guides for many tourist destinations worldwide.

Prentice Hall Press (200 Old Tappan Rd., Old Tappan, NJ 07675; 201-767-4970; 800-223-2348) distributes Gousha "Fastmaps," four-panel city maps that fold to a convenient 5 1/4″ X 11″. On the reverse is an enlarged inset of the downtown area and a street index. Fastmaps are laminated, so routes, directions, and notes can be written on and wiped off later. Fastmaps ($3.95 each) are available for Atlanta, Boston, Chicago, Orlando, Minneapolis/St. Paul, Seattle, New York, Philadelphia, San Francisco, and Washington, D.C.

Raven Maps & Images (24 N. Central, Medford, OR 97501; 503-773-1436; 800-237-0798) produces the "Vineyards & Wineries Map of California" ($35; $65 laminated), a 41 1/2″ X 52 1/2″ colorful map of California wine country, with a detailed inset focusing on the famous Napa-Sonoma region. The map illustrates all bonded wineries and production levels and includes an inset showing the state's 56 viticultural areas.

Square One Map Co. (P.O. Box 1312, Woodinville, WA 98072; 206-485-1511) publishes "Yakima Valley Winery Map and Guide," a two-color map with descriptive text and information on all wineries in this region of Washington state ($1.50).

Tennessee Valley Authority. TVA publishes several lake recreation maps, available for 75 cents each. Also available from TVA is an 1838 lithograph of an "Aboriginal Tourist Map of Tennessee" ($1; 453 G 77), which shows steamboat routes, towns, county outlines, and drainage. The print measures 15″ X 21″.

U.S. Geological Survey. Many USGS map products, especially topographic and general reference maps, include major highways and roads. These aren't necessarily the best to take on a driving trip, but they can help in planning one. The USGS National Atlas map "Highways" (1987), shows the complete interstate system, with route markers, and selected other major highways ($3.10; US05630-38077-AZ-NA-07M-00). See "Topographic Maps" for more information on other USGS maps depicting highways.

University of Hawaii Press (University of Hawaii at Manoa, 2840 Kolowalu St., Honolulu, HI 96822; 808-948-8255) publishes "Reference Maps of the Islands of Hawai'i" ($2.95 each). The topographic shaded-relief maps are available for "Hawai'i," "Kaua'i," "Maui," "Moloka'i/Lana'i," and "O'ahu."

World Impressions (1493 Beach Park Blvd., Ste. 312, Foster City, CA 94404; 415-571-8859) publishes colorfully illustrated sports and travel theme maps of the United States, available in paper or laminated, or as "Coloring ArtMaps," black-and-white line art with pens to color in the map. Among the many sport titles available are:

■ "The Official Major League Baseball ArtMap" (24″ X 36″) features all American and National league teams and stadiums, along with Hall of Famers and baseball trivia.
■ "Raceways" (24″ X 36″) illustrates 70 famous North American motorsport raceways and sanctioning bodies featuring official logos, telephone numbers, track descriptions, and historical trivia.
■ "Airshows of North America" (24″ X 36″) depicts the most popular airshow acts and

A Guide to the Guides

Travel guides are one of the fastest-growing parts of the book publishing business. Whereas there used to be only a couple of guidebooks from which to choose for a given area, there now can be as many as a dozen or more titles for some tourist hotspots. And just as the quality of these books' writing and information varies widely, so, too, do the maps they contain. Astonishingly, some books have no maps at all — or at least no detailed maps. The importance of detailed, translated maps was underscored by the results of a 1988 survey by Berlitz Guides U.S.A. Respondents to the survey called good maps of "dire importance" and "absolutely essential" when selecting a travel guide.

To help you wend your way through the guidebook jungle, here are five examples of guides that contain good maps:

Howard W. Sams & Co. (Box 7092, Indianapolis, IN 46206; 317-298-5400; 800-428-7267) distributes Berlitz travel guides. Among its

many guides, two are notable for their full-color road atlases—the "Blueprint" series and the "More for the Dollar" series. Blueprint Guides, available for France and Italy, contain 1:1,000,000-scale 30-page Hallwag road atlases, as well as 40 itineraries with route maps. The illustrated guides are packed with information on what to see and do, how to get around, restaurant and hotel recommendations, and a brief history of the country ($16.95 each). "More for the Dollar" hotel and restaurant guides to France and Italy ($19.95 each), also have a 30-page Hallwag road atlas, and hotel and restaurant coupons worth more than $5,000.

Langenscheidt Publishers Inc. (46-35 54th Rd., Maspeth, NY 11378; 718-784-0055; 800-432-6277) publishes the "Self-Guided" series of guidebooks emphasizing touring information, with detailed maps for driving and walking tours. Each 368-page book contains more than 100 maps, photos, and illustrations. "Self-

performers, along with illustrated action scenes, airshow trivia, and a listing of more than 200 airshows around North America.

Travel ArtMap titles available include:

■ "The Yosemite Adventure" (16″ X 24″) colorfully celebrates Yosemite's 100th anniversary, with a detailed trail and services map on the reverse side.
■ "The Oregon Adventure" (16″ X 24″) depicts and describes many of the wonders and points of interest of Oregon's coastline, rivers, mountains, towns, and special events.
■ "The Gibraltar Adventure" (16″ X 24″) illustrates more than 100 points of interest and various highlights of the Rock's 400,000 year history. Comes with a street map and index on the reverse.
■ "San Francisco" (24″ X 35″), a 63-color hand-painted ArtMap, illustrates the architec-

tural personality of more than 50 well-known San Francisco neighborhoods.

Yankee Books (Depot Sq., Peterborough, NH 03458; 603-563-8111; 800-872-9265) publishes *Yankee Magazine's Travel Maps of New England* ($5.95), containing 25 full-color pages of the enlarged and divided "National Survey Map of New England." The softcover book includes street maps of 11 major cities, mileage charts, population figures, and a city and town index.

ASIA

American Map Corp. (46-35 54th Rd., Maspeth, NY 11378; 718-784-0055; 800-432-6277) distributes Hallwag maps, including:

■ "Country Road Map" ($7.95) of Israel.

Guided" titles are available for "Alaska," "Canada," "Caribbean," "Egypt," "England," "France," and "Italy" ($12.95 each).

Prentice Hall Press (200 Old Tappan Rd., Old Tappan, NJ 07675; 201-767-4970; 800-223-2348) distributes American Express Pocket Guides, pocket-size guides featuring a section of full-color highway maps, general maps of major cities in the area, and general two-color "Orientation Maps," showing towns and major roads. The handy guides include travel advice, hotel and restaurant recommendations, and a calendar of interesting events. Titles available ($8.95 each) are "California," "England/Wales," "Florence/Tuscany," "Greece," "Hong Kong/Singapore/Bangkok," "London," "Mexico," "New York," "Paris," "Rome," "Spain," "Venice," and "Washington, D.C."

It also publishes Baedeker's Guides, well-known for their map coverage. Each guide features a large fold-out map of the entire area, street maps of cities, small area maps inset in the text, and layout illustrations of museums, galleries, and monuments. The guides are full of information on history, population, religions, transportation, culture, and commerce and industry, and include listings of hotels and restaurants. Among the 22 country and regional titles available ($16.95 each) are "Caribbean," "Egypt," "Germany," "Israel," "Japan," "Mediterranean," "Mexico," "Provence/Cote d'Azur," "The Rhine," "Scandinavia," and "Yugoslavia." There are also 23 titles available in the city-specific Baedeker Guides, including "Athens," "Bangkok," "Copenhagen," "Hong Kong," "Jerusalem," "Moscow," "New York," "Paris," "Tokyo," and "Vienna" ($10.95 each).

Prentice Hall also distributes Mobil Travel Guides, including full-color fold-out city maps, with walking tours shown on smaller maps, along with plenty of information on sightseeing, motels and hotels, transit systems, shopping and entertainment, and a listing of rated restaurants. Regional titles ($9.95 each) include "California & The West," "Great Lakes," "Northeast," "Northwest & Great Plains," "Southeast," "Southwest & South Central," "Major Cities." Five city guides are available: Boston, Chicago, New York, San Francisco, and Washington, D.C. ($5.95 each).

■ "Regional Road Map" ($7.95) of Turkey/Near East.
■ "City Maps" ($5.95 each), including three-dimensional sketches of prominent landmarks, for Istanbul and Tokyo.
■ "Physical Reference Maps" ($7.95 each), showing the physical contours and boundaries of the terrain, available for Southeast Asia/China and the Soviet Union.

Central Intelligence Agency. The CIA produces a surprising number of city and tourist-related maps. Selections include "Moscow, Downtown" ($8.95; PB 88-928355), "Moscow Street Guide" ($18.95; PB 88-928332), "USSR: Travel Restrictions on Foreigners" ($11.95; PB 88-928336), "Clothing Recommendations for Travel in China" ($10; PB 86-928343), "Shanghai Street Guide" ($15.50; PB 87-928303), "Beijing Street Guide" ($21.50; PB 86-928305), "Leningrad, Downtown" ($10; PB 86-928363),

and the ever-useful "Standard Time Zones of the World" ($8.95; PB 88-928358). CIA maps can be ordered through the National Technical Information Service (see Appendix B).

Hammond Inc. (515 Valley St., Maplewood, NJ 07040; 201-763-6000; 800-526-4953) distributes Bartholomew's World Travel Map series ($9.50 each), featuring full-color reference maps for most countries of the world. The maps come in a variety of sizes and scales. Titles include "Arabian Gulf," "Arabian Peninsula," "Southeast Asia," "China/Mongolia," "The Indian Subcontinent," "Israel/Jordan," "Japan," "Singapore/Malaysia," "Thailand," "Iran," and "Eurasia."

Hunter Publishing (300 Raritan Center Pkwy., Edison, NJ 08818; 201-225-1900) distributes Kümmerly & Frey "World Travel Maps" ($6.95 to $9.95), fully indexed maps that are updated annually. Maps are available for Asia, China, the Middle East, and Turkey. Hunter also distributes "Hildebrand Travel Maps" ($6.96 to $7.95) containing travel information on climate, customs, currency, shopping, and more, in the margins, as well as detailed city maps. Titles available are "China," "Hong Kong," "India," "Israel," "Korea," "Philippines," "Soviet Union," "Sri Lanka," "Taiwan," "Thailand/Burma/ Malaysia," and "Turkey."

National Geographic Society (17th & M Sts. NW, Washington, DC 20036; 202-921-1200) offers several maps of Asian countries, including:

■ "Asia" ($4; 02812; 37 1/2" X 31 1/2"), in full color.
■ "Middle East" ($4; 02296; 37" X 23").
■ "Soviet Union" ($4; 02396; 37" X 23").
■ "Japan and Korea" ($4; 20022; 23" X 30").
■ "Mount Everest/Himalaya" ($4; 20033; 23 1/2" X 36").
■ "South Asia" ($4; 02294; 28 1/4" X 22 3/4").

The Talman Co. Inc. (150 Fifth Ave., New York, NY 10011; 212-620-3182) distributes Ravenstein Verlag's "International Road Map" series of detailed, large-scale maps ($7.95 each) depicting topographic features, roads, towns, picturesque locales, cultural sights, and camping grounds, with legends in four to six languages. "South & East Asia," "Western Soviet Union," "Mideast," "Israel," "Turkey," "Eastern Mediterranean," and "Near East" are among the 55 titles available.

INDIVIDUAL COUNTRY MAPS
China
Cypress Book (U.S.) Co. Inc. (Paramus Place, Ste. 225, 205 Robin Rd., Paramus, NJ 07652; 201-967-7820) distributes a "Map of the People's Republic of China" ($5.95), published in China, with text in both Chinese and Roman characters.

Interarts Ltd. (15 Mt. Auburn St., Cambridge,

MA 02138; 617-354-4655) distributes a "China Map" ($18.95 folded, $25 flat) drawn at a scale of 1:4,000,000 and measuring 46" X 62".

Japan
Prentice Hall Press (200 Old Tappan Rd., Old Tappan, NJ 07675; 201-767-4970; 800-223-2348) sells a Baedeker Road Map of Japan ($6.95), depicting roadside attractions, motels, scenic routes, and speed limits, among other useful information.

Teikoku-Shoin Co. Ltd. (29, Jimbocho, 3-chome, Kanda, Chiyoda-ku, Tokyo 101, Japan) publishes maps and atlases of Japan, such as the *Complete Atlas of Japan* and the "Map of Japan." Its maps are available in the U.S. through **Map Link**, 529 State St., Santa Barbara, CA 93101; 805-965-4402.

EUROPE

Aaron Blake Publishers (1800 S. Robertson Blvd., Ste. 130, Los Angeles, CA 90035; 213-553-4535) publishes colorful literary maps, illustrating the homes and hangouts of famous or important authors, and the locations for various books. These maps are useful for planning a day of literary sightseeing. European titles available are: "Literary Map of Paris," "The Sherlock Holmes Mystery Map," and "The Jane Austen Map of England." The "Ernest Hemingway Adventure Map of the World" features inset maps of Paris, Italy, and Spain, while the "Ian Fleming Thriller Map" identifies the locales where James Bond encountered adventure, including many in Europe. The maps measure approximately 20" X 27" and are available rolled ($9.70 each) or folded ($5.75 each).

American Map Corp. (46-35 54th Rd., Maspeth, NY 11378; 718-784-0055; 800-432-6277) publishes the wire-bound "Travel Atlas" series, with each atlas featuring full color, detailed, large-scale driving maps of Europe. Atlas titles include "Great Britain/Ireland," "Germany,"

"Italy," "France/Benelux," "Switzerland/Austria," "Spain/Portugal" ($8.95 each), and "Europe" ($12.95).

American Map distributes Hallwag European maps, including:

■ "Regional Road Maps" ($7.95 each), available for all European countries.
■ "Country Road Maps" ($7.95 each), available for all European countries.
■ "City Maps" ($5.95 each) of most major European cities, which include three-dimensional sketches of prominent landmarks.
■ "Special Holiday Maps" ($7.95 each), featuring enlarged inset maps of cities, ruins, and other points of interest, historical background and commentary, plus a quick-reference index for finding hotels and restaurants. Available for Corfu, Costa Brava, Cyprus, Peloponnesus, Balearics, Malta, and Rhodes.
■ "Physical Reference Map" of Europe ($7.95), showing the physical contours of the terrain and physical boundaries.
■ *Europa Road Atlas* ($19.95), a 238-page atlas showing the complete road system of Europe, including distances, car ferries, and campgrounds, along with 64 city and town maps.
■ *Euro-Guide* ($29.95), an all-in-one guide to European travel. This 800-page book offers travel information and includes the complete road network of Europe, city maps, hotel and restaurant listings, and other information.

Europe Map Service/OTD Ltd. (1 Pinewood Rd., RD 7, Hopewell Junction, NY 12533; 914-221-0208) distributes a European Topographic map series in various scales, including 1:25,000 and 1:1,000,000. The series consists of travel and recreation maps, hydrographic maps, water-resources maps, and administrative maps, for Austria, Belgium, Denmark, Finland, France, Germany (East and West), Ireland, Luxembourg, the Netherlands, Norway, Sweden, and Switzerland. Also available are reproductions of the official administrative World War II maps showing borders of the Third Reich from 1937 to May 1945.

Hammond Inc. (515 Valley St., Maplewood, NJ 07040; 201-763-6000; 800-526-4953) distributes Bartholomew maps, including:

■ "World Travel Map" series ($9.50 each), featuring full-color reference maps for most countries of the world. The maps, in a variety of sizes and scales, are of frameable quality. Map titles include "British Isles," "Central Europe," "Eastern Europe," "Western Europe," "France," "Germany," "Greece," "Italy," "Scandinavia," and "Spain/Portugal."
■ "Tourist Route Maps" ($9.50 each), showing roadways, service areas, golf courses, beaches, scenic spots, and other points of interest. Titles available include "England North," "England South," "Scotland," "Britain," "Ireland," and "Europe."
■ "Easy-Fold World City Plans" ($8.95 each), laminated maps with comprehensive indexes, for Florence, Venice, and Athens.
■ "Holiday Route Planner" series ($12.95 each), showing all types of transportation routes and including tourist information. Titles include "Belgium/Luxembourg," "Holland," "Channel Ports to the Hull," and "Europe."
■ "Handy Map" ($8.50), a color-coded map of Europe.

Hippocrene Books (171 Madison Ave., New York, NY 10016; 212-685-4371) distributes various foreign-produced maps of Europe, including Hildebrand maps ($5.95 each). Titles available are: "Balearics," "Central Europe," "Corsica," "Cyprus," "Grand Canary," "Gulf of Naples," "Majorca," "Peloponnese," "Spanish Coast," and "Teneriffa."

Hunter Publishing (300 Raritan Center Pkwy., Edison, NJ 08818; 201-225-1900) distributes Kümmerly & Frey "World Travel Maps" ($6.95 to $9.95), fully indexed maps which are updated annually. Maps are available for most countries and regions, including "Austria," "Finland," "Greece," "Hungary," "Rumania/Bulgaria," and "Yugoslavia." Also, Kümmerly & Frey's *Euro-Atlas*, a 174-page atlas of road maps from Britain to Moscow, is available for $15.95.

Hunter also distributes "Hildebrand Travel Maps" ($6.96 to $7.95), containing travel information on climate, customs, currency, shopping, and more, in the margins, as well as detailed city maps. Among the 19 European titles available are "Algarve," "Balearic Islands," "Corsica," "Crete," "Gran Canaria," "Majorca," "Peloponnese," "Sardinia," "Tenerife," "Yugoslav Coast North," and "Yugoslav Coast South."

Interarts Ltd. (15 Mt. Auburn St., Cambridge, MA 02138; 617-354-4655) distributes vibrant Esselte maps of Scandinavia, including a "Sweden Map" ($19.95), measuring 24" X 46" and drawn at a scale of 1:1,500,000, and "Scandinavia Folded Maps" ($8.95 each), which include "Stockholm Map and Guide," "Scandinavia," and "Sweden Tourist Map."

Michelin Travel Publications (P.O. Box 3305, Spartanburg, SC 29304; 803-599-3305; 800-423-0485) publishes "Main Road Maps" of Europe ($5.95 each), including "Europe" (1:3,000,000-scale), "Greece" (1:700,000), "Scandinavia/Finland" (1:1,500,000-scale), and "Yugoslavia" (1:1,000,000-scale). Also available is *Road Atlas of Europe* ($19.95).

National Geographic Society (17th & M Sts. NW, Washington, DC 20036; 202-921-1200) offers maps of European countries, including:

- ■ "British Isles" ($4; 02022; 23" X 30").
- ■ "Europe" ($4; 02020; 30" X 40").
- ■ "Germany" ($3; 02814; 23" X 30").
- ■ "Greece and the Aegean" ($4; 20023; 30" X 23").
- ■ "Spain and Portugal" ($4; 02284; 30" X 23").
- ■ "Traveler's Map of the Alps" ($4; 20003; 36" X 22 3/4"), a decorative map with descriptive text.
- ■ "Traveler's Map of Britain and Ireland" ($4; 20018; 23" X 30").
- ■ "Traveler's Map of Spain and Portugal" ($4; 02028; 36" X 27 1/8").

Also available is "Shakespeare's Britain" ($3; 02535), a 19" X 25" sheet based on a map drawn three centuries ago and illustrating 45 sites from the Bard's plays.

Passport Books/NTC Publishing Group (4255 W. Touhy Ave., Lincolnwood, IL 60646; 312-679-5500; 800-323-4900) publishes travel maps and guides, including:

- ■ *Passport's European Atlas for Travelers* ($24.95), containing comprehensive maps of European roads, along with codes for road conditions, points of interest, and detailed city maps. Each map is in the language of the country to correspond with the language of road signs.
- ■ "Passport Maps" ($7.95), showing all highways and secondary road systems, with scenic routes highlighted. Available for Europe, Scandinavia, Italy, Britain, France, Germany, Benelux, Spain/Portugal, and Switzerland/Austria/The Alps.

Prentice Hall Press (200 Old Tappan Rd., Old Tappan, NJ 07675; 201-767-4970; 800-223-2348) sells Baedeker Road Maps ($6.95 each), the cream of the crop of international driving maps. Roadside attractions, motels, scenic routes, and speed limits are all detailed. Titles include "Alps," "Austria," "Belgium," "Denmark," "Europe," "France," "Germany," "Great Britain," "Italy," "Spain/Portugal," "Switzerland," and "Yugoslavia."

Rand McNally (P.O. Box 7600, Chicago, IL 60680; 312-673-9100) publishes the Cosmopolitan map series ($1.95 to $2.95) for Europe. It also publishes the International map series, 52" X 34" maps of the U.S. and the world ($2.95).

Schiffer Publishing Ltd. (Box E, Exton, PA 19341; 215-696-1001) distributes Laura McKenzie's "Travel Notes," 17" X 22" maps that fold to pocket-size. One side is a detailed, color-coded map with an index, while the other side includes useful information on traffic, weather, restaurants, shopping, the media, and some tips by McKenzie ($1.95 each). Travel Notes are available for London and Rome.

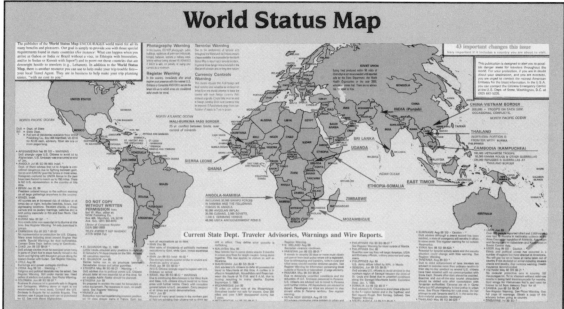

Going somewhere? Want to be prepared for just about anything?

WSM Publishing Co. (Box 466, Merrifield, VA 22116; 301-564-8473) publishes a "World Status Map," a bimonthly-produced map loaded with useful advisories for international travelers. WSM gathers information from 150 sources, including the U.S. Department of State, the World Health Organization, the United Nations, news services, the Centers for Disease Control in Atlanta, and the governments of each country. The map highlights danger areas for travelers, and includes travel warnings and advisories, passport, visa, and vaccination requirements, and a riot report. It also provides valuable information on the special, and sometimes surprising, requirements and penalties found in many countries, such as what happens to travelers arriving in Gabon, Brazil, or India without a visa; in Sudan and Kuwait with liquor; or in Ethiopia with binoculars. A single copy of the map is $6, while a one-year subscription is $36.

WSM Publishing also offers an "International Travel Briefing Service," a computer-disk-based data base that is updated monthly, containing specific travel advisories and information for every country in the world. The disks may be used in any IBM-compatible computer. Contact WSM for further information.

The Talman Co. Inc. (150 Fifth Ave., New York, NY 10011; 212-620-3182) distributes Ravenstein Verlag's "International Road Map" series of detailed, large-scale maps ($7.95 each) depicting topographic features, roads, towns, picturesque locales, cultural sights, and camping grounds, with legends in four to six languages. Maps for Norway, Czechoslovakia, Greece, Ireland, and the Benelux countries are among the 55 titles available. The Talman Co. also distributes Ravenstein Verlag's "European Pocket Map" series of full-color detailed maps of 22 European countries. The topographic maps show cities and towns, railroad routes, roads, scenic routes, camping grounds, tourist sights, and other points of interest. Titles include "Europe," "Austria," "Italy," "Yugoslavia," "Hungary," "France," "Poland," and "Rumania/Bulgaria" ($3.95 each).

VLE Limited (Box 547, Tenafly, NJ 07670; 212-580-8030; 201-567-5536) publishes "Walks

Through," a series of European city maps designed especially for "map illiterates." Visitors start reading at the bottom of the page and walk straight ahead, following a dotted line to the top of the page, turning where indicated. Photos or drawings of landmarks are positioned along the dotted line, as they appear along the route, and short descriptions of each site are also included. "Walks Through" maps ($3 each) are available for Paris, Rome, Vienna, Zurich, Munich, Stuttgart, Madrid, Barcelona, Seville, and London. VLE Limited also produces EUR*oad* ($5.95), a 64-page guide to motoring in Europe.

INDIVIDUAL COUNTRY MAPS

France

Michelin Travel Publications (P.O. Box 3305, Spartanburg, SC 29304; 803-599-3305; 800-423-0485) publishes "Detailed Maps" a series of over 30 road maps drawn at a 1:200,000 scale, covering all regions in France ($2.95 each); larger "Regional Maps," road maps drawn on a 1:200,000 scale for all French regions ($4.95 each); and various maps of Paris, including *Paris Atlas* ($8.95), arranged by *arrondissements* (neighborhoods), and "Paris Transports" ($2.95), showing bus, metro, and rail routes, as well as major roads. Michelin also publishes the "Road of History" series, including "Valley of the Kings" ($5.95) and "Treasure Houses of the Sun King" ($5.95).

Germany

Central Intelligence Agency. The CIA produces a surprising number of city and tourist-related maps, including "West Berlin, Transportation Systems" ($11.50; PB 82-928044), and the ever-useful "Standard Time Zones of the World" ($8.95; PB 88-928358). CIA maps can be ordered through the National Technical Information Service (see Appendix B).

Europe Map Service/OTD Ltd. (1 Pinewood Rd., RD 7, Hopewell Junction, NY 12053) distributes a *Gazetteer of the Federal Republic of Germany*, a detailed publication packed with

information on geographical names and features, official map series and their numbers, and various administrative offices ($88; 794 pages). Also available is *Gazetteer of the German Democratic Republic* ($43; 352 pages).

United Kingdom

Hammond Inc. (515 Valley St., Maplewood, NJ 07040; 201-763-6000; 800-526-4953) distributes Bartholomew maps, including:

■ "Leisure Maps" ($9.95 each), large-scale maps with layer contouring, an index of leisure activities available in each area, and marked points of interest, such as historic houses, ancient monuments, nature reserves, museums, and sailing and fishing areas. Maps are available for most areas in the U.K.
■ "Handy Maps" ($8.50 each), color-coded maps for Britain/Ireland, London, Ireland, North England, Scotland, South England, and The Royal Mile.

England/Wales
■ "Tourist Map of Britain" ($8.95), showing every tourist information center and describing each one's services.
■ "Cathedrals and Abbeys" ($9.50).
■ "London Big Ben" ($3.95), a pocket-size map showing tube stations, bus routes, theaters, cinemas, shops, hotels, and restaurants.

Scotland
■ "Scotland Touring Map" ($5.95), a driving map with an index to places of interest, such as castles, historical houses, golf courses, and picnic spots.
■ "Scottish Kitchen Map" ($9.50), with regional recipes.
■ "Tartan Map" ($9.50).
■ "Whiskey Map" ($9.50).
■ "Loch Ness and the Monster Map" ($9.50).
■ "Scotland Motorways and Main Roads" ($7.95).
■ "Historical Clan Map" ($9.50).
■ "The Clans of Scotland" ($8.95), a wall map describing the events and development of Scotland's culture and its clans.

■ City plans and street guides for major metropolitan areas in Scotland.

Ireland
■ "Dublin Central Street Plan" ($3.95).
■ "Historical Map of Ireland" ($9.50).
■ "Irish Family Names Map" ($9.50).
■ "Irish Kitchen Map" ($9.50), with regional recipes.
■ "Ireland Touring Road Map" ($9.50).

Hunter Publishing (300 Raritan Center Pkwy., Edison, NJ 08818; 201-225-1900) distributes numerous tourist maps, including:

■ *London* A-Z ($8.95; 132 pages), a street atlas of London with full-color, highly detailed maps. Underground stations, railway lines and stations, hospitals, fire and police stations, and post offices are also shown.
■ "A-Z Maps of Britain," full-color regional road maps and tourist area maps, are available for many areas of England, Wales, and Scotland ($5.95 to $7.95).

International Specialized Book Services Inc. (5602 NE Hassalo St., Portland, OR 97213; 503-287-3093; 800-547-7734) distributes the British government's Ordnance Survey maps ($8.95 each), including "Landranger Maps," popular "all-purpose" maps for driving, walking, and cycling tours in England; "Motorway Maps," showing major roadways in and around cities, with a gazetteer of place names—available are "M25 & London/Manchester," "Sheffield," "Leeds," and "York"; "Great Britain Routeplanner Map," covering the entire British Isles, with inset maps of major towns and park areas; "RouteMaster Maps," nine maps covering Britain, showing ground relief, road distances, and tourist information; and "Town & City: London Central," including transportation systems, car parks, cinemas, sports grounds, and other points of interest.

Random House (201 E. 50th St., New York, NY 10022; 301-848-1900; 800-733-3000) publishes Fodor's "Flashmaps Instant Guides," a series of

guides containing single-subject maps that are color-coded and cross-indexed to listings of addresses and phone numbers. The "Flashmap Instant Guide to London" ($5.95) includes maps on "Arts and Antiques," "Blue Plaque Houses," "Colleges and Universities," "Day Outings by Rail," "Hyde Park," "National Trust Houses," "Royal Residences," "Literary England," "Bond Street," "Cambridge," "Cathedrals and Churches," "Cinemas," "Museums," "Shopping," "Theatres," "The Underground," "Architectural Highlights," "Ferry Information," "Embassies," "Historical Sites," "Mansions and Houses," "Hotels," and "The British Isles."

Travel Graphics International (1118 S. Cedar Lake Rd., Minneapolis, MN 55405; 612-377-1080) produces colorfully illustrated maps and posters of popular vacation areas. The 17″ X 23″ "Pocket Maps of London" ($2.75) is a simplified set of maps showing tourist points of interest. Colorful posters ($6.50 each), although not useful as road or street guides, illustrate activities and things to see; they are available for London and Britain (including England, Scotland, Wales and Ireland). Prices are discounted for orders of more than one map or poster.

TripBuilder (30 Park Ave., Ste. 5-0, New York, NY 10016; 212-603-0042; 800-255-4433) produces an England TripBuilder ($29.95), a 100-page guide custom-designed according to the buyer's interest. Pick 10 categories of interest from a list of 23 (including "Country Gardens," "Historic Castles," "Royalty Watching," "Georgian England," and "Imbiber's England"), and you'll receive a loose-leaf book with a base map of England and a transparent overlay for each category with corresponding information sheets on sites of interest.

OCEANIA

Hammond Inc. (515 Valley St., Maplewood, NJ 07040; 201-763-6000; 800-526-4953) distributes Bartholomew's World Travel Map series featuring full-color reference maps for most countries of the world. The maps, in a variety of sizes and scales, are of frameable quality. Maps available are: "Australia" and "The Pacific," including inset maps of the islands ($9.50 each).

Hunter Publishing (300 Raritan Center Pkwy., Edison, NJ 08818; 201-225-1900) distributes "Hildebrand Travel Maps" ($6.96 to $7.95), with travel information on climate, customs, currency, and shopping, in the margins, as well as inset detailed city maps. A Hildebrand Travel Map is available for Western Indonesia.

National Geographic Society (17th & M Sts. NW, Washington, DC 20036; 202-921-1200) offers two maps of Australia:

■ "Australia" ($4; 20002; 23 1/2" X 30").
■ "Traveler's Map of Australia" ($4; 20028), a 20 1/2" X 27 1/8" map printed on both sides.

Schiffer Publishing Ltd. (Box E, Exton, PA 19341; 215-696-1001) distributes Laura McKenzie's "Travel Notes," 17" X 22" sheets that fold to pocket size. One side is a detailed, color-coded map with an index, while the other side includes useful information on traffic, weather, restaurants, shopping, the media, and some tips from McKenzie ($1.95

each). Travel Notes are currently available for Hawaii, Kauai, Maui, Oahu, and Sydney.

The Talman Co. Inc. (150 Fifth Ave., New York, NY 10011; 212-620-3182) distributes Ravenstein Verlag's "International Road Map" series of detailed, large-scale maps ($7.95 each) depicting topographic features, roads, towns, picturesque locales, cultural sights, and camping grounds, with legends in four to six languages. Maps for New Zealand and Australia are available.

INDIVIDUAL COUNTRY MAPS
Australia
Australia In Print Inc. (110 W. Ocean Blvd., Ste. 537, Long Beach, CA 90802; 213-432-2223) distributes "Gregory Maps" of Australia, including State Road Maps ($4.95 each) for New South Wales, Victoria, Queensland, South Australia, Northern Territory, Western Australia, and Tasmania; City Pocket Maps ($2.95 each) for Sydney and Melbourne; and City Maps ($4.50 each) for Sydney, Brisbane, and Adelaide.

■ **See also: "Business Maps," "County Maps," "Foreign Country Maps," "Recreation Maps," "Selected Atlases," "United States Maps," and "Urban Maps and City Plans."**

MAPS OF SPECIFIC AREAS

County Maps

William Faulkner's Yoknapatawpha County may have been fictional, but it embodied the regional distinctiveness and closely knit culture many American counties possess. County maps are both useful and fascinating reflections of these national dividing lines that are smaller than states but bigger than cities.

The boundaries of U.S. counties usually are subdivided into townships, cities, or villages. There are some states, especially those in the West, that do things a bit differently, however: New Mexico's counties are divided into election precincts and two Indian reservations, for example, and Wyoming's counties are divided into election districts, although a large portion of the state is given over to Yellowstone Park, which has no county affiliation. Louisiana has no counties at all; it is divided into parishes.

Some counties have personalities of their own. New York's Westchester County, for example, is known as one of the richest in the nation. California's ultra-hip Marin County has been parodied in the movie "The Serial." Harlan County, Kentucky, was the title and subject of a well-known documentary film about coal miners. As a microcosm of American life, the county often represents a portrait of its region or the resources around which it has grown.

County maps both illustrate and mirror the variety of county life and land in America. There are basic county maps showing the subdivisions, highways, parks, and industrial areas. County census maps show a myriad of demographic characteristics. County congressional district maps provide a political reference for an area. County recreational maps show rivers, lakes, parks, and forests open for public use.

The federal government creates a number of topographic, agricultural, geologic, and recreational maps of counties. State and county governments also create county maps, many of which are distributed free to the public. County and regional historical societies and map collections are good places to find older county maps for research purposes. And there are a number of commercial map companies specializing in the production of county and regional maps.

GOVERNMENT SOURCES

Census Bureau. The Census Bureau creates maps of counties from statistics it gathers during its surveys. For more information on these maps, see "Census Maps."

Tennessee Valley Authority. TVA produces inexpensive blueline and lithographic prints of both contemporary and historical maps of the region, several of which detail counties and county outlines. Include 50 cents for postage when ordering these maps. Titles include:

■ "Railroad and County Maps of Tennessee" (50 cents; 453 D 754-7), a blueline print of a map published by E. Meoendenhall in 1864.
■ "County Map of Virginia and North Carolina" ($1.25), a lithographic reproduction from a mid-1800s atlas that includes some counties in West Virginia.
■ "County Map of Kentucky and Tennessee" ($1.50), a reproduction from the same atlas as the above map.
■ "Map of Tennessee" ($1; 453 K 315), a blueline print of an 1824 map showing county outlines and county seats.

U.S. Geological Survey. USGS has an ongoing program to produce county-formatted topographic maps at scales of 1:50,000 or 1:100,000. Published maps vary in size and are available in a limited number of states at this time ($4).

Portion of a USGS base map of the United States showing county names and boundaries.

COMMERCIAL SOURCES

The "City Maps" chapter lists companies specializing in local and regional maps. Most of the companies listed also produce county maps. Others include:

ADC (6440 General Green Way, Alexandria, VA 22312; 703-750-0510) publishes county street maps for most counties in Virginia and Maryland, as well as some counties in North Carolina, Georgia, Pennsylvania, and Delaware. Prices are around $8.95.

Alfred B. Patton Inc. (Swamp Rd. & Center St., Doylestown, PA 18901; 215-345-0700) publishes county road maps and atlases for

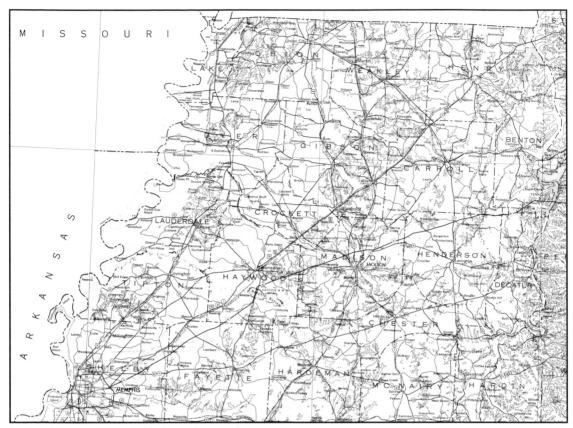

Portion of Tennessee Valley Authority county map of Tennessee.

Pennsylvania and New Jersey counties. Prices vary; a catalog is available upon request.

Champion Map Corp. (200 Fentress Blvd., Daytona Beach, FL 32014; 800-874-7010; 800-342-1072 in Fla.) produces "City/County Wall Maps" ($121.50) detailed maps showing county boundaries as well as roads and highways, subdivisions, schools, parks, cemeteries, etc. Maps are available for most counties in most states. Champion also produces custom maps.

Clarkson Map Co. (1225 Delanglade St., P.O. Box 218, Kaukauna, WI 54130; 414-766-3000) sells the "Wisconsin County Map Book" ($10.95), featuring county maps that show roads, wildlife information, public hunting-ground locations, and lake size, depth, and fish types. It also distributes individual maps for Wisconsin counties.

Compass Maps Inc. (P.O. Box 4369, Modesto, CA 95352; 209-529-5017) produces street and road maps for nearly all counties in California.

Dolph Map Co. Inc. (430 N. Federal Hwy., Ft. Lauderdale, FL 33301; 305-763-4732) publishes county maps for most parts of Florida. Dolph also publishes county street atlases for Broward County, Dade County, and Palm Beach County.

Geographia Map Co. (231 Hackensack Plank Rd., Weehawken, NJ 07087; 201-867-4700; 212-695-6585) distributes county road maps of New York, New Jersey, and Pennsylvania.

Hammond Inc. (515 Valley St., Maplewood, NJ 07040; 201-763-6000) publishes county outline maps for business use. Each 8 1/2″ X 11″ map depicts a single state with county outlines, cities, towns, state capitals, and county seats ($4 per pkg. of 10).

Marshall-Penn-York Co. Inc. (538 Erie Blvd. W., Syracuse, NY 13204; 315-422-2162) publishes county maps for areas in Connecticut, New York, and Pennsylvania.

Map Works (2125 Buffalo Rd., Ste. 112, Rochester, NY 14624; 716-426-3880; 800-822-6277 in New York) produces county highway maps that include parks, railroads, and towns. Maps ($2.25 to $2.95) are available for Licking County, Ohio, and the following New York counties: Monroe, Niagara, and Orange.

Phoenix Mapping Service (2626 W. Indian School Rd., Phoenix, AZ 85017; 602-279-2323) produces maps for counties in Arizona, including Maricopa County, Pinal County, and Eastern Pima County.

Pittmon Map Co. (732 SE Hawthorne Blvd., Portland, OR 97214; 503-232-1161; 800-547-3576; 800-452-3228 in Oregon) publishes "Pittmon Book Maps" of counties in the Pacific Northwest ($15 to $19.95), as well as "Pittmon Recreational County Maps" for all counties in Oregon and Washington ($2.99 each).

Rockford Map Publishers (P.O. Box 6126,

Rockford, IL 61125; 815-399-4614) has been producing county plat books since 1944, with more than 600 County Land Atlases and Plat Books currently in publication. The Land Atlas and Plat Books measure 8 1/2" X 11" and are wire-bound, with black-and-white maps of cities and population centers, as well as a black, red, and white county highway map in the center of the volumes. Rockford has County Land and Plat Books available for counties in Alabama, Florida, Idaho, Illinois, Indiana, Iowa, Minnesota, Mississippi, Missouri, New York, Oklahoma, Pennsylvania, South Dakota, West Virginia, and Wisconsin. Prices range from $18 to $30.

Square One Map Co. (P.O. Box 1312, Woodinville, WA 98072; 206-485-1511) produces the following two-color maps: "Southwest Snohomish County" ($1.95) and "Western Whatcom County" ($1.50).

Thomas Bros. Maps (17731 Cowan, Irvine, CA 92714; 714-863-1984; 800-432-8430 in Calif.) publishes "Thomas Guides" for California, Washington, and Oregon counties.

■ **See also: "Business Maps," "Census Maps," "Congressional District Maps," "Topographic Maps," and "Urban Maps and City Plans."**

Foreign Country Maps

Not all maps of other lands were designed to get you from here to there. Some maps of foreign countries are intended for use by students, scientists—or spies. While most foreign country maps can be found under "Tourism Maps," here is a selection of both general and specialized maps of our neighbors around the globe. For more specialized maps of specific countries—geologic or energy maps, for example—you should contact the appropriate government agencies of those countries directly (see Appendix C). Maps also may be available from countries' tourism offices or from their embassies or consulates.

GOVERNMENT SOURCES

Central Intelligence Agency. The CIA, one of the federal government's more prolific mapmakers, has more than 100 maps of foreign countries available to the public. Some are not very detailed, so it's best to get a description of each map before ordering. Examples of CIA foreign country maps, available through the National Technical Information Service, are:

■ "Africa" ($8.95; PB 88-928359).
■ "German Democratic Republic ($11.95; PB 88-928337).
■ "Terrain Map of Iceland" ($10; PB 87-928321).
■ "Mainland Southeast Asia" ($8.95; PB 88-928327).
■ ""Terrain Map of Lesotho" ($10; PB 88-928311).
■ "United Arab Emirates" ($11; PB 88-928317).
■ "Uruguay" ($11; PB 88-928319).

Defense Mapping Agency. DMA produces topographic, hydrographic, and related maps and charts of foreign countries. DMA map series of foreign countries include the following: (Stock numbers for individual sheets in

multisheet series are listed in DMA's "Public Sales Catalog"; see Appendix B).

■ "Area Outline Maps" ($5 per sheet; Series 1105), a series of black-and-white planimetric maps, drawn at a scale of 1:20,000,000, that delineate international and major subdivision boundaries, national capital cities of major importance, and water drainage patterns. The entire world is covered in this series of 27 sheets, each of which is about 14″ X 11″.
■ "Europe" ($5 per sheet; Series 1209), based on a British topographic map of Europe created by the British Mapping and Charting Establishment Royal Engineers. This series of six sheets shows international boundaries, major civil subdivisions and administrative boundaries, city and town populations, road classifications (by importance and weatherability), operable railways and airports, and other key topographic features. The sheets are drawn at a scale of 1:2,000,000 and are designed to fit together to form a 3 1/2′ X 5′ wall map. The average size of each sheet is 43″ X 62″.
■ "Middle East Briefing Map" ($5; Series 1308; 1308XMEBRMAP), measuring 34″ X 38″ and drawn at a scale of 1:1,500,000. This multicolored physical map shows armistice demarcation lines and international boundaries, populations of significant towns, roads, railways, airfields, and oil pipelines.
■ "Africa" ($4 per sheet; Series 2201), a multicolored topographic map series comprised of 36 sheets drawn at a scale of 1:2,000,000. Shown are international and major administrative boundaries and prominent topographic features, including cities, towns, transportation routes, and vegetation. The average size of each sheet is 29″ X 26″.
■ "Administrative Areas of the USSR" ($5; Series 5103; 5103XADMAUSSR) is a multicolored planimetric, political administrative map showing boundaries, towns, transportation and

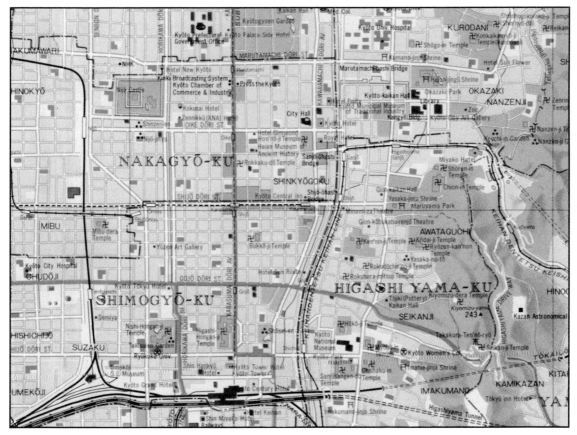

Portion of map of Kyoto, from Teikoku–Shoin Co. Ltd.'s Complete Atlas of Japan.

hydrographic features, and pipelines. A glossary and administrative list with abbreviations are also included.

A description of DMA's hydrographic products can be found under "Nautical Charts."

Government Printing Office. The State Department's Bureau of Public Affairs produces several atlases of foreign countries, available from GPO, including:

■ *Atlas of the North Atlantic Treaty Organization* ($1.75; S/N 044-000-02039-4) contains basic information about NATO, as well as 19 maps displaying NATO's membership and structure, military strength, members' roles in world affairs, and relations with the USSR and Warsaw Pact countries.

■ *Atlas of the Caribbean Basin* ($1.50; S/N 044-000-02022-0) includes maps and charts illustrating the Caribbean Basin's economic and political features.
■ *Atlas of U.S. Foreign Relations* ($5; S/N 044-000-02102-1), a 1985 atlas containing basic information about U.S. foreign relations, divided into six sections dealing with various aspects of foreign relations. There are 90 maps and charts.

COMMERCIAL SOURCES

Many publishers of tourist and road maps of foreign countries can be found in "Tourism Maps." A sampling includes:

Cypress Book (U.S.) Co. Inc. (Paramus Place, Ste. 225, 205 Robin Rd., Paramus, NJ 07652;

201-967-7820) distributes maps published in the Republic of China, with text in both Chinese and Roman characters. Titles available are "Map of the People's Republic of China" ($5.95), a poster-sized 59″ X 45″ map featuring both English and Chinese characters, and a smaller "Map of the United States" ($2.95).

Europe Map Service/OTD Ltd. (1 Pinewood Rd., RD 7, Hopewell Junction, NY 12533; 914-221-0208) distributes a European topographic map series in various scales, including 1:25,000 and 1:1,000,000. The series consists of travel and recreation maps, hydrographic maps, water resources maps, and administrative maps for Austria, Belgium, Denmark, Finland, France, Germany (East and West), Ireland, Luxembourg, the Netherlands, Norway, Sweden, and Switzerland. Also available are reproductions of the official administrative World War II maps showing the borders of the Third Reich from 1937 to May 1945.

Gabelli U.S. (11500 W. Olympic Blvd., Ste. 475, Los Angeles, CA 90064; 213-312-4546) distributes the brightly colored French-made

"Editions Géographiques et Touristiques" Gabelli maps. Their vivid "Montiscolor Picture of the Earth" maps, showing the depths of the oceans and geographic relief of the land, are available for each continent. The maps are printed on either paper or flexible plastic, with sizes and prices varying from $4.95 to $15.75. Titles include "Europe," "Africa," "Asia," "Oceania," "North America," and "South America."

Geoscience Resources (2990 Anthony Rd., P.O. Box 2096, Burlington, NC 27216; 919-227-8300; 800-742-2677) sells a wide variety of foreign-country maps, from topographic to geologic maps to tourist maps. A catalog is available.

Hippocrene Books (171 Madison Ave., New York, NY 10016; 212-685-4371) is the U.S. distributor for several European publishers with foreign map series, including Hildebrand Maps ($5.95 each), for more than 20 countries and regions such as Cuba, Haiti/Dominican Republic, Indonesia, Israel, Malta, Sri Lanka, Tunisia, and the U.S.; and Falk Country Maps

CIA map of Tanzania showing agricultural, manufacturing, and mining regions.

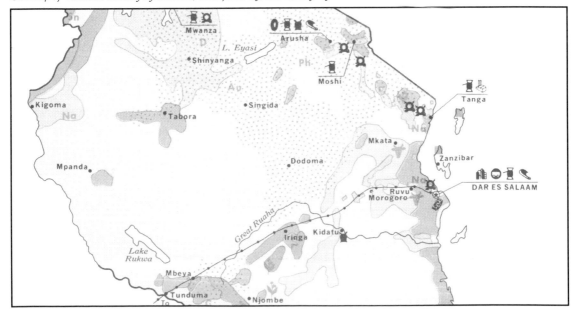

($8.95 each) for most western European countries. Hippocrene also sells maps for other countries and cities (see "Tourism Maps").

Hunter Publishing (300 Raritan Center Pkwy., Edison, NJ 08818; 201-225-1900) distributes Kümmerly & Frey "World Travel Maps" ($6.95 to $9.95), fully indexed maps that are updated annually. Maps are available for most countries and many regions, including "Austria," "Canada," "Finland," "Greece," "Hungary," "Rumania/ Bulgaria," "South America," and "Yugoslavia," as well as world political and world physical maps. Also available are Recta/Foldex International Maps for "Belgium/Luxembourg," "France," and "Sardinia" ($6.95 each).

Interarts Ltd. (15 Mt. Auburn St., Cambridge, MA 02138; 617-354-4655) distributes a "China Map" ($18.95 folded, $25 flat) drawn at a scale of 1:4,000,000 and measuring 46" X 62". They also distribute Esselte maps of Scandinavia, including a "Sweden Map" ($19 95), measuring 24" X 46" and drawn at a scale of 1:1,500,000, and "Scandinavia Folded Maps" ($8.95 each), which include "Stockholm Map and Guide," "Scandinavia," and "Sweden Tourist Maps."

International Map Services Ltd. (P.O. Box 2187, Grand Cayman, Cayman Islands, British West Indies; 809-94-94700) produces maps of the Caribbean and Central America. Prices vary.

National Geographic Society (17th & M Sts. NW, Washington, DC 20036; 202-921-1200) offers several full-color maps of foreign countries, including:

- "Australia" ($4; 20002; 23 1/2" X 30").
- "Soviet Union" ($4; 02396; 37" X 23").
- "Greece and the Aegean" ($4; 20023; 30" X 23").

- "Spain and Portugal" ($3; 02284; 30" X 23").

Rand McNally (P.O. Box 7600, Chicago, IL 60680; 312-673-9100) publishes the "Cosmopolitan" map series ($1.95 to $2.95) for Canada, Mexico, Europe, South America, Africa, the West Indies/Caribbean, and Alaska. It also publishes the "International" map series, 52" X 34" maps of the U.S. and the world ($2.95).

Teikoku-Shoin Co. Ltd. (29, Jimbocho, 3-chome, Kanda, Chiyoda-ku, Tokyo 101, Japan) publishes maps and atlases of Japan, such as the *Complete Atlas of Japan* and the "Map of Japan." Its maps are available in the U.S. through **Map Link**, 529 State St., Santa Barbara, CA 93101; 805-965-4402.

Warren Communications (P.O. Box 8635, San Diego, CA 92102; 619-531-0765) distributes topographic maps of Mexico published by the Mexican government at scales of 1:50,000, 1:250,000, and 1:1,000,000, as well as bathymetric maps drawn on a scale of 1:1,000,000. Coverage includes most of the country; individual copies of maps Warren Communications imports can be ordered from **The Map Centre**, 2611 University Ave., San Diego, CA 92104; 619-291-3830.

World Eagle Inc. (64 Washburn Ave., Wellesley, MA 02181; 617-235-1415) sells "Would You Believe?", a 24" X 31 1/2" outline map of Africa with the superimposed boundaries of Argentina, China, Europe, India, New Zealand, and the U.S., shown fitting within the continental borders of Africa.

See also: "Political Maps," "Tourism Maps and Guides," and "World Maps."

Native American Land Maps

A legacy of government mapping was born in 1804, when Thomas Jefferson sent Meriwether Lewis and William Clark in search of the mysteries of what later became the western part of the United States. And so began the federally recorded history of the absorption of Indian territory into the body of the fledgling country. Within a short time, the mapping of Indian lands became routine.

The Lewis and Clark maps showed general locations of Indian tribes, bands, and villages, as well as the number of tents, lodges, and "souls." Later maps reflected the growing conflict between Indians and whites: They stated the number of warriors in a tribe, for example, or the favorite haunts of raiding parties. Although the War Department was charged with duties that included those "relative to Indian affairs," Uncle Sam learned quickly that burying the hatchet was a better tactic in winning land from Indians. So, in 1824, the Bureau of Indian Affairs was established within the War Department to handle the nonmilitary aspects of Indian affairs. The Bureau (transferred to the Department of the Interior in 1849), with its various divisions covering everything from Indian forestry to population statistics, created the need for maps of Indian land and life.

Still more maps were produced to redefine boundaries each time a treaty was drawn up between the government and an Indian tribe or nation. Westward expansion and development led to the creation of specialized maps of Indian lands, showing railroad rights-of-way across Indian territories and hydroelectric facilities near reservations, among other things.

Sometimes the process of mapping itself spurred the erosion of Indian land ownership. The General Allotment Act of 1887 charged the government with surveying Indian reservations to establish land values for agricultural purposes. If agricultural worth was significant, the land was allotted to qualifying Indians, the catch being that any surplus allotted lands were usually bought up by the government. Legislation in the 1930s prohibited future sales of Indian land, but by that time the Indians had lost some 100 million acres of treatied land to the government's allotment program.

Aside from depicting the steady loss of territory over the years, Indian land maps can be helpful in studying irrigation, crop rotation, and other land use, as well as the social structure and community planning of various tribes. Although the major wave of Indian-land-mapping abated at the beginning of the 20th century, the federal government includes reservations in many of its overall mapping projects (census, topography, and utilities, for example), and several thousand earlier maps have been carefully preserved in map libraries. Indian maps are also produced by a number of map companies as American history teaching tools and by organizations involved in the study and understanding of Native American cultures.

GOVERNMENT SOURCES, UNITED STATES

Bureau of the Census. The decennial census of 1860 was the first to treat the Indian population as a separate race, but it wasn't until 1890 that the census counted the number of Indians living in their own territories or on reservations. (The Bureau of Indian Affairs Statistics Division, abolished in 1947, surveyed the Indian population long before the Census Bureau got around to the task.) Some BIA population maps are included in the holdings of the National Archives, as are early census maps (see "Census Maps").

Tennessee Valley Authority. Not only does the TVA produce power from its battery of transmission stations, but it is also one of Uncle Sam's more prolific map-makers. The TVA has two Indian-related maps available:

■ "A Draft of the Cherokee Country" (50 cents plus 50 cents postage; 11" X 17"). This print of Henry Timberlake's 1762 map shows the Little Tennessee River, Indian governors, Indian villages, and forts.

■ "Cherokee 'Nation' of Indians 1884" ($1.50 plus 50 cents postage; 12" X 15").

U.S. Geological Survey. USGS, part of the Interior Department, produces maps that include BIA information. USGS topographic and other maps depicting boundaries illustrate Indian lands according to BIA categorization: tribal lands (reservations owned by an entire tribe that are subject to tribal laws and are intact); Indian lands allotted in part (reservations, including both tribal lands and tracts of land allotted to individual Indians); and allotted and open (reservations with individually owned land tracts, other tracts to the public, and any remaining tribal lands). Former Indian reservations generally are not noted.

As part of the National Atlas Program, the USGS also produces a general map of Indian lands, titled "Early Indian Tribes, Culture Areas, and Linguistic Stock" ($3.10). This multicolored map, created in 1967, includes explanatory text on the back.

Library of Congress and **National Archives.** The map collections of both the Library of Congress and the National Archives are full of old Indian land maps. The maps in the Library's Geography and Mapping Division date back to the colonial explorers who set out to diagram the wilds of the New World for their sovereign nations. Most of these maps are general exploration studies that note Indian villages and territories along with other interesting features of the land.

More comprehensive and specific is the enormous collection of Indian maps in the Cartographic and Architectural Branch of the National Archives, composed of several collections from various federal mapping agencies. Maps of Indian lands and life made by the BIA, the Census Bureau, the Corps of Topographic Engineers, and the War

Portion of the "Guide to Indian Country," showing Indian reservations in southern New Mexico. The map is available from the Southwest Parks and Monuments Association. ©1989 *by the Automobile Club of Southern California. Reproduced by permission.*

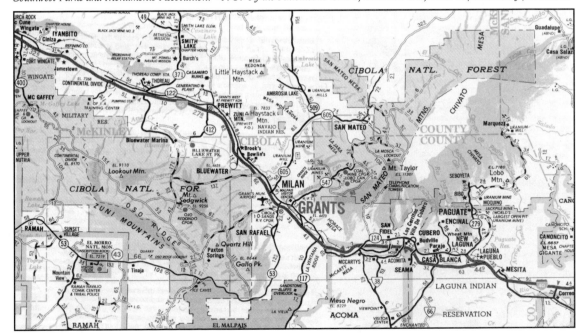

Department's Office of Explorations and Surveys, to name a few, are included in the Archives holdings. Some maps are as recent as 1950, although most date from 1781 to 1883. Included in the collections are such maps as:

■ "Sectional Map From the Coast of Maryland, Virginia, and North Carolina From Cape Henlopen." This 1781 map was drawn by John Purcell under the instructions of Lt. Col. Thomas Brown, Superintendent of Indian Affairs in the Southern District. The map is in six sections and measures 73 1/2" X 75".
■ "A Map of Lewis and Clark's Tract Across the Western Portion of North America, From the Mississippi to the Pacific Ocean, by Order of the Executive of the United States in 1804, 5 & 6." This map, in manuscript on tracing paper, was copied by Samuel Lewis from the original drawing of William Clark. The map includes the positions of numerous Indian tribes, villages, and bands, as well as some population information.
■ "Map Illustrative of the Route of H.R. Schoolcraft Ind|ian| A|gent| between L. Superior & Mississippi R. in the Summer of 1831. By D. Houghton, Surgeon 7 Nat|uralist| to the Exp|edition|." This map illustrates parts of the Chippewa, Sioux, Menomincc, and Winnebago Territories, including names of Indian chiefs and population numbers for Indian villages along the route and on area lakes and rivers.
■ "Map of the Indian Reservations Within the Limits of the United States." This 1883 map, drawn by Paul Brodie, shows Indian reservations by color and includes population figures.

Reproductions of maps from both the National Archives and Library of Congress may be obtained through the reproductions services of each collection. For further information, see "Uncle Sam's Treasure Chest."

GOVERNMENT SOURCES, CANADA
Canada Map Office (615 Booth St., Ottawa, Ontario K1A 0E9; 613-952-7000) distributes Canadian Indian land maps as part of the

National Atlas of Canada, 5th Edition, published in 1980 ($5.50 Canadian each). Maps available are "Indian and Inuit Communities and Languages" (MCR 4001), "Indian and Inuit Population Distribution" (MCR 4031), and "Native Peoples of Canada, 1630" (MCR 4054). Indian land maps from the *National Atlas of Canada, 4th Edition,* published in 1973, are also available, ($2 Canadian each), including: "Indian Lands and Languages" (MCR 1170) and "Indian and Eskimo Population" (MCR 1171). All maps from the *National Atlas* are available with either English or French text.

Maps Alberta (Land Information Services Div., Main Fl., Brittania Bldg., 703 6th Ave. SW, Calgary, Alberta T2P 0T9; 403-297-7389) also distributes the above maps from the *National Atlas of Canada, 5th Edition,* as well as "Indian and Inuit Communities, Prairie Provinces" (MCR 4028).

COMMERCIAL SOURCES
George F. Cram Co. (P.O. Box 426, Indianapolis, IN 46206; 317-635-5564) publishes an American History Series, which includes "Indians During Early Explorations and Settlements" ($63 to $72, depending on mounting), a color map depicting the location and physical features of American Indian tribes at the time of the white man's arrival and settlement.

Modern School (425 E. Jackson St., Goshen, IN 46525; 219-533-3111) produces "The Basic Series" of history maps. One example is "Indian Tribes and Cultures" ($244 to $336 for set of 30 maps, depending on mounting), which illustrates the locations, migrations, and languages of Indian tribes at the time of Columbus's arrival in Northern America.

Nystrom (3333 Elston Ave., Chicago, IL 60618; 312-463-1144; 800-621-8086) produces a set of colorful American history maps as part of its social studies series. Two relate to Indians: "Indian Tribes and Settlements in the New World 1500-1700," illustrating the locations of natural resources and Indian cultures, as well as Cortez's conquest of Mexico and Pizarro's conquest of Peru; and "The Roots of American Culture: Westward to the Mississippi" (same prices and size as above), depicting the old Southwest and Northwest and showing sites of major Indian battles. Each is $43, 50″ X 38″, folded and eyeletted for hanging; $66 on spring roller.

Rand McNally (P.O. Box 7600, Chicago, IL 60680; 312-673-9100) produces two series of American history maps for schools that include Indian maps: American History Maps for Intermediate Grades, "Homelands of the American Indians—North America" ($61; 50″ X 50″), a brightly colored markable and tear-resistant map depicting the homelands of the major Indian nations before the arrival of Columbus; and Our America Series, "Early Indians and Their Culture" ($59; 44″ X 38″), a full-color map showing seven major and 18 other linguistic areas, as well as pictorials of food sources, products, habitats, and culture.

Other producers of Indian maps and charts include:

Automobile Club of Southern California (P.O. Box 2890, Terminal Annex, Los Angeles, CA 90051) produces "Guide to Indian Country" ($3.95 plus $1 postage), an annually revised map featuring the "Four Corners" region—Arizona, New Mexico, Colorado, and Utah—and highlighting the region's Indian reservations, national parks and monuments. It also contains detailed information on points of interest in the region, local Indian tribes and events, recreational opportunities, wilderness trips, and campgrounds. The guide is available at most national parks and monuments in the area, or by mail from **Southwest Parks and Monuments Association**, 157 Cedar St., Globe, AZ 85502; 602-622-1999.

Celestial Arts (P.O. Box 6326, Berkeley, CA 94707; 415-524-1801; 800-841-2665) sells a wall chart, "The Aztec Cosmos" ($9.95 plus $2 shipping and $.65 tax for Calif. residents), illustrating the cultural and artistic accomplishments of the Aztec civilization and religion. A 32-page explanatory booklet is included.

Facts On File Inc. (460 Park Ave. S., New York, NY 10016; 212-683-2244; 800-322-8755; 800-443-8323 in Canada) publishes *Atlas of the North American Indian* ($16.95 paperback; $29.95 hardcover), a 288-page atlas by Carl Waldman that covers the history, culture, and location of native Americans from ancient times to the present, with more than 120 maps.

National Geographic Society (17th & M Sts. NW, Washington, DC 20036; 202-921-1200) produces several Indian maps, including: "Indians of North America" ($3; 02816; 32 1/2″ X 37 1/2″), a full-color ethnological map created in 1982; "Indians/Archaeology of South America" ($3; 02846; 37″ X 23″), a two-sided map including illustrations, chronology, notes, and text; "North America Before Columbus" ($3; 02817; 32 1/2″ X 37 1/2″), an archaeological map that includes an inset chart highlighting the "prehistory" of North America; and "Visitor's Guide to the Aztec World" ($4; 02722; 25″ X 20″), illustrating the Valley of Mexico on one side and Mexico City on the reverse.

Phoenix Mapping Service (2626 W. Indian School Rd., Phoenix, AZ 85017; 602-279-2323) publishes an Arizona topographic map showing, among other things, Indian reservations in the state ($7.95 paper, $31.95 laminated; 48″ X 54″).

■ **See also: "Boundary Maps," "Census Maps," "Land-Ownership Maps," and "Topographic Maps."**

State and Provincial Maps

Like national and county maps, state and provincial maps encompass much of the cartographic spectrum. In addition to state road maps and tourism maps, there are land plat maps, geologic maps, topographic maps, recreation maps, boundary maps, and many other charts, surveys, and maps depicting resources or features. All are described in their respective chapters elsewhere in this book.

The wide variety of available state and provincial maps is reflected in the number of producing companies and agencies. Among the many kinds of state maps that Uncle Sam creates are the **U.S. Geological Survey's** topographic, geologic, and natural resource investigations maps; the **National Oceanic and Atmospheric Administration's** hydrologic maps; aerial photomaps available from the USGS ESIC offices and **EROS Data Center**; and the **Census Bureau's** demographic maps.

For purposes of getting from here to there, the best source of free, up-to-date road maps are the state tourism offices (see Appendix A). Most are straightforward road maps, although a few are spiced with interesting graphics or themes. (Indiana, for example, distributes an illustrated "Adventure Map," with state activities and attractions geared toward entertaining and educating children.) But that's just the beginning of maps published by state governments. Most states produce and distribute one or more of the following: geologic maps, soil maps, natural resource maps, recreation maps, and maps related to travel, land use, and industry within their borders. Local commercial publishers produce state atlases, road maps, travel guides, recreational and business maps, and basic wall maps for use in schools and businesses.

Older state maps can be found in the collections of the **National Archives** and the **Library of Congress**. A 1794 map of Maryland, created by Dennis Griffith, and an 1851 map, "A New Constructed and Improved Map of the State of California," by J.B. Tassin, can be found in the Archives. Another source for older state maps are the holdings of state, county, and local historical societies or agencies.

Here are selected sources of state maps:

GOVERNMENT SOURCES

The federal government produces many types of state maps, all covered more fully in other chapters of this book:

■ The **U.S. Geological Survey** produces a series of state topographic maps, covering an entire state or specific areas; scales vary by coverage. Other USGS products that encompass state maps include the National Atlas Program, geologic and energy investigations maps, seismicity maps, and Bureau of Land Management land plat maps for public lands (see "Emergency Information Maps," "Energy Maps," "Geologic Maps," "Land-Ownership Maps," "Natural Resource Maps," "Selected Atlases," and "Topographic Maps").

■ The **Census Bureau's** *Congressional District Atlas*, regularly updated, contains state-by-state maps of congressional districts. The Census Bureau also publishes a wide range of other maps (see "Business Maps," "Census Maps," and "Congressional District Maps").

■ The **National Oceanic and Atmospheric Administration** creates numerous state maps related to lakes, rivers, and waterways, as well as aeronautical charts that cover certain states (see "Aeronautical Charts," "River, Lake, and Waterway Maps," and "Recreation Maps").

■ The **Defense Department** produces an annual *Atlas/State Data Abstract for the United States* ($6), available from the Government Printing Office, containing state-by-state maps and information on military bases, personnel, and defense contracts (see "Military Maps").

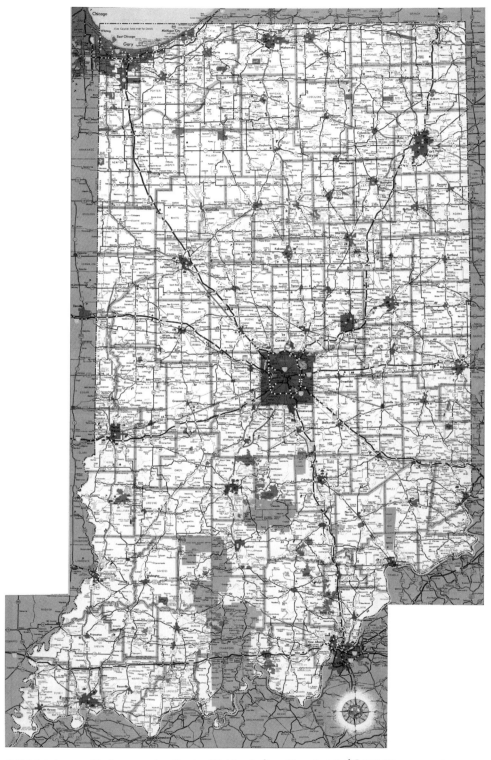

Official state map of Indiana, courtesy Tourism Division, Indiana Department of Commerce.

■ The **Tennessee Valley Authority** has topographic, geologic, and utilities maps for states in its region (see "Energy Maps," "Geologic Maps," "Natural Resource Maps," "Topographic Maps," and "Utilities Maps").

■ The **Federal Emergency Management Agency's** Flood Insurance Administration publishes state maps of flood plains (see "Emergency Information Maps").

COMMERCIAL SOURCES

Virtually every major map company produces some kind of state or provincial map. Any map store and most bookstores carry state maps, at least local ones. Other sources include most of the publishers listed under "Urban Maps and City Plans," as well as those listed in Appendix D. Here are some examples:

American Automobile Association (8111 Gatehouse Rd., Falls Church, VA 22047; 703-222-6000) has free state travel maps available to AAA members only. The maps may not be ordered through the mail, but must be obtained at local AAA offices.

Facts On File Inc. (460 Park Ave. S., New York, NY 10016; 212-683-2244; 800-322-8755; 800-443-8323 in Canada) produces "State Maps on File" ($345, $465 Canadian), a seven-volume compendium of reproducible state maps. There are an average of 20 different maps for each state, including political, historical, environmental, cultural, economic, and natural resources maps.

MapArt/Peter Heiler Ltd. (72 Bloor St. E., Oshawa, Ontario, Canada L1H 3M2; 416-668-6677) distributes maps of many Canadian provinces.

The Maps Place (700 E. Benson Blvd., Country Village Mall, Anchorage, AK 99503; 907-274-6277) sells a map of Alaska superimposed over the U.S., illustrating the true size of the state. Interesting facts and details are included on the map ($9.50).

Rand McNally (P.O. Box 7600, Chicago, IL 60680; 312-673-9100) produces a series of state road maps ($1.50 each), as well as a set of mural wall maps of 28 states ($59.95 each).

Raven Maps & Images (24 N. Central, Medford, OR 97501; 503-773-1436; 800-237-0798) produces shaded-relief maps of western states. The maps ($20 paper; $45 laminated) show towns, roads, and railroads, as well as geographical features. Maps are available for Alaska, Arizona, California, Colorado, Hawaii, Idaho, Montana, Nevada, New Mexico, Oregon, Utah, Washington, and Wyoming.

University of Hawaii Press (Univ. of Hawaii at Manoa, 2840 Kolowalu St., Honolulu, HI 96822; 808-948-8255) publishes "Reference Maps of the Islands of Hawai'i" ($2.95 each).

University of New Mexico Press (Albuquerque, NM 87131; 505-277-7564) publishes *New Mexico In Maps*, a 409-page book featuring maps and accompanying essays, graphs, tables, and charts on 131 topics such as "Groundwater Pollution," "Mining and Stagecoaching, 1846-1912," "Physician Distribution," "Ranching and Rangeland," "Alternative Communities," and "Presidential Elections." This impressive book is a good source of otherwise hard-to-find facts on this state ($24.95; 2nd Edition published 1986).

■ **See also: "Tourism Maps and Guides."**

United States Maps

"It was wonderful to find America, but it would have been more wonderful to miss it," wrote Mark Twain in *Pudd'nhead Wilson's Calendar*. For many, finding America—in all its many cartographic forms—is nearly a national pastime. From tourists to teachers, biologists to bus drivers, nearly everyone uses maps of the United States in their quest to find America. And there's a U.S. map for nearly everyone, too.

When Thomas Jefferson authorized purchase of the Louisiana Territory from Napoleon in 1803, the size of America more than doubled overnight. Little was known about the new land, and even less was known about the trails and passages that were purported to exist throughout this vast wilderness. There was believed to be a single, small mountain range running across the center of the territory, through which passage to the Pacific Coast could be easily maneuvered. But when the Lewis and Clark expedition, ordered by Jefferson to find this "Pacific Passage," discovered instead the nearly impenetrable Rocky Mountains, the mysteries of the new territory became evident.

The Lewis and Clark team brought back maps that illustrated everything from the number of "souls" in Indian villages to the placement of tributaries along the West's major rivers. It would take many subsequent explorations before the extent, treachery, and wealth of the land was fully understood and mapped.

Today, the spectrum of U.S. maps is vast and comprehensive. The agencies of the federal government produce hundreds of maps, ranging from general reference works to specific thematic maps of U.S. history, resources, transportation, agricultural, industrial, military, and recreational areas—and anything else that can be mapped. Commercial mapmakers also produce thousands of general and thematic U.S. maps, in forms ranging from atlases to classroom wall maps to small, pocketsize road maps.

The vaults of map libraries, historical societies, and local land offices hold a wealth of U.S. maps within their protective care. Although the originals of some of these maps may be examined only on the premises, the national collections—the **Library of Congress** and the **National Archives**—as well as many of the smaller ones, provide reproduction services for a fee (see "Uncle Sam's Treasure Chests").

Whether one is seeking maps of America's soil or soul, its parks or pipelines, they likely exist in abundance. Most chapters of this book contain some kind of U.S. map, although there are others that defy simple classification. Here is a sampling of U.S. maps not included elsewhere in this book:

GOVERNMENT SOURCES

Every map-producing agency of the federal government creates maps of the United States in one form or another. The National Archives and the Library of Congress have extensive collections of U.S. maps that are available to the public for study or reproduction through the collections' facilities (see "Uncle Sam's Treasure Chests"). Most sources of U.S. maps fall under other headings in this book. Other federal sources include:

Tennessee Valley Authority. TVA has several inexpensive blueline and lithographic reproductions of aboriginal and historical maps of the United States, including:

■ "Aboriginal Map of Eastern U.S." ($1; G MD 453 G 552), a lithoprint of a 17″ X 22″ map made by a Frenchman in 1718 that shows towns and drainage.
■ "Map of Eastern U.S. Made to Accompany 'History of American Revolution'" ($1.50; G MD

453 K 701), a blueline print from a map reprinted in 1811.
■ "North West Section of Map of the United
States" ($1.50; 453 K 274), a blueline print of a
1784 map.
■ "United States and Mexico" ($2; G MD 453 P
754-4), a blueline print of an 1860 map.

When ordering these maps, include 50 cents
for postage and handling.

U.S. Geological Survey. USGS produces and
distributes many U.S. maps, far too many to list
in detail here; a free brochure, "United States
Maps Available from the U.S. Geological
Survey," is available. A sampling of USGS maps
includes:

■ National Atlas Program's "U.S. General
Reference Map" ($3.10; 00438, Code 38077-
AA-NA-07M-00), an all-purpose map suitable
for use as a basic reference tool. The map
measures 19″ X 28″ and is drawn at a scale of
1:17,000,000. Several thematic U.S. maps are
also available from the National Atlas Program,
many of which are mentioned elsewhere in this
book.
■ "Routes of the Explorers" (Sheet 8-Λ), a
colorful historical map tracing the paths of
principal explorers of North America from
1501 to 1844, measuring 18″ X 25″.
■ Base, contour, outline, and physical division
maps of the United States, available in various
scales, colors and sheet sizes. Prices range
from 70 cents to $6.10. Many of these are
mentioned in other chapters; see "Geologic
Maps," "Natural Resource Maps," and "Topographic Maps."

COMMERCIAL SOURCES

There are hundreds of U.S. map and atlas
publishers. All major companies produce maps
of the U.S. for a wide range of purposes, from
business maps to travel maps, general refer-

ence maps, and historical maps, among many
others. Most of these are mentioned elsewhere
in the book. Here are a few more, intended for
general reference use:

American Automobile Association (8111
Gatehouse Rd., Falls Church, VA 22042; 703-

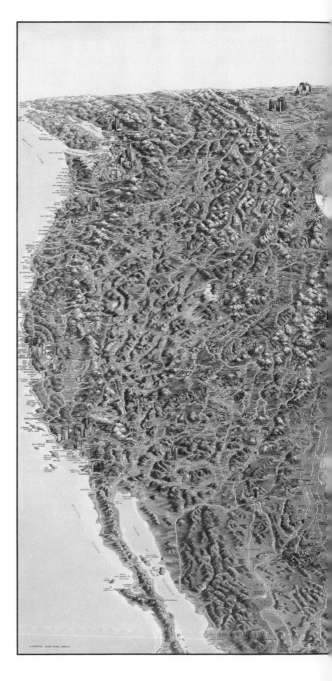

*Right: "Portrait USA," illustrating attractions and skylines of
major cities. © Meridian Graphics.*

222-6000) produces road maps of the United States that are free to AAA members but are not for sale to the general public. The maps may be obtained at local AAA offices.

The Chart House Travel Map Co. Inc. (13263 Ventura Blvd., Studio City, CA 91604; 818-986-

1866; 800-322-1866) produces "Personalized Travel Maps," U.S. maps designed to record one's travels. The back of the map features a "Charting Log" to keep a detailed record of each trip, showing dates, destinations, carriers, hotels and points of interest. Each map comes with a "Charting Kit," consisting of map pins, a

route charting curve, and a pen for drawing routes. The full-color maps come in three sizes, standard (30″ X 20″), deluxe (33″ X 22″), and king size (51″ X 33″), and most are laminated and trimmed in walnut-finish wood. The maps retail for $38 to $98. Personalized World Travel Maps are also available. (See "World Maps.")

Cypress Book (U.S.) Co. Inc. (Paramus Place, Ste. 225, 205 Robin Rd., Paramus, NJ 07652; 201-967-7820) distributes a novel "Map of the United States," published in the Republic of China, with text in both Chinese and Roman characters ($2.95).

Meridian Graphics (421 Bell St., Milton, Ontario, Canada L9T 4R6; 416-876-2260) publishes "Portrait USA," a pictorial U.S. map illustrating tourist attractions and the skylines of major cities. The maps are available in the U.S., both wholesale and retail, from **Latitudes**, 3180 Presidential Dr., Ste. N, Atlanta, GA 30340; 404-455-4234.

National Geographic Society (17th & M Sts. NW, Washington, DC 20036; 202-921-1200) has several maps and atlases of the U.S., including:

■ "United States Political" ($4; 02003),a full-color 41 1/2″ X 29 1/2″ map. An index is available ($1.50; 02004). The map is also available enlarged to 69 1/2″ X 48 1/2″ ($6; 02008).

■ "United States Physical" ($4; 20024), measuring 25 1/8″ X 35 1/8″.
■ "Portrait U.S.A." ($4; 02842), a 42 1/2″ X 29 1/2″ photomosaic of Landsat satellite imagery.
■ "United States/Territorial Growth" ($4; 20025), measuring 29 1/2″ X 42 1/2″.
■ "Wild and Scenic Rivers of the U.S.," including descriptive notes and insets on the reverse side. Available in paper ($3; 02244) or plastic ($4; 02344).

Rand McNally (P.O. Box 7600, Chicago, IL 60680; 312-673-9100) produces a "United States Map" ($2.95; 52″ X 34″), part of the "Cosmopolitan" map series. The map is finely detailed and subtly colored, making it ideal for decorating an office or boardroom. Rand McNally publishes another "United States Map" ($2.95; 52″ X 34″), showing physical features in graphic detail. Both are available laminated for $14.95 each.

■ **See also: "Agriculture Maps," "Antique Maps," "Boundary Maps," "Business Maps," "Census Maps," "Congressional District Maps," "Energy Maps," "Geologic Maps," "History Maps," "Military Maps," "Native American Land Maps," "Natural Resource Maps," "Railroad Maps," "Recreation Maps," "Space Imagery," "Topographic Maps," "Tourism Maps and Guides," and "Wildlife Maps."**

Urban Maps and City Plans

"All cities are mad: but the madness is gallant. All cities are beautiful: but the beauty is grim," writes Christopher Morley in *Where the Blue Begins*. All cities are confusing, he might have added, which can be both maddening *and* grim. Getting lost at least once in a new city is almost a given, but it needn't become habit. A good map can be a valuable guide to the gallant madness and grim beauty of any modern metropolis. City maps are as useful to the native urbanite as they are for the urban neophyte; some cities can take a lifetime to learn.

Basic city maps come in a variety of sizes, shapes, forms, and detail. There are simple pocket and glove-compartment maps showing major roads and official buildings, there are wall maps showing most streets, there are indexed atlases with detailed maps of every neighborhood and district. There's more: land plats showing the location and use of buildings and parks, tourism maps pinpointing locations of award-winning eateries. There are even aerial and satellite photomaps of most American cities.

Large cities like New York or Los Angeles appear on countless maps by both national and local map-makers, while many smaller cities and towns have been mapped only by a local chamber of commerce, city hall, or bank. There are probably hundreds of city maps made for internal use by local, state, and federal governments, from sewer maps to taxation maps, but these usually aren't helpful to the typical visitor. The number, type, and quality of maps available for any city is usually directly related to the size of its tourist trade.

Hundreds of commercial map-makers specialize in creating city maps for a particular region or state. Many such companies do not sell directly to the public, but sell promotional maps that banks, Realtors, and other businesses distribute to customers. There are also companies specializing in certain kinds of city maps. **Historic Urban Plans** (P.O. Box 276, Ithaca, NY 14850; 607-273-4695), for example, produces full-color reproductions of antique maps of American and European cities.

The best sources for free city maps are a local tourist bureau, chamber of commerce, or city planning office. State and county governments produce general tourism and highway maps of their regions; morever, they often can advise you on where to find local maps. Although the federal government includes city outlines on some of its general-use maps, it does not produce detailed, street-by-street maps of American cities. (The CIA does, however, produce maps and guides to various foreign cities—see "Foreign Country Maps".)

More detailed street maps and atlases, or specialized theme maps, probably will be more expensive. They may be purchased at a local bookstore, travel store, or through the producing companies themselves. Many aerial and Landsat photomaps of cities are available from the federal government through its **EROS Data Center** and other distributing agencies. There are also several commercial producers of Landsat images and numerous commercial aerial photo companies that can provide photomaps of cities (see "Aerial Photographs" and "Space Imagery").

For older or antique city maps, the best national resources are the map collections of the **Library of Congress** and the **National Archives**. The Archives collection includes city plans and plats mapped by the government up to 1950. The Library of Congress maps include thousands of commercially-made city maps, as well as those made by federal, state, county, and local governments. One of the Library's most prized possessions is its set of Sanborn city fire insurance maps, which trace the history of hundreds of American towns from the mid-19th century (see "Emergency Information Maps"). Also in the Library of Congress is a large collection of panoramic city views.

Local map collections, university libraries, and historical societies are also good resources for finding older city maps and plans.

GOVERNMENT SOURCES, UNITED STATES

Tennessee Valley Authority. TVA has a series of inexpensive lithographic reproductions of pictorial historical maps for four cities in its region. The lithographs, made from artists' sketches of the cities from an oblique aerial view, are suitable for framing. Available lithographs (include 50 cents postage for each map when ordering) are:

■ "Pictorial Map of Knoxville" ($1.25; 455 K 94; 22" X 29").

■ "Pictorial Map of Chattanooga" ($1.25; G MD 455 K 558; 23" X 30").

■ "Pictorial Map of Nashville in the 1880s" ($1.25; 24" X 41").

■ "Pictorial Map of Memphis" ($1.25; 16" X 23").

GOVERNMENT SOURCES, CANADA

Canada Map Office (615 Booth St., Ottawa, Ontario K1A 0E9; 613-952-7000) distributes "Military City Maps," full-color street maps of Canadian areas, compiled by the Mapping and Charting Establishment of the Department of National Defence. The maps, drawn at a scale of 1:25,000, include street, road, and major building indexes. They are available for major metro areas in all provinces ($3.50 Canadian each). A map index is free upon request.

COMMERCIAL SOURCES

Allmaps Canada (390 Steelcase Rd. E., Markham, Ontario, Canada L3R 1G2; 416-477-8480) publishes city maps for metropolitan

Three views of New York City: Portion of "New York City and Region," from Unique Media (below), ©1989, Unique Media Inc., Box 4400, Don Mills, Ontario, Canada M3C 2T9; 416-924-0644; "Streetwise Manhattan," from Streetwise Maps Inc. (top right); and portion of "Aerial View of Lower Manhattan," from David Fox (bottom right), © David Fox, 1986.

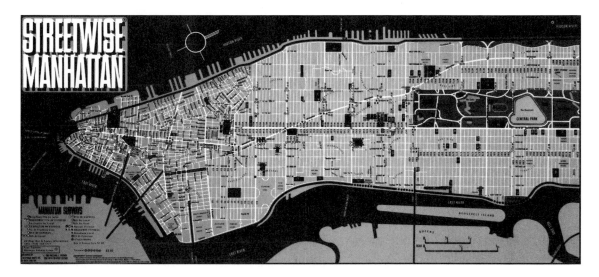

areas in most provinces, including Brampton/
Georgetown, Calgary, Edmonton, Kingston,
Mississauga, Montreal, and Toronto.

American Map Corp. (46-35 54th Rd., Maspeth, NY 11378; 718-784-0055; 800-432-6277)
distributes the following Hallwag and Guia Roji
maps:

■ Hallwag city maps ($5.95 each), with three-
dimensional sketches of important landmarks.
Cities available include Amsterdam, Athens,
Barcelona, Berlin, Berne, Budapest, Copen-
hagen, Florence, Istanbul, Rome, Tokyo,
Vienna, and Zurich. There are also maps of the
London, Munich, and Paris subway systems.
■ Guia Roji pocket maps ($3.95 each) for the

Mexican cities of Guadalajara, Monterrey, Acapulco, Oaxaca, Mazatlan, Mexico City, and Puerto Vallarta.

■ Guia Roji Atlases of Monterrey ($7.95), Mexico City ($7.95), and San Diego/Tijuana ($22.95).

Arrow Maps (Myles Standish Industrial Park, 25 Constitution Dr., Taunton, MA 02780; 508-880-2880; 800-343-7500) produces maps for cities in Massachusetts, New Hampshire, Rhode Island, Maine, Connecticut, Georgia, Delaware, and Virginia.

Champion Map Corp. (200 Fentress Blvd., Daytona Beach, FL 32014; 904-258-1270; 800-874-7010; 800-342-1072 in Fla.) produces folding city maps for many U.S. cities, including Washington, D.C.; Provo, Utah; Topeka, Kansas; and Champaign-Urbana, Illinois.

Creative Sales Corp. (1350 Michael Dr., Ste. B, Wood Dale, IL 60191; 312-350-0770) publishes the *Chicagoland Atlas* ($34.95), a fully indexed, spiral-bound street atlas of the Chicago area. Also available are "Chicagoland Street Maps" (36″ X 25″; $1.95 each), a series of 12 pocket maps of the area.

David Fox (P.O. Box 533, Narberth, PA 19072; 215-667-2136) produces axionometric, four-color, three-dimensional "aerial view" maps of Boston, Chicago, New York, New Orleans, Philadelphia, and Washington, D.C. (24″ X 38″; $20 each). Another map is of the Boston financial district (24″ X 28″; $24).

Dolph Map Co. Inc. (430 N. Federal Hwy., Ft. Lauderdale, FL 33301; 305-763-4732) publishes city maps for most metropolitan areas in Florida, as well as Alabama, Georgia, Kansas, Louisiana, Maryland, Mississippi, New Mexico, Ohio, Pennsylvania, South Carolina, Tennessee, Texas, and Virginia.

Geographia Map Company, (231 Hackensack Plank Rd., Weehawken, NJ 07087; 201-867-4700; 212-695-6585) publishes street maps and atlases for New York City, Buffalo, Chicago, Cincinnati, Gary, Pittsburgh, and Rochester. Geographia also distributes Interarts' "Streetwise" city maps and DeLorme maps and atlases, among other brands.

Geoscience Resources (2990 Anthony Rd., P.O. Box 2096, Burlington, NC 27216; 919-227-8300; 800-742-2677) distributes maps for more than 3,000 cities worldwide. Geoscience Resources stocks more than 500 foreign city maps and has access to more than 2,500 other city maps through affiliated foreign publishers, including maps for the capital cities of nearly every country, most tourist-oriented cities, as well as hard-to-find maps of cities in developing countries. Prices vary.

Hagstrom Map Co. (46-35 54th Rd., Maspeth, NY 11378; 718-784-0055; 800-432-6277) sells pocket maps for various eastern cities, including New York, Stamford, Bridgeport, Buffalo, Philadelphia/Camden, Rochester, and Boston/Cambridge. Prices range from $2.50 to $3.95.

Hammond Inc. (515 Valley St., Maplewood, NJ 07040; 201-763-6000) distributes Bartholomew's "Easy Fold World City Plans" ($8.95 each) for Athens, Cairo, Florence, Istanbul, Rome, and Venice. Each map gives 11 different "spreads" of the city.

Hippocrene Books (171 Madison Ave., New York, NY 10016; 212-685-4371) is the U.S. distributor of several foreign publishers, including:

■ Cartographia city maps ($4.95 each), artistic maps of mostly East and West European cities, including Amsterdam, Belgrade, Berlin, Dubrovnik, Florence, Istanbul, Krakow, Leningrad, Madrid, Reykjavik, Sofia, Thessaloníki, and Warsaw.
■ Falk city maps ($7.95 to $8.95 each) of most major cities worldwide, such as Athens, Berlin, Cairo, The Hague, Jakarta, Peking, and Seoul.
■ Numerous travel guides with maps inside, including Bonechi city/museum guides.

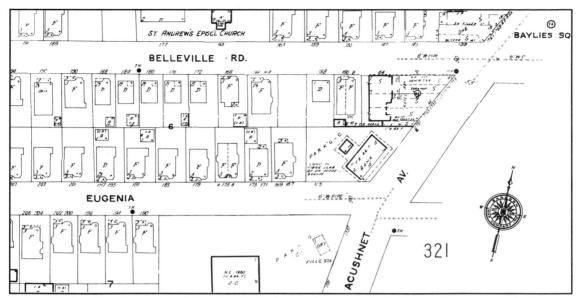

Portion of Sanborn Map Company's land plat of New Bedford, Massachusetts, showing building numbers, streets, and other specific features of the city.

Hunter Publishing (300 Raritan Center Pkwy., Edison, NJ 08818; 201-225-1900) distributes Recta-Foldex "Grandes Villes du Monde" maps ($5.95 each; 32″ X 31″) for more than 30 major cities worldwide, including Amsterdam, Berlin, Cairo, Hamburg, Jerusalem, New York, Prague, Singapore, and Venice.

Interarts Ltd. (15 Mt. Auburn St., Cambridge, MA 02138; 617-354-4655) offers "Streetwise City Maps" ($3 each), laminated guides titled "Boston," "Manhattan," "Mid-Manhattan," "Art-Wise Manhattan," "Philadelphia," "Washington, D.C.," and "San Francisco." Also available are a "Streetwise Manhattan Bus & Subway Map" ($1.50), and a "Streetwise City Map of Chicago" ($4).

Maptec International Ltd. (5 S. Leinster St., Dublin 2, Ireland; 353-1-766266) produces city maps made from enhanced satellite images for various foreign cities, as well as 15 U.S. cities. The maps are available in the U.S. through Rand McNally ($3.95 each). Among the areas Maptec maps are New York, Chicago, Los Angeles, Puget Sound, Toronto, Ireland, Zurich, Corfu, and the U.K.

Marshall-Penn-York Co. Inc. (538 Erie Blvd. W., Syracuse, NY 13204; 315-422-2162) publishes city maps and atlases of New York, Pennsylvania, and Southern New England under the Visual Encyclopedia trademark.

Metro Graphic Arts Inc. (P.O. Box 7035, Grand Rapids, MI 49510; 616-245-2271) produces wall maps, street guides, and pocket maps of various cities in Alabama, Georgia, Indiana, Iowa, Kentucky, Michigan, Missouri, Nebraska, Ohio, Tennessee, Texas, West Virginia, and Wisconsin.

Ozark Map Co. Inc. (RR 2, Box 700-B, Gravois Mills, MO 65037; 314-374-6553) produces city maps for Kansas City, St. Louis, and Springfield, MO ($1.75 to $2.25 each), and a "Greater Kansas City Street Atlas" ($9.95).

Perigee Books (200 Madison Ave., New York, NY 10016; 201-933-9292; 800-631-8571) distributes Van Dam's "America Unfolds" map series of American cities. These colorful pocket-size street maps ($5.95 each) unfold to 12 times the original size and refold automatically. Each map includes two section maps, an address locator, neighborhood profiles, and bus and subway maps. Maps are available for Atlanta, Boston, Chicago, Dallas, Detroit, Los Angeles, Manhattan, Orlando, Philadelphia, San Francisco, Seattle, and Washington, D.C.

Pierson Graphics Corp. (899 Broadway, Denver, CO 80203; 303-623-4299) publishes the "Birds-Eye View Series": four-color axionometric maps of Chicago, Dallas, Denver, Houston, and the West Loop of Houston ($2.95 to $27.95). Also available are 4″ X 9″ street maps of Colorado metropolitan areas ($1.95 to $3.95), and David Fox Studios axionometric maps of other cities.

Prentice Hall (200 Old Tappan Rd., Old Tappan, NJ 07675; 800-223-2348) distributes Gousha city maps ($1.95 each) for more than 100 American cities, including Akron, Albuquerque, Baton Rouge, Cleveland, Gainesville, Las Vegas, Little Rock, Oklahoma City, Pensacola, Salt Lake City, and Winston-Salem. Gousha "Metro Maps" each show a major city and its suburbs, with inset maps of downtown areas. They are available for more than 30 metropolitan areas in the U.S.

Random House (201 E. 50th St., New York, NY 10022; 212-751-2600; 800-726-0600) distributes Fodor's "Flashmaps" ($4.95), 80-page single-subject guides to various cities in the U.S. and abroad. Included in the London guide are maps of "Historical Sites," "National Trust Houses," "Literary England," "Architectural Highlights," "Major Shopping Areas," "Royal Residences," "Blue Plaque Houses," "Treasures," "Art Galleries," "Antiques," "Transportation," and "Museums." Flashmaps are available for: Boston, Chicago, Dallas-Fort Worth, Los Angeles, San Francisco, Washington, D.C., and London. Maps of European cities are planned.

Streetwise Maps Inc. (P.O. Box 2219, Amagansett, NY 11930; 516-267-8617) produces handy-size city maps of New York City, including "Artwise Manhattan" ($3), "Address Map" ($1.50), "Brooklyn" ($3), "Bus/Subway"

($1.75), "Manhattan" ($3), "Mid-Manhattan" ($4), "Downtown Manhattan" ($2), "Transitwise" ($3.50), and "Zipwise Manhattan" ($1.50). Also available are Streetwise maps for Boston, Chicago, London, Paris-Central City, Philadelphia-Central City, Rome, San Francisco, and Washington, D.C.

Sunset Books (Lane Publishing Co., 80 Willow Rd., Menlo Park, CA 94025; 800-227-7346; 800-321-0372 in Calif.) produces *The Frequent Traveler's City Atlas* ($9.95), 240 pages of detailed road and street maps covering major U.S. metropolitan areas. Also available is the *State & City Atlas* ($14.95), 352 pages of colorful maps and travel information, with more than 160 metropolitan and downtown street maps, as well as full-color state maps.

T.R. Map Co. (16403 W. 126th Terr., Olathe, KS 66062; 913-782-6168) publishes wall maps of metropolitan areas in Arkansas, Iowa, Kansas, Missouri, Nebraska, and Oklahoma.

Unique Media Inc. (Box 4400, Don Mills, Ontario, Canada M3C 2T9; 416-924-0644) publishes "picture maps" featuring a bird's-eye perspective of the physical terrain, landmarks, and individual buildings of cities in Canada and the U.S. ($4.50 each, or $15 for laminated copies). They include "Greater Los Angeles," "Downtown Los Angeles," "San Francisco," "San Diego," "San Antonio," "Pittsburgh," "New York City & Region," "Metro Toronto & Region," "Downtown Toronto," "North York," "Canada" (Ottawa on backside), and "Canada" with Vancouver, B.C. on the backside.

VanDam Inc. (430 W. 14th St., New York, NY 10014; 212-929-0416; 800-321-6277) publishes "The World Unfolds" series of pop-up city maps that open to six times their original size, then refold automatically. These unique maps hold a U.S. patent as kinetic sculpture. Maps are available for Amsterdam, Atlanta, Boston, Chicago, Dallas, Detroit, Hawaii, Hong Kong, London, Los Angeles, Miami, New Orleans, New York, "NYC Subway," Orlando, Paris,

Rome, San Francisco, Tokyo, Washington, D.C., and Zurich ($6.95 each).

COMMERCIAL SOURCES—LOCAL

Other sources for cities in their regions include:

Alaska
Funmap Inc.
705 W. 6th Ave.
Anchorage, AK 99501
907-272-6773

The Maps Place
P.O. Box 91975
Anchorage, AK 99509
907-284-6277

Arizona
Phoenix Mapping Service
2626 W. Indian School Rd.
Phoenix, AZ 85017
602-279-2323

California
Global Graphics
2819 Greentop St.
Lakewood, CA 90712
213-429-8880

Colorado
Macvan Productions Inc.
809 N. Cascade Ave.
Colorado Springs, CO 80903
303-633-5757

Connecticut
Mail-a-Map Street Maps
P.O. Box 282
Guilford, CT 06437
203-453-5525

District of Columbia
Mino Publications
9009 Paddock Ln.
Potomac, MD 20854
301-294-9514

Florida
Trakker Maps
12027 SW 117th Ct.
Miami, FL 33186
305-255-4485; 800-432-3108

Louisiana
New Orleans Map Co. Inc.
3130 Paris Ave.
New Orleans, LA 70119
504-943-0878

Maine
DeLorme Mapping Co.
P.O. Box 298
Freeport, ME 04032
207-865-4171

Massachusetts
Yankee Books
Depot Sq.
Peterborough, NH 03458
603-563-8111; 800-423-2271

Butterworth Co. of Cape Cod Inc.
476 Main St.
Harwichport, MA 02646
508-432-8200

Nevada
Front Boy Service Co.
3340 Sirius Ave.
Las Vegas, NV 89102
702-876-7822

New York
Map Works Inc.
2125 Buffalo Rd., Ste. 112
Rochester, NY 14624
716-426-3880; 800-822-5277

Oregon
Pittmon Map Co.
732 SE Hawthorne Blvd.
Portland, OR 97214
503-232-1161; 800-452-3228

Washington
Square One Map Co.
P.O. Box 1312
Woodinville, WA 98072
206-485-1511

Canada
Perly Toronto Inc.
1050 Eglinton Ave. W.
Toronto, Ontario M6C 2C5
416-785-6277

World Maps

It's simply not possible to count the number of world maps that exist—even those currently available. There are thousands of maps, dating from the birth of map-making. There are maps of the world when only one continent was known to exist and maps of the world today, with every inch of the globe plotted with astounding accuracy. There are maps of world vegetation, rainfall, mineral reserves, soil types, and industrial strongholds. There are, in short, as many maps as there are ways to interpret and comprehend the modern world.

The uses of world maps are as varied as the maps themselves. Geologic maps of the world are vital to the work of earth scientists; maps showing the amount of annual sunshine that falls on various spots around the world are key to planning energy resources. Maps showing the growth of socialism, the decline in arable land, the shifts of industrial wealth, and the flows of maritime commerce all have their respective professional constituencies.

World maps are created by most carto-graphic government agencies and private companies. The federal government publishes several general, as well as specific, world maps, ranging from "mosaics" of Landsat images to a map of illegal drug-growing regions and smuggling routes.

Commercial companies also make a vast range of world maps, as do various geographic, political, and earth science societies around the world.

Some of the most beautiful world maps are also the oldest, drawn in the days when artists moonlighted as map illustrators. Sources for antique world maps include the collections of the **Library of Congress** and the **National Archives**, as well as a number of dealers in antique maps, globes, and atlases specializing in world maps; some may be willing to search for a particular map (see "Uncle Sam's Treasure Chests" and "Antique Maps").

Here is a sampling of what's available:

GOVERNMENT SOURCES, UNITED STATES

Central Intelligence Agency. The CIA has several world maps, available through the National Technical Information Service, including:

■ "Political Map of the World" ($11; PB 88-928323).
■ "World Map" ($11.95; PB 88-928325).
■ "Standard Time Zones of the World" ($8.95; PB 88-928358).
■ "U.S. Foreign Service Post and Department of State Jurisdictions" ($10; PB 84-928009), published in 1984, so information may now be slightly out of date.

National Geophysical Data Center. The NGDC, part of the National Oceanic and Atmospheric Administration, produces a three-sheet set of world maps, "Relief of the Earth's Surface." The maps are: "Computer-Generated Shaded Relief," a gray monochrome with shaded relief depicting the Earth's geographi-cal features; "Computer-Generated Color-Coded Shaded Relief"; and "Hemispheric Images," a color map made from simulated views of the Earth from various angles. The maps measure 36" X 46" each and are avail-able as a set for $20.

U.S. Geological Survey. USGS has several large world maps available, the most popular being the "International Map of the World" ($3.60), a multicolor reference Mercator projection map showing borders, capital cities, elevation tints and shaded relief, and other key features to delineate the nations of the world. This map is available in three scales: 1:30,000,000 ($4.30), a single sheet measuring 42" X 56"; 1:22,000,000 ($10.80), consisting of three sheets that assemble to 55" X 74"; and 1:14,000,000 ($36), consisting of six sheets that assemble to 7'2" X 9'8".

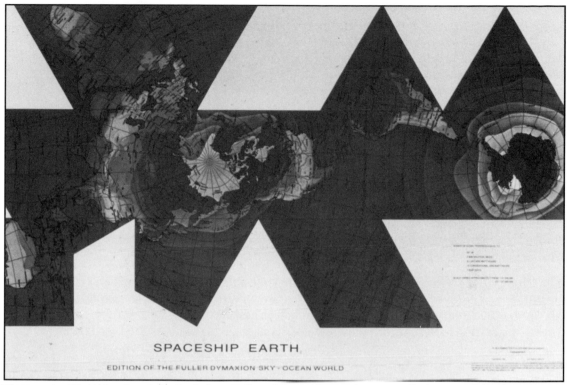

SPACESHIP EARTH

EDITION OF THE FULLER DYMAXION SKY · OCEAN WORLD

The Dymaxion World Map, invented by Buckminster Fuller, shows our planet without any visible distortion of the relative shapes and sizes of the land and sea areas, and without any breaks in the continental contours.

Also available from USGS are:
■ "World Seismicity Map" ($3.10), showing different areas of seismic activity on an international level, using a Mercator projection centered on the Americas.
■ "Political Map of the World" ($3.10), which shows nations, dependencies, and island groups, drawn with the Miller Cylindrical projection centered on the Prime Meridian.
■ "Outline Map of the World," based on the Van der Grinten projection centered on the Americas, a two-color map showing political boundaries, country names, capitals, and selected cities, available in two sizes: 48" X 33" ($3.10) and 25" X 18" ($1.70, with no cities named).

GOVERNMENT SOURCES, CANADA
Maps Alberta (Land Information Services Div., Main Fl., Brittania Bldg., 703 6th Ave. SW,

Calgary, Alberta T2P 0T9; 403-297-7389) publishes "Alberta in the World" ($3.50 Canadian; 28" X 30"), an azimuthal projection of the world from a perspective above Edmonton, showing political boundaries in eight colors.

COMMERCIAL SOURCES
World maps are the backbone of many map companies, the standard maps that "never go out of style," so finding a map of the world is generally as easy as walking into a local travel, book, or map store. There are thousands of world maps available; many of a thematic nature are listed elsewhere in this book. Here is a small sampling of available world maps:

American Geographical Society (156 Fifth Ave., Ste. 600, New York, NY 10010; 212-242-0214) distributes two world maps: "World Map" ($4; 14" X 14"), reproduced from the small "Under Heaven Map Book," published in Korea in the late 17th or early 18th century; and "The World" ($11 per sheet), a 17-sheet series using the Miller oblated stereographic projection system, showing land and water physical features, bathymetry, political boundaries, towns, roads, railroads, airports, and other features. Eight sheets, drawn on 1:5,000,000 scale, are available from the AGS; size varies.

American Map Corp. (46-35 54th Rd., Maspeth, NY 11378; 718-784-0055; 800-432-6277) distributes Hallwag's "The World" physical reference map (6″ X 9 1/2″), showing physical contours of the terrain and physical boundaries ($7.95; Mercator projection).

The Chart House Travel Map Co. Inc. (13263 Ventura Blvd., Studio City, CA 91604; 818-986-1866; 800-322-1866) produces "Personalized Travel Maps," designed to record one's travels. The legend of the map has space for personalizing, while the back of the map features a "Charting Log" to keep a detailed record of each trip, showing dates, destinations, carriers, hotels and points of interest. Each map comes with a Charting Kit, consisting of map pins, a route-charting curve, and a pen for drawing routes. The full-color maps come in three sizes, standard (30″ X 20″), deluxe (33″ X 22″), and king size (51″ X 33″), and most are laminated and trimmed in walnut-finish wood. They range in price from $38 to $98 and are drawn on the Mercator projection centered on the Americas. Personalized U.S. maps are also available (see "United States Maps").

Class Publications Inc. (237 Hamilton St., Hartford, CT 06106; 203-951-9200) distributes several four-color world maps, all priced at $5, plus $3 shipping: "Map of the World," with national flags (27″ X 39″; Mercator projection centered on Africa); "World Relief Map," showing geographical features, a poster produced by Dutch Verkerke (24″ X 36″; 26699; Molleweide projection); "Political Map of the World," also published by Verkerke of The Netherlands (24″ X 36″; 26654; Molleweide

projection); and "Old World Map," a reproduction of a 1636 map of the world (27″ X 39″; C3030).

Environmental Graphics Inc. (15295 Minnetonka Blvd., Minnetonka, MN 55345; 612-938-1300; 800-328-3869) produces a 8′8″ X 13′ World Map Photomural ($51.95), a brightly-colored map drawn with the Mercator projection.

Buckminster Fuller Institute (1743 S. La Cienega Blvd., Los Angeles, CA 90035; 213-837-7710) sells copies of the Dymaxion map in several forms: "Spaceship Earth Edition" ($9.95), a 34″ X 22″ wall map; "Spaceship Earth Greeting Card" (95 cents each), a formal-size card with a full-color Dymaxion map on the cover, with quotes from Fuller inside; "Raleigh Edition Dymaxion Map" ($7.95), a reprinting of Fuller's first edition of the Dymaxion map, in very subtle colors; "Dymaxion Sky-Ocean Globe" ($5), a 22″ X 14 1/2″ map on heavy stock that folds into a 5 1/2″ icosahedron; and postcards ($1.95 each).

Gabelli U.S. (11500 W. Olympic Blvd., Ste. 475, Los Angeles, CA 90064; 213-312-4546) distributes two vibrant French-made Gabelli maps (drawn with the Mercator projection centered on Africa):

■ "World Political," available in paper or flexible plastic in a number of sizes and prices, from $3.30 for a 25″ X 16″ paper map to $18.75 for a 60″ X 37″ plastic map.
■ "World Montiscolor," showing the continents and oceans in colorful relief, available for the same prices as the World Political map.

Gabelli's world maps are also available on notebooks, desk blotters, folders, wastepaper baskets, pencil holders, and traveling bags (see "Map Stuff").

Geochron Enterprises Inc. (899 Arguello St., Redwood City, CA 94063; 415-361-1771) produces the "Geochron" time indicator, a

The World Game Map

Imagine that you are 2,000 miles high, standing on the Earth looking down. The space shuttle would be in orbit at your ankles, the sun would seem to be a mere 25 miles away, the length of your foot would stretch 300 miles. It would take four human steps to traverse the United States. Assuming you took one step per second, your speed would reach the equivalent of 2.5 million miles per hour. At this speed, you could circumnavigate the Earth 109 times per hour.

Clearly, you would be viewing the planet Earth from a whole new perspective.

Those are precisely the intentions of The World Game, an innovative playground of learning created by the nonprofit **World Game Institute** (University City Science Ctr., 3508 Market St., Philadelphia, PA 19104; 215-387-0220), which describes itself as "a peace research and education organization developing tools and solutions for global and local problems."

The World Game features what its creators call "the world's largest and most accurate map of the whole earth," a 40-by-70-foot map (its creators have dubbed it "Big Map") painted onto the asphalt or concrete of a playground, parking lot, or neighborhood plaza. The map, at a scale of 1:2,000,000, is based on Buckminster Fuller's Dymaxion projection (see "Map Projections"). This gigantic map becomes the basis for a variety of presentations, games, and workshops, focusing on such issues as economics, hunger, energy, population, nuclear war, education, resources, technology—and, of course, geography. In the World Game Sessions, for example, a two- to three-hour event using Big Map, participants involve themselves in simulations dealing with the structure of global problems, resource distribution, and political interconnections. Other programs deal with global food, energy, and economics; the "greenhouse effect"; and Africa's unique problems. All of these actively involve participants.

Fees vary according to location, length of program, and other factors. For more information, contact the World Game Institute.

Mercator projection world map that moves slowly from left to right, one inch per hour. With Geochron, it is possible to see where it is day and where it is night at any moment. The Geochron retails for $1,265 to $2,465, depending on model.

Hansen Planetarium (1098 S. 200 W., Salt Lake City, UT 84101; 801-538-2242) sells "Earth at Night" ($6), a map made from a montage of satellite images showing what the Earth looks like at night. It reveals gas flares in the Middle East, slash-and-burn agriculture in Southeast Asia, and urbanization in Europe, Japan, and North America. This fascinating map measures 23″ X 35″ and comes with a reference guide.

Interarts Ltd. (15 Mt. Auburn St., Cambridge, MA 02138; 617-354-4655) distributes a number of world maps, including:

■ World Political Maps drawn with the Van der Grinten projection are: 30″ X 53″, available in Tyvek ($20) or laminated ($40); 43″ X 54″, available in paper ($15.95) or laminated ($35); and 17″ X 27″, available in paper ($7.95), laminated ($20), or as a desk blotter ($14.95).
■ World Political Maps drawn with the Mercator projection are: 27″ X 39″, with country flags, available varnished ($14.95) or laminated ($30); 22″ X 39″, in Tyvek ($14.95) or laminated ($30); and 17″ X 27″, available in paper ($7.95), varnished ($9.95), or laminated ($20).

■ World Environmental Maps drawn with the Mercator projection, available in the following sizes: 27″ X 39″ ($9.95 paper, $30 laminated), 17″ X 27″ ($7.95 paper, $20 laminated), and a 9′ X 12′ eight-sheet wall map ($79.95, paper).

Interarts also distributes "Wearin' The World" sports jackets made from the Tyvek World Map (see "Map Stuff").

MapArt/Peter Heiler Ltd. (72 Bloor St. E., Oshawa, Ontario L1H 3M2; 416-668-6677) publishes a 24″ X 36″ political/physical world map, drawn with the Peters projection ($4.95 paper, $9.95 laminated).

Nystrom (3333 Elston Ave., Chicago, IL 60618; 312-463-1144; 800-621-8086) produces various world maps, including the educational "Readiness Map of the World" made with either the Van der Grinten or the Robinson projection ($98 each). A special Canadian edition shows adjacent provinces in contrasting colors. "Pacific Rim and the World" is a 76″ X 52″ map drawn with the Eckert IV Equal Area projection centered on the Pacific Rim area, with a shaded-relief inset map showing the Pacific region's "Ring of Fire" volcanoes and earthquake areas.

Rand McNally (P.O. Box 7600, Chicago, IL 60680; 312-673-9100) Rand McNally distributes Wenschow Maps, large, German-made, full-color relief maps drawn with the Molleweide Homographic projection. Rand McNally also offers Wenschow's series of spring-mounted, 50″ X 35″ cultural geography world maps, including:

■ "Population Density" ($100; 114-10741-6), showing population density by continent. The map measures 50″ X 35″.
■ "Languages" ($100; 114-10742-4), showing distribution of 30 of the world's language groups. The map measures 50″ X 35″.
■ "Religions" ($100; 114-107432-2), showing areas of dominance for 14 religions throughout the world. The map measures 50″ X 35″.
■ "Cultural Regions and Migrations" ($100; 114-10744-0), which indicates the directions of population migrations throughout the world since 1500. The map measures 50″ X 35″.
■ "World Cultural Geography" ($228; 114-10740-8), which consists of all four maps in the set, mounted on one cloth-backed sheet on a spring roller.

There is also a Wenschow "World/Ocean Relief" ($266; 111-10203-0), a 96″ X 65″ full-color map with hypsometric tinting depicting the ocean floor in three dimensions. The map, drawn at a scale of 1:15,000,000, is mounted on a plastic rod but is available on a spring roller mounting for $281.

■ **See also: "Antique Maps," "Foreign Country Maps," "Globes," "Ocean Maps," "Political Maps," "Selected Atlases," and "Space Imagery."**

BOUNDARY MAPS

Boundary Maps

When it comes to drawing the line, nobody does it better than Uncle Sam. The federal government has worked diligently to draw and maintain the intricate boundaries that separate it from Canada and Mexico, as well as those dividing the contiguous 48 states from one another. The United States isn't alone in this endeavor: The surveyors of most other countries have long strived to divide conquered kingdoms and define lands given as gifts in royal marriages or annexed after a war; so, too, have those who govern cities, counties, states, and provinces around the world. Throughout the world, boundary maps are helpful to police, border guards, and customs agents, as well as to geographers, landowners, developers, historians, and statesmen.

Though they are perhaps the simplest and most direct maps in the world—they merely show where one country, state, county, city, town, or land plat ends and another begins—there is a certain romance to a boundary map. To appreciate the mythical proportions some boundaries take, one need only recall Jean Renoir's classic film about World War I, *The Grand Illusion*. In the final scene, German soldiers cease their fire at escaping Frenchmen who have just made it over the Swiss border. There are less dramatic scenes daily at borders, as it is necessary to determine where to build, whom to sue, how to tax, and other issues that are trivial to all but a few interested parties.

The men and women who survey boundaries follow a long tradition that includes such renowned American surveyors as Daniel Boone and George Washington. From drawing out the lines of a cattle ranch or sheep farm to setting the borders of growing metropolises, the history of boundary mapping is one of expansion and intrigue. Surveyors like Charles Mason and Jeremiah Dixon, who tackled the 18th-century American wilderness to pinpoint the North-South dividing line that bears their names, did so with a battery of axmen in the lead to battle the vegetation, and often the Indians.

Boundary maps, however, are not the last word on borders. They are merely references drawn to illustrate the border lines set forth in treaties, annexations, or other agreements, and they draw lines accordingly. If this sounds easy, consider the border specifications—simplistic by today's standards—of the first charter of Virginia as declared by England's King James in 1606:

> ...situate, lying, or being all along the Sea Coasts, between four and thirty degrees of Northerly Latitude from the Equinoctial Line and five and forty degrees of the same Latitude, and in the main Land between the same four and thirty and five and forty Degrees and the Islands therunto adjacent, or within a hundred miles of the coast thereof.

Sometimes, the process of correcting boundary lines that were carelessly mapped can require more words than the original boundary specification. In an 1875 document, the North Carolina Geological Survey described the incorrect mapping of the state's borders:

> ...it appears from the South Carolina geographical State survey of 1821-1825 that the course from the starting point is N. 47 degrees 30' W., and instead of pursuing the parallel of 35 degrees, it turns west about ten miles south of that line, and then, on approaching the Catawba River, turns northward, pursuing a zigzag line to the forks of the Catawba River...

The report estimated that such mapping errors caused the state to lose between 500 and 1,000 square miles of territory. The course of boundary drawing may not run smoothly, but the importance of the process is boundless.

Boundary maps are available through a number of federal agencies, as well as through local and state land management offices and the cartographic offices of foreign countries.

GOVERNMENT SOURCES, UNITED STATES

The lines drawn on boundary maps produced by the federal government contain lines that are either solid (if their accuracy is absolutely established) or broken (if they are believed, but not certain, to be accurate). Monuments such as engraved stones or monoliths built as border markers are also shown along boundary lines, usually with a small square symbol. Lines within states are usually mapped by local or state agencies, and these boundaries are then incorporated into federal maps.

International Boundary Commission, United States and Canada (425 I St. NW, Rm. 150, Washington, DC 20001; 202-632-8058) has boundary maps that show the detailed border locations separating the United States and Canada. The maps, drawn in 1922, are printed on heavy paper and show the vegetation, major waterways, and some topography in shades of blue, green, and brown. The commission's 226 maps that represent the entire U.S.-Canadian border are divided into seven subsections:

■ Source of the St. Croix River to the Atlantic Coast (18 maps).
■ Source of the St. Croix River to the St. Lawrence River (61 maps).
■ Northwesternmost Point of Lake of the Woods to Lake Superior (36 maps).

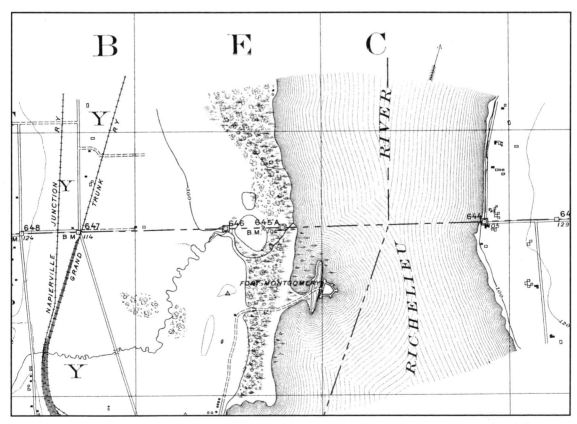

Portion of an International Boundary Commission map, sheet 3 in the St. Lawrence River–St. Croix River series, which illustrates the U.S.–Canadian boundary between New York and Quebec.

■ Gulf of Georgia to Northwesternmost Point of Lake of the Woods (59 maps).
■ 49th Parallel to the Pacific Ocean (one map).
■ Tongass Passage to Mount St. Elias (13 maps).
■ 141st Meridian from the Arctic Ocean to Mount St. Elias (38 maps).

Each map is $3. Write to the IBC for complete listings of maps within each subsection to help determine which maps you need.

International Boundary and Water Commission, United States and Mexico (The Commons, Bldg. C, Ste. 310, 4171 N. Mesa, El Paso, TX 79902). U.S. boundary maps are available for the Lower Rio Grande Project (Falcon Dam to the Gulf of Mexico) and the El Paso Project (Elephant Butte Dam to below El Paso). The maps vary in cost from $4 to $20. Price information is available upon request.

Bureau of the Census. The Census Bureau produces two editions of U.S. county outline maps:

■ Black Edition ($3.50) shows state, county, and county equivalent boundaries as of Jan. 1, 1980, in black, with state boundaries drawn with heavier lines (30″ X 42″).
■ Black and Green Edition ($3.50): Same as Black Edition, but shows state boundaries in black, county boundaries in green (30″ X 42″).

Most maps produced by the **U.S. Geological Survey** include boundary lines for the area mapped. Topographical quadrangle maps include boundary lines drawn by local or state authorities. Only the maps of West Virginia contain boundaries established by the USGS at the request and compliance of the state (see "Topographic Maps" for more information on topographic quadrangles).
    USGS also publishes an Alaska Boundary Map Series at 1:250,000 scale (1° X 2°), showing boundaries of federal lands ($2.50 each) and color photoimage quadrangle-size maps along the U.S.-Mexico border and along

part of the U.S.-Canada border from Massena, NY, to East Richford, VT. These 20″ X 22″ maps center approximately on the border and sell for $2.50 each. See the USGS *Catalog of Maps* for ordering information.

    The **Central Intelligence Agency** publishes hundreds of maps of foreign countries that include internationally accepted boundary lines. Some CIA maps are created specifically to illustrate border areas, such as "China-India Border Area" ($10; PB 86-928343) and "China-Vietnam Boundary Markers" ($10; PB 87-928335). The maps are available from the National Technical Information Service (see Appendix B).
    Both the **Library of Congress** and the **National Archives** have boundary maps in their vast collections. The Library's collection consists of both American and international boundary maps, some dating to the 1300s. The Archives collection, which spans the two centuries between 1750 and 1950, contains American boundary maps commissioned by the federal government. While many of the best boundary maps are created privately by hired surveyors and are rarely made public, the Library and the Archives have collected and made available many such maps of both domestic and foreign origin. Reproduction services are available from both collections (see "Uncle Sam's Treasure Chests").

State and Local Boundary Maps. These may often be obtained through local land management or survey departments. Most cities have a records office where early land-ownership agreements and boundary maps may be available for study; local historical societies are another good source for boundary maps of significance to an area. Some state cartographic offices provide maps or can direct you to good resources (see Appendix A).

Foreign Boundary Maps. These often are available through each country's cartographic agency (see Appendix C). Another good source may be foreign tourism bureaus in the U.S.

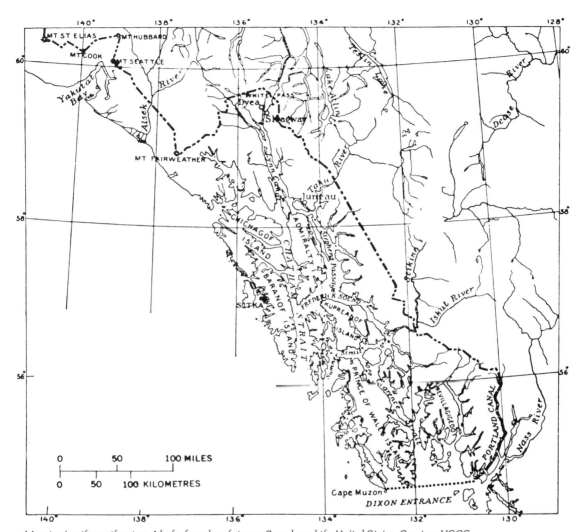

Map tracing the southeastern Alaska boundary between Canada and the United States. Courtesy USGS.

COMMERCIAL SOURCES

Clarkson Map Co. (1225 Delanglade St., P.O. Box 218, Kaukauna, WI 54130; 414-766-3000) distributes Canadian Boundary Waters maps ($13 each), including maps for Kabetogama Lake, Rainy Lake, and Lake of the Woods.

Hammond Inc. (515 Valley St., Maplewood, NJ 07040; 201-763-6000) distributes "Daily Telegraph Maps" of foreign areas. The maps feature information on international bounda-

ries, as well as shipping routes and road and rail networks. Maps ($8.95 each) are available for Africa, Australia, Europe, India/Pakistan/Far East, Great Britain/Northern Ireland, Middle East, New Zealand, North America, South Africa, South America, and the world.

■ **See also: "Foreign Country Maps," "Land-Ownership Maps," "Nautical Charts and Maps," "State and Provincial Maps," and "United States Maps."**

Congressional District Maps

Congressional district maps tell many tales. The history of congressional representation can be found in the shadings and the boundaries that outline the voting realms—as well as the history of political deal-making and redistricting, also known as "gerrymandering."

The game of gerrymandering is one of the oldest in politics: redrawing the boundaries of a legislative district to create an imbalance of power, giving one political party an advantage at the voting booth. At times, congressional district boundaries meander around and through an area so sinuously as to include only the voters on the right side of one street, and only the voters on the left side of the street a block away. To look at maps of these districts, one might think the boundaries were astrological symbols—sea horses or snakes slithering across the land with no apparent method to their madness. Without maps to accurately define the district lines, few would believe the crazy-quilt patterns to be legal divisions of voting boundaries.

The term "gerrymander" was coined when Elbridge Gerry was elected governor of Massachusetts in 1812 with the aid of a bit of creative state-wide redistricting. A clever cartoonist, noting the serpentine shape of the new boundaries, drew a caricature of the "gerrymander," a salamander-like namesake of the governor.

Gerrymandering has helped more than a few politicians win elections in areas where they normally would have had little support. The Supreme Court sought to abolish the practice in a 1963 ruling, which established the "one man, one vote" precedent. Now, legislative districts must have relatively equal populations, effectively putting an end to the days when 500 bankers could elect three members of Congress, while 5,000 farmers elected just one. But even within these constraints, state legislatures still have the right to carve up voting districts as their political leaders see fit.

In most states, the party in power in the legislature has the right to draw up district boundaries. In the end, the district maps have the final say: once the maps are drawn, voters and politicians must follow their guidelines until a shift in power sparks the next bout of redistricting.

Some older, gerrymandered congressional district maps are works of art, colored and intricately drawn to include just the right citizenry. More recent maps are a bit less ornate—and less obvious about the purposes of redistricting—and though they may not be as pretty to look at, they paint an accurate portrait of the U.S. population's physical and political distribution.

Congressional district maps are useful to more than just members of Congress: businesspeople use them to study population concentration; fund-raisers use them to pinpoint areas where certain political loyalties may create a donation base; teachers use them to educate students about American history and government. Anyone with an interest in U.S. political make-up and population distribution will find congressional district maps enlightening, even entertaining.

The federal government creates (and updates with each new Congress) a number of congressional district maps and atlases, as do several commercial map producers. Older congressional district maps are available from map dealers specializing in American history maps, or in reproduction form from map libraries, including the **Library of Congress** and the **National Archives**.

GOVERNMENT SOURCES

Government Printing Office. The following congressional district maps and atlases created by the Census Bureau are available from GPO:

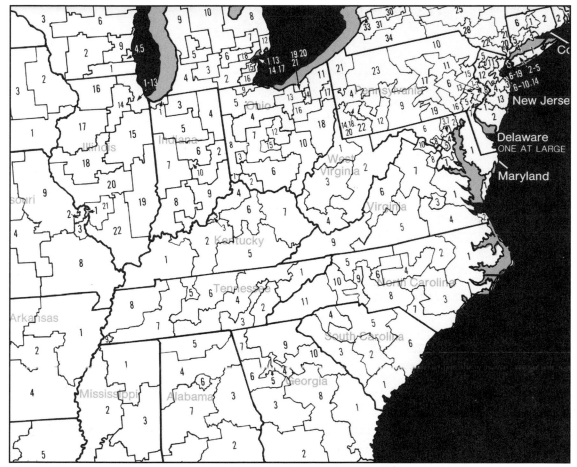

Portion of the district boundary map of the 101st Congress, from Congressional Quarterly.

■ *Congressional District Atlas: Districts of the 100th Congress* ($33; S/N 003-024-06234-8). This 696-page atlas contains maps of all congressional districts during the 100th Congress.

■ *Congressional District Atlas: Districts of the 99th Congress* ($15; S/N 003-02406132-5). 583 pages.

■ "Congressional Districts of the 100th Congress of the United States" ($4.75; S/N 003-024-06228-3), a 36″ X 47″ two-sided map. One side shows district boundaries for the 100th Congress; the other shows the districts for the first, 25th, 50th, 75th and 100th Congresses.

U.S. Geological Survey. The National Atlas program produced a colorful 1987 map, "Congressional Districts for the 100th Congress" ($3.10; US00446-38077-AT-NA-07M-00).

COMMERCIAL SOURCES

Congressional Quarterly (Book Distribution, One Capitol Dr., Ridgely, MD 21685; 800-543-7793) sells a U.S. map of all the congressional districts, updated every two years. The 101st Congress map is $14.95 prepaid, plus $1.95 postage and handling.

Western Economic Research Co. Inc. (8155 Van Nuys Blvd., Ste. 100, Panorama City, CA 91402; 818-787-6277) sells a Congressional District Map of the Los Angeles five-county area ($20; 30″ X 42″).

■ **See also: "Political Maps."**

Land-Ownership Maps

From land-grant maps to county plat maps, the portrayal of lands bought, granted, or inherited paints a picture of our perpetual need to own real estate. The detail with which some of these maps are drawn—every house, street corner, park, and fire hydrant in town diagrammed and labeled—preserves on paper one moment in the history of America's urbanization. By studying older ownership maps, one may locate the first house where immigrant grandparents lived. By studying current ones, one might ascertain information about the landholdings of neighbors and others.

The mapping of land ownership dates back to the Babylonian cadastral maps that delineated individual land-holdings. Somewhat later, according to legend, Ramses II established a cadastral survey in the 13th century B.C., but little of the papyrus on which the maps were made has survived. By the time of the Renaissance, the mapping of European estates was commonplace, and, by the 18th century, maps of counties were being published in England, with specific houses and residents named.

In the 19th century, the development of American county "land plats"—simple but detailed grid maps showing buildings, public lands, roads, and other features—became a necessity for the rapidly expanding nation. The importance of diagrams showing exactly who owned what and who lived where was immeasurable in a country where land-grant acts and squatters' rights were hotly debated. Today, county land plats, updated regularly by both commercial and government mappers, still serve as valuable land-planning tools; older plats serve as portraits of an area's past.

Everyone from genealogists to geographers uses land-ownership maps. Planners, builders, and investors can outline prospective developments by examining land-ownership blueprints. These maps sometimes reveal lands granted to, then taken from, native Indians, or the planta-

tions of slave owners. And the exploding growth of cities, towns, and industrial centers is all recorded on land-ownership maps.

Boundary maps (see "Boundary Maps") are one form of land-ownership maps, but most land-ownership maps do more than merely illustrate property lines. While boundary maps designate where one land plot ends and another begins, many land-ownership maps disclose to whom those lands belong.

Current land-ownership maps are available from several federal agencies, among them the **USGS**, the **Army Corps of Engineers**, and the **National Park Service**. Older boundary, land-grant, and plat maps may be found in the collections of the **Library of Congress** and the **National Archives**. State, county, and local mapping departments create their own land-ownership maps for taxation purposes, and some are available to the public. Land plat books and other ownership maps also are published by commercial map and blueprint companies around the country.

GOVERNMENT SOURCES

Uncle Sam owns a lot of land, and the surveying of this wealth of real estate is a job undertaken zealously by the federal government. Land-ownership maps are produced by several agencies or departments to diagram the continuing use of the land.

The **Bureau of Land Management** (BLM) is the largest federal source for plats of townships surrounding public land. In many cases, BLM's surveys were the first to establish boundary lines, and subsequent maps have detailed the growth of many American towns. The plats are simple black-and-white representations of an area, noting streets, houses, lakes, and parks. Other features, such as cultural centers and noteworthy topography, are included when possible. BLM no longer distributes its own land plats, but they can be obtained through two sources: BLM township plats of Illinois, Indiana, Iowa, Kansas, Missouri, and Ohio are available through the **National Archives and Record Service**; township plats of other public-land states can be obtained from the Eastern States Office of the Bureau of Land Management (350 S. Pickett St., Alexandria, VA 22304; 703-343-5717).

The **National Park Service** keeps track of the purchase and development of federally protected parks. Maps showing the extent of each park's land-holdings are available at the parks themselves or from the Park Service (see "Recreation Maps").

The **Army Corps of Engineers** produces maps showing public water-recreation areas. Land-ownership maps depict the boundaries of

Land plat of Winnebago County, Illinois. Permission for reproduction granted by Rockford Map Publishers, Inc.

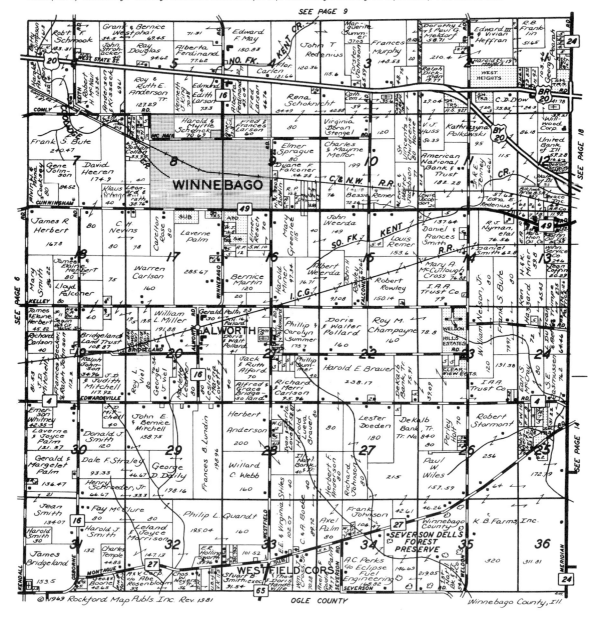

these lands and their potential for safe recreational use. The maps are inexpensive and are available from the Army Corps of Engineers District Offices (see Appendix B).

The **Tennessee Valley Authority** has land maps and reservation property maps available for its 40,190-square-mile region, encompassing parts of Tennessee, Alabama, Georgia, Kentucky, Mississippi, North Carolina, and Virginia. TVA maps show property corner markers, boundaries, bearings, and distances for areas or reservations affected by TVA reservoirs. There are several sheets covering each reservoir area, and free indexes for these sheets are available.

Holdings by the **Library of Congress** of pre-20th century land-ownership maps include 1,449 county maps, representing approximately one-third of all U.S. counties. The Library publishes *Land Ownership Maps: A Checklist* ($5.50), available from the Library or the **Government Printing Office**, which lists these maps and gives a brief history of the collection. Copies of maps listed may be ordered from the Library's Photoduplication Service. The library also has 1,296 county land-ownership maps contained on 105-millimeter microfiche. Individual diazo microfiche copies may be purchased for $2 per fiche ($10 minimum), with the complete set available from the Photoduplication Service. Another microfiche collection is entitled "Ward Maps of United States Cities" and includes 232 pre-20th century ward maps for 25 major cities. They are available from the Photoduplication Service on individual diazo microfiche ($2 each, with a $10 minimum) or as a complete set on either diazo microfiche ($645) or positive silver halide microfiche ($975).

The National Archives map collection includes the surveys of the Land Grant Office and later maps from the Bureau of Land Management. Indexes of these maps are available for study at the Archives' map research facilities in Alexandria, Virginia. Copies may be ordered through the Archives' reproductions service.

State, county, and local governments often produce land-ownership maps for such uses as settling zoning disputes or raising property taxes, and these may be available through local records or land offices. Other local sources include historical societies and long-established land sales companies, which may have ownership maps available for reproduction.

COMMERCIAL SOURCES

Publishing land-ownership maps is one of the oldest sources of income for commercial map companies. No sooner did any town spring up than the mappers arrived to plot boundaries and ownership agreements. Producers of land-ownership maps, especially land plats, may be found in most counties or population centers. Examples of maps being produced by such companies include:

M.A.P.S. Midwestern (City Park Rd., Oelwein, IA 50662; 319-283-3912) produces county/township maps depicting ownership and residency ($15 each). Maps are available for rural areas in Illinois, Minnesota, Missouri, South Dakota, Wisconsin, and eastern Iowa. Also available are books of residency ($15 each) for these areas.

Sanborn Map Co. (629 5th Ave., Pelham, NY 10803; 914-738-1649) began producing detailed city maps in 1866 and by the mid-1950s had diagrammed most U.S. communities with more than 2,500 people. Renowned for its 19th-century Fire Insurance Maps (now a prized collection in the Library of Congress), the Sanborn Map Co. has maps or atlases available for most cities. The maps and atlases, produced in both black-and-white and color, show roads, buildings, parks, fire hydrants, and other features, as well as places labeled by usage and type of structure. Sizes range from 11″ X 13″ to 22″ X 28″, and prices vary. The archival collection of Sanborn Fire Insurance Maps, 1867-1950, held by the Library of Congress, is available on microfilm.

■ **See also: "Boundary Maps."**

Political Maps

There are two types of political maps. In strict cartographic terms, a political map refers to maps that outline the political boundaries of the world. With their color-coded outlines of states and nations, these are the basic maps found in virtually every classroom, library, and home atlas—the ones that taught us early on that Italy looks like a boot "kicking" the island of Sicily, or that Mississippi was on the left and Alabama was on the right. This type of political map is created by practically every map-making company, as well as by most countries' governments. They can be found elsewhere in this book; see especially "Boundary Maps," "Foreign Country Maps," "Globes," "Selected Atlases," "State and Provincial Maps," "United States Maps," and "World Maps."

The other type of political map does more than simply divide nation from nation or state from state—it provides political characteristics, illustrating, for example, a ruling party's influence and role, or tracing a region's political development over time. These are "political" in the truest sense of the word, showing areas of human-rights abuse, for example, or countries under dictatorial rule. Many such maps are nonbiased and factual; some are created by those intending to deliver a message.

One of the best collections of political maps may be found in a book, *The New State of the World Atlas*, by Michael Kidron and Ronald Segal (New York: Simon & Schuster, 1984; $10.95), a paperback chock-full of fascinating four-color maps. Each compares the world's nations in some qualitative way, the choices and titles of which clearly reflect the authors' political leanings. Examples: "Shares in the Apocalypse" (which nations have how many of what kinds of weapons), "The First Slice of the Cake" (the percentage of gross national

"The United States as Seen from Canada." © World Eagle, Inc., 64 Washburn Ave., Wellesley, MA 02181, U.S.A. *Reprinted with permission.*

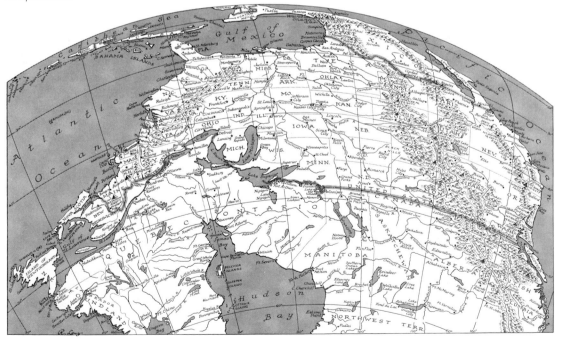

Map of the military balance in the Caribbean, from the State Department's Atlas of Foreign Relations.

■ "Israeli Settlement in the Gaza Strip" ($8.95; PB 88-928353).
■ "Lands and Waters of the Panama Canal Treaty" ($11.95; PB 88-928328).
■ "Registered Afghan Refugees in Pakistan" ($8.95; PB 88-928335).
■ "Major Insurgent Groups of Afghanistan" ($10; PB 85-928040).
■ "Muslim Peoples in the Soviet Union" ($10; PB 86-928358).
■ "China" ($11.50; PB88-928315).

product spent by each nation's government), "Webs and Flows" (each country's share of transnational parent corporations), and "Exploitation" (a ratio of the price of manufactured products compared to the wages of the workers who make them). In the back of the book, the idea and rationale behind each map is explained in simple terms. For those anywhere along the political spectrum, this atlas provides a wealth of thought-provoking maps.

While some commercial map publishers and government agencies produce political maps, most come from other sources, particularly public-interest groups, political organizations, and private foundations. There is no single directory of such organizations or the maps they produce. The gems must be uncovered one by one. Many of the best political maps are contained in atlases (see "Selected Atlases").

GOVERNMENT SOURCES, UNITED STATES

Central Intelligence Agency. The CIA creates several political maps, many of which are available from the National Technical Information Service. For information on ordering CIA publications, see Appendix B. CIA's political maps and atlases include:

Government Printing Office. The GPO distributes "Economic Map of the Persian Gulf" ($3.50; S/N 041-015-00140-0), a full-color 36" X 33" map showing terrain, primary transportation networks, oil and pipelines, tanker terminals, and oil refineries.

U.S. Geological Survey. As part of the National Atlas program, the USGS has published a two-sided map, "Presidential Elections, 1789-1968/1972-1984," that shows the results of the 1789-1968 elections by state, and the 1972-1984 elections by county, and includes U.S. totals ($3.10; US05627-38077-BH-NA-63M-00/38077-AY-NA-20M-00).

GOVERNMENT SOURCES, CANADA

Canada Map Office (615 Booth St., Ottawa, Ontario, Canada K1A 0E9; 613-952-7000) distributes maps of the *National Atlas of Canada* series, including many in the "political geography" section, such as:

■ "Canada—The 32nd Parliament" ($5.50 Canadian; MCR 4045).
■ "Canada—Confederation" ($5.50 Canadian; MCR 4051).
■ "Canada—Results of the 34th Federal

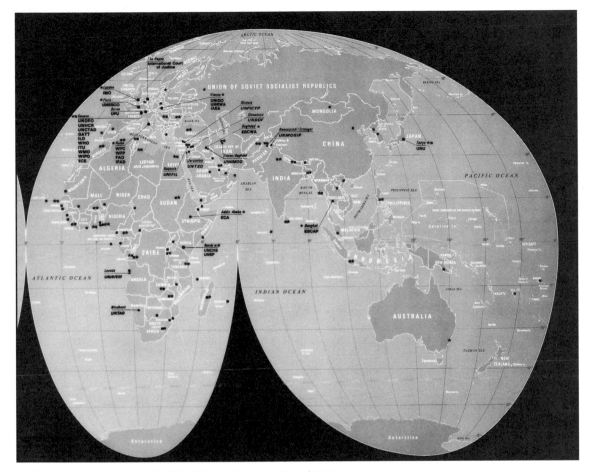

Portion of "Descriptive Map of the United Nations," courtesy United Nations.

Election, Nov. 21, 1988" ($5.50 Canadian; MCR 197).

Listings of National Atlas maps available, from both the 4th and 5th edition, are available upon request.

COMMERCIAL SOURCES

Many map-making companies produce political maps. A sample is:

George F. Cram Co. (P.O. Box 426, Indianapolis, IN 46206; 317-635-5564) publishes various political maps, including "States of Europe," four 87″ X 76″ sheets depicting names of every country in native languages, information about

the North Atlantic Treaty Organization (NATO), the Warsaw Pact, the European Community), and the Comecon ($249).

European Community Information Service (2100 M St. NW, Ste. 707, Washington, DC 20037; 202-862-9500) distributes official publications of the European Community, including a colorful free map, "The European Community," available in 12 languages. Other EC maps are available from **Unipub** (see below).

National Geographic Society (17th & M Sts. NW, Washington, DC 20036; 202-921-1200) offers several maps of foreign countries, including:

■ "Africa: Political Development" ($4; 02311; 23″ X 29″), with illustrations and text on both sides.
■ "Ireland: Political/Historical" ($4; 02843; 14″ X 21″).
■ "Philippines: Political/Historical" ($4; 20010; 20 1/2″ X 27″), printed on both sides.
■ "United States Political" ($4; 02003; 42 1/2″ X 29 1/2″).

Unipub (4611-F Assembly Dr., Lanham, MD 20706; 301-459-7666, 800-274-4888; 800-233-0504 in Canada) distributes several titles from the European Community, including "Political Map of the Community of Twelve" ($5) and "Population Map of the European Community, Present and Future" ($8).

United Nations Publications (Rm. DC2-0853, United Nations, New York, NY 10017; 212-963-8302) publishes "Descriptive Map of the United Nations" ($3; E.87.1.14), a color poster map of the world showing U.N. member states, giving their population and area, as well as pinpointing locations of U.N. offices and information centers. The U.N. publishes more than 6,000 other maps, which are parts of U.N. documents. As they are published, new U.N. maps are listed in UNDOC, a quarterly ($125 annually) that also gives ordering information.

UNDOC can be found in most public libraries.

World Eagle Inc. (64 Washburn Ave., Wellesley, MA 02181; 617-235-1415) publishes "Global Perspective Maps," black-and-white maps showing countries as they appear from certain neighboring countries. Ten titles available are: "U.S. as seen from Canada," "Africa as seen from India," "East Europe as seen from West Europe," "Southeast Asia as seen from People's Republic of China," "North Africa as seen from Mideast," "People's Republic of China as seen from Japan," "Central Africa as seen from South Africa," "Soviet Union as seen from Southeast Asia," "Latin America as seen from Cuba," and "Mideast as seen from Israel." The 24″ X 38″ maps are available as a set for $31.95 paper or $25.95 laminated. Also available are "Worldview Posters" ($8.95 to $9.95), world maps depicting themes, including "World Military Expenditures" (24″ X 37″ or 17″ X 23″), "World Energy Situation" (24″ X 37″), "U.S. and World Population" (24″ X 37″), and "World Interdependence" (24″ X 37″). The currency of data varies from map to map.

■ **See also: "Census Maps," "Foreign Country Maps," "Military Maps," "Selected Atlases," and "World Maps."**

Maps ◆ Travel Maps ◆ Bicycle Route Maps ◆ Mass Transit Maps ◆ Railroad Maps
aps ◆ Tourism Maps ◆ World Status Map ◆ County Maps ◆ Foreign Country Map
nd Maps ◆ State Maps ◆ United States Maps ◆ Urban Maps and City Plans ◆ Wo
rld Game Map ◆ Boundary Maps ◆ Congressional District Maps ◆ Land Ownersh
al Maps ◆ Scientific Maps ◆ Agriculture Maps ◆ Geologic Maps ◆ Land Use Maps
rce Maps ◆ Topographic Maps ◆ Wildlife Maps ◆ Antique Maps ◆ Researching C
c Site Maps ◆ History Maps ◆ Military Maps ◆ Treasure Maps ◆ Business Maps ◆
Emergency Information Maps ◆ Energy Maps ◆ Utilities Maps ◆ Water Maps ◆ N
Ocean Maps ◆ River, Lake, and Waterway Maps ◆ Tide and Current Maps ◆ Sky N
ical Charts ◆ Star Charts ◆ Star Magnitudes ◆ Weather Maps ◆ How to Read a W
mages as Maps ◆ Aerial Photographs ◆ Space Imagery ◆ How to Buy an Atlas ◆ A
Geography E

SCIENTIFIC MAPS

Map Stuff ◆ The Turnabout Map ◆ How to Choose a Map ◆ Map Projections N
t Map Skills ◆ Copying Maps ◆ Travel Maps ◆ Bicycle Route Maps ◆ Mass Transit
Maps ◆ Recreation Maps ◆ Tourism Maps ◆ World Status Map ◆ County Maps ◆
Maps ◆ Indian Land Maps ◆ State Maps ◆ United States Maps ◆ Urban Maps and
World Maps ◆ The World Game Map ◆ Boundary Maps ◆ Congressional District M
nership Maps ◆ Political Maps ◆ Scientific Maps ◆ Agriculture Maps ◆ Geologic
Maps ◆ Natural Resource Maps ◆ Topographic Maps ◆ Wildlife Maps ◆ Antique
ng Old Maps ◆ Historic Site Maps ◆ History Maps ◆ Military Maps ◆ Treaure Ma
Maps ◆ Census Maps ◆ Emergency Information Maps ◆ Energy Maps ◆ Utilities
ps ◆ Nautical Charts ◆ Ocean Maps ◆ River, Lake, and Waterway Maps ◆ Tide ar
s ◆ Sky Maps ◆ Aeronautical Charts ◆ Star Charts ◆ Star Magnitudes ◆ Weather
ead a Weather Map ◆ Images as Maps ◆ Aerial Photographs ◆ Space Imagery ◆
tlas ◆ Atlases ◆ Globes ◆ Geography Education Materials ◆ Map Accessories ◆
tions ◆ Map Software ◆ Map Stuff ◆ The Turnabout Map ◆ How to Choose a Map
ns ◆ Learning About Map Skills ◆ Copying Maps ◆ Travel Maps ◆ Bicycle Route N
nsit Maps ◆ Railroad Maps ◆ Recreation Maps ◆ Tourism Maps ◆ World Status M
Maps ◆ Foreign Country Maps ◆ Indian Land Maps ◆ State Maps ◆ United States
aps and City Plans ◆ World Maps ◆ The World Game Map ◆ Boundary Maps ◆ C
strict Maps ◆ Land Ownership Maps ◆ Political Maps ◆ Scientific Maps ◆ Agricul
Geologic Maps ◆ Land Use Maps ◆ Natural Resource Maps ◆ Topographic Maps
◆ Antique Maps ◆ Researching Old Maps ◆ Historic Site Maps ◆ History Maps ◆
Treaure Maps ◆ Business Maps ◆ Census Maps ◆ Emergency Information Maps ◆
Utilities Maps ◆ Water Maps ◆ Nautical Charts ◆ Ocean Maps ◆ River, Lake, and
Tide and Current Maps ◆ Sky Maps ◆ Aeronautical Charts ◆ Star Charts ◆ Star M
er Maps ◆ How to Read a Weather Map ◆ Images as Maps ◆ Aerial Photographs ◆
◆ How to Buy an Atlas ◆ Atlases ◆ Globes ◆ Geography Education Materials ◆ N
ies ◆ Map Organizations ◆ Map Software ◆ Map Stuff ◆ The Turnabout Map ◆
Map ◆ Map Projections ◆ Learning About Map Skills ◆ Copying Maps ◆ Travel N
oute Maps ◆ Mass Transit Maps ◆ Railroad Maps ◆ Recreation Maps ◆ Tourism N
atus Map ◆ County Maps ◆ Foreign Country Maps ◆ Indian Land Maps ◆ State N

Agriculture Maps

Silicon Valley may be America's gold mine, but corn fields and cow pastures are still its heartland.

The history of American farmers is the history of America itself: the Indians who taught white settlers how to grow corn; the gentlemen planters who led the American Revolution and formed the federal government; the antebellum plantation owners, brought to their knees in the Civil War; the sharecroppers, cowboys, and immigrant farmers who helped tame the West. These are a few of the people who created America's agricultural legacy.

A century ago, farming in the United States was an enormous industry. New technology increased productivity, creating a need for greater knowledge of the land and its resources. In 1899, the **U.S. Department of Agriculture** began its soil-survey program. At the time, little was known about the quality or content of the nation's soil. But the methodological survey created a fund of knowledge that has grown over the years to encompass a complex system of soil testing, categorization, treatment, and usage.

The surveys contain soil maps as well as general information about soil quality and use. Since the turn of the century, USDA has published 3,653 surveys, 1,972 of which are still available. Surveys published since 1957 include a number of different interpretations for the various soils mapped in each area. Among these are interpretations of estimated yields of common crops, land capability, range land, soil-woodland, and soil suitability for community or recreational use. The maps in these later surveys are printed on a photo-mosaic base, usually at scales of 1:24,000, 1:20,000, or 1:15,840.

Older surveys are useful to agricultural or land-use historians, but the maps are more general and the interpretations out of date.

More up-to-date soil surveys and other agriculture maps are used by farmers, engineers, land planners, developers, geologists, and agricultural scientists.

Through its **Agricultural Stabilization and Conservation Service (ASCS)**, USDA also produces and maintains a large collection of aerial photographs of agricultural lands. The aerial photography program began in 1935, as a result of a law passed to alleviate the farm crisis brought on by the Depression. The law, designed to establish and maintain a balance between agricultural consumption and production, required the extensive and accurate measurement of the nation's farmlands. But the standard method of mapping since George Washington's time—the surveyor's chain—was too slow for USDA's purposes, so, in the mid-1930s, the agency began using "rectified-to-scale" aerial photomaps. With the use of aerial photomaps and a measuring device called a planimeter, it was possible to make land measurements that were 99 percent accurate. Today, the ASCS has aerial photomaps covering all major U.S. agricultural areas.

USDA's soil-survey and aerial-photomap programs are two parts of an extensive federal agriculture-mapping effort. The USDA's Forest Service creates maps that indicate the use of forest resources (see "Recreation Maps"). Other agriculture-map-producing agencies include the **Central Intelligence Agency**, which produces maps of foreign agriculture, and the **Army Corps of Engineers** and **National Oceanic and Atmospheric Administration**, both of which create agriculture-related maps about water sources (see "River, Lake, and Waterway Maps").

State, county, and local agriculture and conservation departments are good resources for agriculture maps of specific local areas. A few commercial cartographers create maps for classroom use.

GOVERNMENT SOURCES, UNITED STATES

Agricultural Stabilization and Conservation Service (Aerial Photography Field Office, P.O. Box 30010, Salt Lake City, UT 84130; 801-524-5856), part of the U.S. Department of Agriculture, produces aerial photographs of U.S. agricultural land. Products available include Landsat images; Skylab 2, 3, and 4 imagery; photographs taken as part of the National High Altitude Program; and photos taken by the Forest Service and Soil Conservation Service. Prices for copies of aerial photographs range from $3 for a 10″ X 10″ black-and-white paper contact print to $45 for a 38″ X 38″ paper print made from a color positive. A price sheet (ASCS-441A) and order form (ASCS-441) are available upon request, as is an explanatory pamphlet, *Aerial Photography*.

Central Intelligence Agency. The CIA produces hundreds of maps, some depicting agriculture or vegetation in foreign countries, such as "Natural Vegetation in Africa" ($10; 800630), "China/Agricultural Regions" ($10; 800635) and "South Africa: Agricultural Activity" ($10; 800482). CIA maps and a map index are available from the National Technical Information Service (see Appendix B).

Forest Service. The U.S. Forest Service produces several forestry maps and atlases, many of them available from the Government Printing Office (see "Recreation Maps").

Government Printing Office. The following GPO publications contain agricultural maps or surveys and related text:

■ *Soil Survey Manual* ($17; S/N 001-000-00688-6; 503 pages), provides guidelines on survey techniques.
■ "Environmental Trends" ($11; S/N 041-011-

Soil types in the United States, from the U.S. Department of Agriculture's map "Soils of the World."

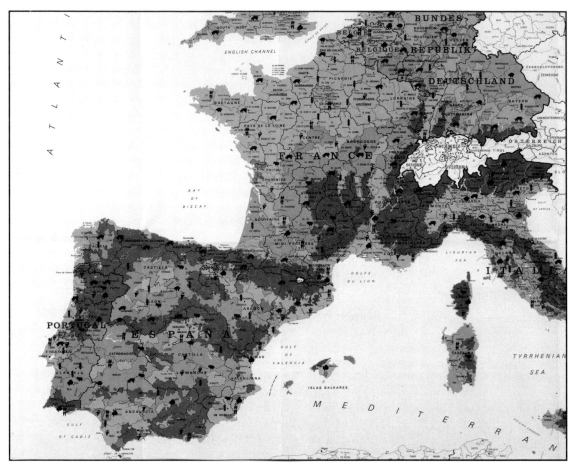

Portion of "The European Community: Farming," one of a series of four popular maps published by the European Communities Commission (distributed by Unipub).

00058-1; 352 pages), contains maps, charts, and text that show key changes in the environment.

■ "Major World Crop Areas and Climatic Profiles" ($8; S/N 001-019-00562-9; 165 pages), includes maps defining the key production regions and pinpointing production concentrations within the regions; histograms and statistical tables are also included.

Soil Conservation Service (Soils Div., Rm. 5105, South Bldg., Washington, DC 20250; 202-447-6385) has soil surveys for almost all of America's farmland. There are 1,972 available surveys, although many are out of date; 1,681 are out of print. The surveys still in print can be obtained at no charge by land-users and by

representatives of state and local offices of the Soil Conservation Survey, county agents, and congressional representatives. Many libraries have surveys on file for study and reproduction; these are good places to start the search for any of the out-of-print surveys. SCS publishes a free *List of Published Soil Surveys*, the most recent edition of which is available from state offices upon request. Also available from the national office and many state offices is "Soils of the World," a colorful and informative map. A list of state offices is available from the Publications Division of the Washington, D.C. office (see address above).

Tennessee Valley Authority. "Soils in the Tennessee Valley" ($2.50 plus $2 for mailing tube), compiled by TVA's Soils and Fertilizer Branch in cooperation with USDA's Soil and Conservation Service, is a colorful map describing in detail a variety of aspects of Tennessee Valley soils.

GOVERNMENT SOURCES, CANADA

Canada Map Office (615 Booth St., Ottawa, Ontario K1A 0E9; 613-952-7000) distributes maps from the *National Atlas of Canada, 5th Edition* series, including: "Agricultural Lands" (MCR 4022), "Soil Capability for Agriculture" (MCR 4023), "Farm Operators" (MCR 4047), and "Farm Types" (MCR 4109). The maps, available with English or French text, are $5.50 Canadian each.

Maps Alberta (Land Information Services Div., Main Fl., Brittania Bldg., 703 6th Ave. SW, Calgary, Alberta T2P 0T9; 403-297-7389) publishes numerous agricultural maps as part of the *Alberta Resource and Economic Atlas*. The "Agriculture Production" series consists of 16 maps, most produced in 1981, depicting the production of the following items: "Barley," "Canola," "Flaxseed," "Oats for Grain," "Tame Hay," "Rye," "Wheat," "Cattle and Calves," "Hogs, Milk Cows," "Poultry," "Poultry Hatcheries," "Sheep and Lambs," "Greenhouses," "Potato Growers," and "Vegetable Growers." The maps are 75 cents Canadian each. A "Soils" map from the "Physical Features" series of the atlas is also available (75 cents).

Maps Alberta also publishes Forage Inventory maps of Alberta's public land, which separate the natural vegetation into recognized plant associations ($3.50 Canadian each; Code 232; 30″ X 36″). They also publish Canada Land Inventory Agriculture maps for Alberta, available in booklets ($1.50 to $3.50 Canadian; Series CLI/ALI). Write or call for latest catalog.

GOVERNMENT SOURCES, ENGLAND

Ordnance Survey (Romsey Rd., Southampton S09 4DH; 0703-792398) publishes "European Community Maps," two of which are agriculture-related: "Farming" colorfully shows agricultural zones and products for each community region and includes an inset map with statistics for each member country and the entire European Community; "Soil Map" is a seven-sheet map showing the soils in the Community. It comes with two sheets of legend and an explanatory booklet.

COMMERCIAL SOURCES

Although most agriculture maps are created by government cartographers, there are a few commercial maps created as classroom tools that depict the development of agriculture in America; among them:

American Geographical Society (156 Fifth Ave., Ste. 600, New York, NY 10010; 212-242-0214) distributes A.W. Kuchler's map, "Potential Natural Vegetation of the Conterminus U.S." ($26; 61″ X 40″), which illustrates 116 types of vegetation, as well as states, major cities, water bodies, and topographic features.

George F. Cram Co. (P.O. Box 426, Indianapolis, IN 46206; 317-635-5564) produces a vibrant agricultural map as part of its American History Series, "Agricultural Regions of the United States" ($63 to $72, depending on mounting).

Geoscience Resources (2990 Anthony Rd., P.O. Box 2096, Burlington, NC 27216; 919-227-8300; 800-742-2677) distributes soil maps of many countries, acquired from various sources. Maps are available for countries such as Burkina Faso, Cameroon, Central African Republic, Great Britain, Italy, New Zealand, and the Soviet Union. Also available is a "World Soils Map" created by the United Nations Educational, Scientific, and Cultural Organization (UNESCO) and a vegetation map of China.

Librarie Pédagogique Mtl. Inc. (7957 St. Michel Blvd., Montreal, Quebec, Canada H1Z 3C9; 514-729-3844) produces an agriculture map of the European Community ($99 Canadian; M-1036-MP), showing the agriculture and formation of the European Community on one side, and participant countries' strongest economic characteristics on the other. A colorful "World Vegetation Map" ($79 Canadian; G-1777) is available, as well as thematic maps of Canada ($99 Canadian; MA-1244-MP), the U.S.S.R. ($99 Canadian; M-1165-MP), Asia ($99 Canadian; M-1065-MP), and China ($99 Canadian; M-1195-MP), with a physical/political country map on one side, and four themes on the other: agriculture, population, natural resources, and industry. All are available with French text; the European Community, World Vegetation, and Canada thematic map are also available in English.

Nystrom (3333 Elston Ave., Chicago, IL 60618; 312-463-1144; 800-621-8086) publishes a 50" X 80" brightly-colored teaching map titled "Patterns of Agriculture in the United States" ($43 to $66; QJ29), which depicts the products of America's major agricultural regions.

Rand McNally (P.O. Box 7600, Chicago, IL 60680; 312-673-9100) publishes agriculture maps in two of its classroom series. The "American History Maps" series for intermedi-

ate grades includes "Grain Growers on the Great Plains-Western North America" ($61; 114-12584-8), illustrating farming development in the West around 1912. Its "American Studies" Series includes "Farm, Factory, and Forest (1900)" ($46; 112-12521-2), showing the impact agricultural and industrial development have had on the land. Both measure 50" X 50".

Unipub (4611-F Assembly Dr., Lanham, MD 20706; 301-459-7666, 800-274-4888; 800-233-0504 in Canada) distributes official publications of the European Community, including the following colorful maps: "The European Community: Farming" ($5); "The European Community: Forests" ($5); "Soil Map of the European Community," a seven-sheet map with an explanatory booklet ($72 flat, $66 folded); "Map of Natural Vegetation of the Member States of the European Community and the Council of Europe," a four-sheet map with an illustrated explanatory booklet ($59); and "Forests in the European Community" ($5).

■ **See Also: "Aerial Photographs," "Land-Use Maps," "Natural Resource Maps," "Recreation Maps," "Selected Atlases," and "Weather Maps."**

Geologic Maps

More than any other type of map, geologic maps are portraits of the earth. Like any good portrait, a geologic map shows not only the face of its subject but also something of its inner nature.

Geologic maps show the earth's face by delineating the characteristics and distribution of exposed rocks and surface materials. And they make it possible to infer the earth's "inner nature"—the size, shape, and position of rock masses and mineral deposits they may contain—by means of symbols. These qualities are often interpreted with the help of slices, or cross-sections, made by combining surface observations with whatever subsurface information may be at hand from drill holes, mine workings, caves, or geophysical measurements.

Geologic maps are models of both space and time. The geologic history of an area can usually be reconstructed using a geologic map because the relative ages of rocks can often be determined, and even their ages in years estimated, if they contain fossils or certain naturally radioactive elements that decay at known rates. Geologic maps have been used to plot the pathways of meltwaters from glaciers that disappeared tens of thousands of years ago, leading, for example, to the discovery of groundwater reserves. Reserves of petroleum and natural gas are also found with the help of geologic maps.

Because they carry so much information, geologic maps are considerably more complex than topographic maps, which simply trace the surface contours of the earth (see "Topographic Maps"). Using a topo map requires mastering only two ideas beyond those needed to use a standard road map: the contour line and the interval between lines. Geologic maps, in contrast, convey a wealth of additional information. Some idea of their complexity may come from realizing that rocks, once formed—on the ocean floor, on river flood plains, beneath glaciers, or around and below volca-

noes—can be eroded; be lifted or lowered thousands of feet; be folded into troughs, basins, arches, domes, and far more intricate shapes; or be broken by great fractures and dragged or pushed hundreds of miles from

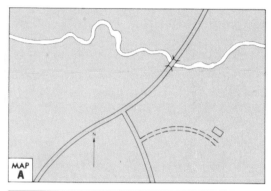

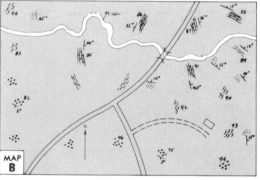

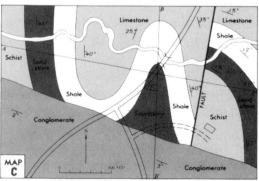

The development of a geologic map, from base map (top), to field map (center), to finished product (bottom).

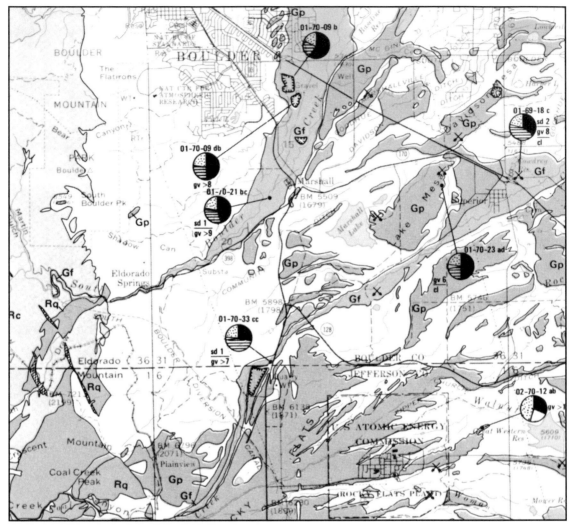

Map illustrating gravel and aggregate resources around Boulder, Colorado.

their place of origin. The earth has been actively making, deforming, eroding, and remaking rocks for more than four-and-a-half billion years. This entire history can be shown through symbols, colors, and patterns on geologic maps.

A geologic map is important for two reasons:

■ It is the most effective means of recording observations about rocks in a way that shows their spatial relationships to each other.
■ Along with the cross-sections that can be drawn from it, it is a device for study and analysis of many kinds of geologic features and processes, such as sequence and thickness of rock formations, their geologic structure, and their history.

Scientists and engineers in many fields use geologic maps as basic tools. Because we know, for example, that certain kinds of rock or geologic structures are associated with certain kinds of mineral deposits, geologic maps can help an exploration geologist find oil, coal, copper, uranium, and many other minerals.

Hydrologists use geologic maps to locate sources or movement paths of groundwater.

Geologic maps can be used for these and other purposes, whether or not the geologist who made them had such purposes in mind. A map prepared initially to solve a specific geologic problem may later help an engineering geologist choose between potential construction sites; a map made as part of a program of petroleum exploration may turn out, ultimately, to have far greater value in a search for uranium and potash.

GOVERNMENT SOURCES, UNITED STATES

The **U.S. Geologic Survey** makes many kinds of geologic maps as part of the continuing program to ". . . examine the geologic structure . . . of the national domain," as its mission is described. These maps may be published singly or in one of several series:

■ **Geologic Quadrangle Maps** are multicolor geologic maps on topographic bases in 7 1/2- or 15-minute quadrangle units, showing bedrock, surficial, or engineering geology. They include text and some maps by structure and columnar sections.

■ **Geophysical Investigations Maps** are based on topographic or planimetric maps showing the results of surveys using geophysical techiques, such as gravity, magnetic, seismic, or radioactivity, which reflect subsurface structures that are of economic or geologic significance. Many are correlated with the geology.

■ **Miscellaneous Investigations Series Maps**, drawn on topographic or planimetric map bases, include a wide variety of format and subject matter, including 7 1/2-minute quadrangle photogeologic maps showing geology as interpreted from aerial photographs. The series includes maps of Mars and the moon.

■ **Coal Investigations Maps** show bedrock geology, stratigraphy, and structural relations in certain coal-resource areas; they are based on topographic or planimetric maps.

■ **Miscellaneous Field Studies Maps** are color or black-and-white maps of quadrangle or irregular areas. Pre-1971 maps show bedrock geology in relation to specific mining or mineral-deposit problems; post-1971 maps are preliminary black-and-white maps on various subjects such as environmental studies or wilderness mineral investigations.

■ **Mineral Investigations Resources Maps** show geographic distribution and grade of mineral-resource commodities.

■ **Hydrologic Investigations Atlases** are color or black-and-white maps based on topographic or planimetric maps, showing a wide range of geohydrologic data.

■ **State Geologic Maps** for 14 states: Alaska, Arizona, Arkansas, Colorado, Kentucky, Massachusetts, Montana, Minnesota, New Hampshire, New Mexico, North Dakota, Oklahoma, South Dakota, and Wyoming.

■ **Special Maps**, including "Geologic Map of the United States," "Coal Map of North America," "Subsurface Temperature Map of North America," and "Basement Rock Map of the United States, Exclusive of Alaska and Hawaii."

Geologic maps also may be published as folded sheets in envelopes bound with book-type series, such as "Bulletins," "Water-Supply Papers," and "Professional Papers," the texts of which contain descriptive and interpretive material that the maps alone cannot provide.

There are also several full-color 19″ X 28″ thematic maps from USGS's authoritative National Atlas Program, including "Geology" (1966), which shows the distribution of sedimentary, volcanic, and intrusive rock types; "Surficial Geology" (1979), showing the distribution of transported, untransported, and other deposits for the United States; "Classes of Land-Surface Form" (1964) and "Tectonic Features" (1968) covering the lower 48 states; "Land Surface Form" (1968) and "Tectonic Features" (1968) covering Alaska. Each is $3.10, available from USGS.

One of the best information resources at USGS is the **Geologic Inquiries Group** (GIG, USGS, 907 National Ctr., Reston, VA 22092;

703-648-4383). GIG provides specific responses to all geology-related questions from the public, other government agencies, and within the USGS. GIG specialists maintain up-to-date reference lists on popular geologic subjects, listing USGS publications (as well as those of other organizations) on frequently-asked-about subjects. These subjects include the geology and resources of each state; prospecting and mineral economics; oil and gas maps of the United States; earthquakes and volcanoes; state geologic maps; rocks, minerals, and gemstones; geology of the National Parks; plate tectonics; careers in the geosciences and environmental technology.

GIG also publishes Geologic Map Indexes (GMI's) for each state, and maintains the GEOINDEX data base. GMI's show outlines of published geologic maps more detailed than the state geologic maps. Publishers of geologic maps shown include the USGS, other federal agencies, state geoscience agencies, universities, professional societies, and private publishers. GIG helps inquirers find the geologic map for their area of interest, and directs them to the appropriate publisher. GEOINDEX is the bibliographic data base maintained by GIG that can be queried to answer most geologic map questions.

Two additional free publications available from GIG are *State Geologic Maps*, listing maps available for each state, and *Geologic Maps of the United States, North America, and Large Regions of the United States*, listing maps available from USGS and other sources.

Also available from GIG are two packets of earth-science teaching aids, which differ according to grade level and geographic location. The packets include lists of reference materials, various maps and map indexes, and a selection of general interest publications. The packets are free to teachers who mail requests to GIG on school letterhead.

A complete description of USGS-produced maps can be found in the free publication *Guide to Obtaining USGS Information* (Circular 900), available from USGS Books and Open Files, 25425 Federal Center, Denver, CO 80225.

USGS-published geologic maps range widely in scale depending on the type of information to be portrayed. A single mine or landslide may be drawn on a scale of 1:400, for instance, while the National Atlas series features geologic maps drawn on a scale of 1:7,500,000. However, most USGS maps come in one of five standard scales: 1:24,000, 1:48,000, 1:62,500, 1:100,000, or 1:250,000. The smaller-scale maps are used for general planning and resource evaluation over large regions. Larger-scale maps may be used for detailed planning, zoning, site selection, and resource evaluation. Geologic maps at 1:250,000 or larger scales are available for nearly 85 percent of the United States. The entire country has been mapped by USGS at 1:2,500,000 scale; only about fifteen percent of the United States has been mapped at 1:24,000.

Geologic maps are available directly from USGS and its Earth Science Information Centers (see Appendix B). USGS maintains free geologic map indexes by state, which are available from USGS Books & Open-File Reports (BORFOR) Centers in Denver, Colorado and Fairbanks, Alaska, as well as from Earth Science Information Centers.

GOVERNMENT SOURCES, STATE

Regional USGS Earth Science Information Center offices also sell maps of local areas and books of local and general interest. In addition, most states have a Geological Survey, Bureau of Mines, or Department of Natural Resources that publishes and sells geologic maps. The **New Mexico Bureau of Mines and Mineral Resources** (Publications Rm., Socorro, NM 87801; 505-835-5334), for example, publishes a map "Geology of Veteado Mountain quadrangle," which sells for $5.50; **Oklahoma's Geological Survey** (100 Boyd, Rm. N-131, Norman, OK 73019; 405-325-3031) sells a "Geologic map of Oklahoma" for $7.30. There are countless others. See Appendix A for names and addresses of state geoscience agencies.

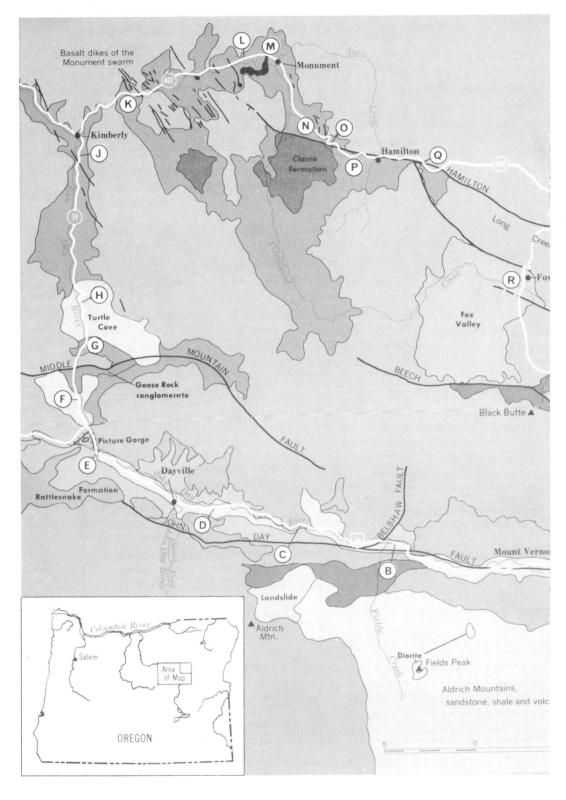

Government Sources, Canada

Canada Map Office (615 Booth St., Ottawa, Ontario K1A 0E9; 613-952-7000) distributes maps from *The National Atlas of Canada, 5th Edition*, including "Geomorphology Relief — Canada" ($5.50 Canadian; MCR 4097).

Maps from the 4th edition of the *National Atlas* are also available ($2 Canadian each), including: "Geology" (MCR 1123), "Geological Provinces" (MCR 1124), "Tectonics" (MCR 1125), "Retreat of the last ice sheet" (MCR 1126), "Glacial Geology" (MCR 1127), and "Post-Glacial Rebound" (MCR 1128). All maps from *The National Atlas* are available with either French or English text.

Maps Alberta (Land Information Services Div., Main Fl., Brittania Bldg., 703 6th Ave. SW, Calgary, Alberta T2P 0T9; 403-297-7389) distributes geology maps produced by the Geological Survey of Canada. The color maps ($7.50 Canadian each) focus on various aspects of Canadian geology. Available maps are "Principal Mineral Areas of Canada" (900A), "Geological Map of Canada" (1250A), "Tectonic Map of Canada" (1251A), "Mineral Deposits of Canada" (1252A), "Glacial Map of Canada" (1253A), and "Physiographic Regions of Canada" (1254A). Maps Alberta also distributes maps from the *Alberta Resource and Economic Atlas* ($7.50 Canadian each), including "Bedrock Geology" (D6).

Commercial Sources

AAPG Bookstore (P.O. Box 979, Tulsa, OK 74101; 918-584-2555) is affiliated with the American Association of Petroleum Geologists and distributes a series of "highway maps" that include compilations of state and regional geologic data with an overlay depicting major highways and landmarks. The entire country is covered in twelve maps ($5 each) at a scale of one inch per thirty miles. AAPG's "Map of the World Project" consists of maps produced by the International Union of Geological Sciences. The six titles include "Tectonic Map of South America" ($28), "Tectonic Map of Europe" ($18), "Tectonic Map of South and East Asia" ($85), "Metamorphic Map of Africa" ($18), "Metamorphic Map of Europe" ($54), and a *Geologic World Atlas* ($285), a large-format atlas with 22 sheets designed to cover world geology. Also available is "Geological Map of Sector of Gondwana" ($24). AAPG stocks a wide variety of other geologic maps.

Circum-Pacific Map Project (345 Middlefield Rd., MS 952, Menlo Park, CA 94025; 415-329-4002) is an international cooperative effort between the Circum-Pacific Council for Energy and Mineral Resources and the USGS to assemble and publish new geological and geophysical maps of the Pacific Ocean and surrounding areas. The widely acclaimed maps are intended to show the relation of geology, tectonics, and crustal dynamics to known energy and mineral resources, and also to aid in exploration for new resources. The area has been divided into six regions (Northeast Quadrant, Northwest Quadrant, Southeast Quadrant, Southwest Quadrant, Antarctic Region, and "Total Area"). Geographic, base, plate-tectonic, geodynamic, and mineral resources maps have been completed for all areas ($12 each; set of 6 maps, $40); "Tectonostratigraphic Terranes" and "Manganese Nodule/Sediments" maps have been produced for the "Total Area" ($12 each). The maps are distributed through the AAPG bookstore (see address above).

Crystal Productions (Box 2159, Glenview, IL 60025; 312-657-8144; 800-255-8629) sells four "Geologic Map Portfolios" ($39.95 for set): "Portfolio 1—Physical Geology" ($14.50), including eight full-color geologic maps with 23 color cross-sections and a set of the same sections partially unfinished for student completion; "Portfolio 2—Historical Geology" ($14.50), containing 13 color maps showing major geologic provinces of the U.S.; "Portfolio 3—Physical Geology" ($15.50), including bed-

(continued on page 133)

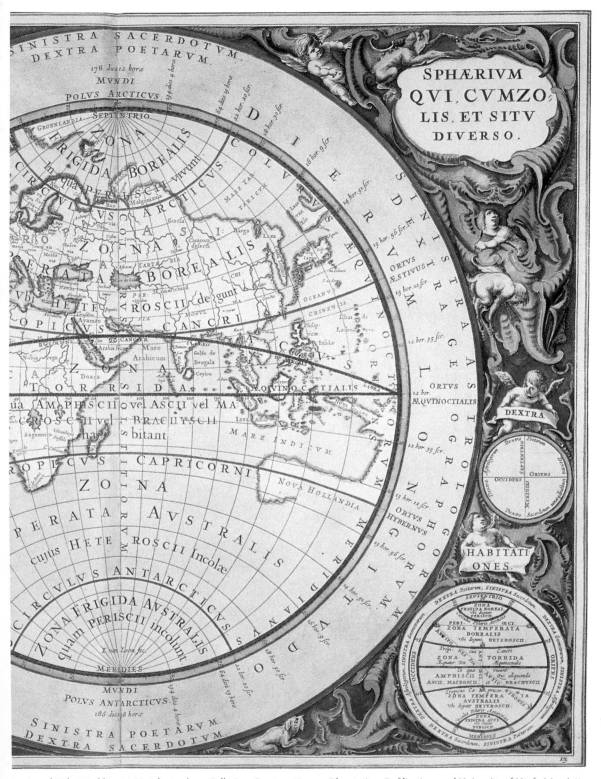

Portion of "The World in 1660," by Andreas Cellarius. Courtesy Hansen Planetarium Publications and University of Utah, Marriott Library Special Collections.

Aerial photograph of Wilmington, Delaware. Courtesy U.S. Geological Survey.

Space imagery of Grand Canary Island by the French satellite Spot. ©1988 CNES. Provided courtesy SPOT Image Corporation, Reston, Virginia.

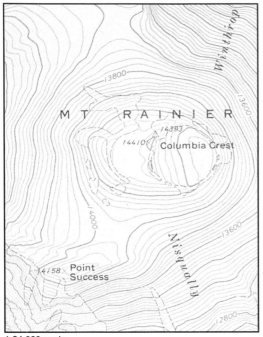

1:24,000-scale

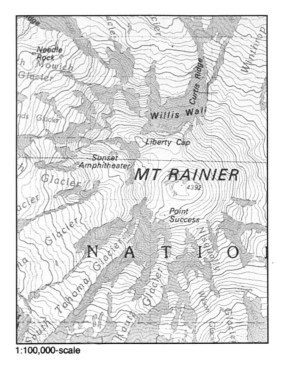

1:100,000-scale

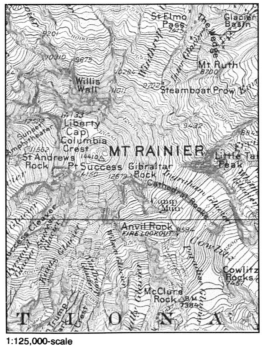

1:125,000-scale

Topographic maps of Mt. Rainier, Washington, illustrating three perspectives of the same region at different scales. Courtesy U.S. Geological Survey.

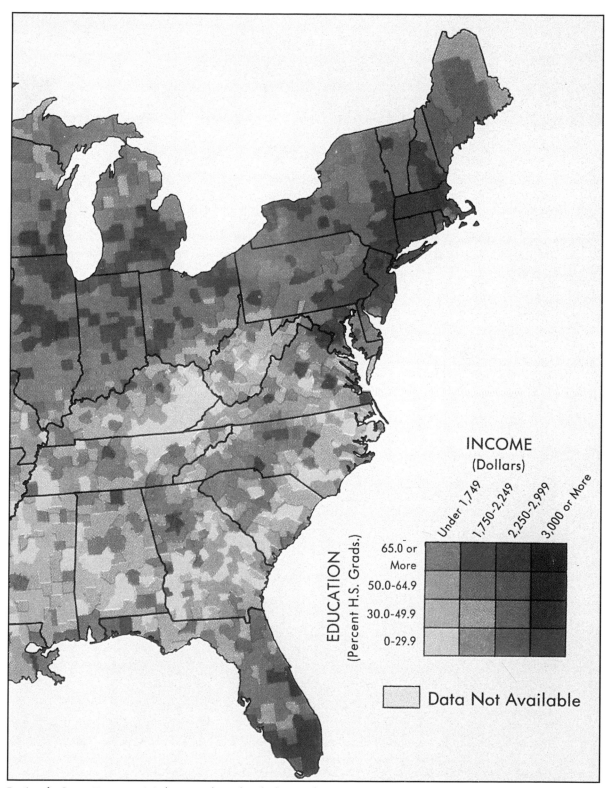

INCOME
(Dollars)

Under 1,749

1,750-2,249

2,250-2,999

3,000 or More

EDUCATION
(Percent H.S. Grads.)

65.0 or More

50.0-64.9

30.0-49.9

0-29.9

Data Not Available

Portion of a Census Bureau statistical map, "Relationship of Educational Attainment to Per Capita Income, 1969."

"Geology and Active Faults in the San Francisco Bay Area," from the Pt. Reyes National Seashore Association, showing the
fault lines and adjacent geology, described in terms of relative stability during an earthquake.

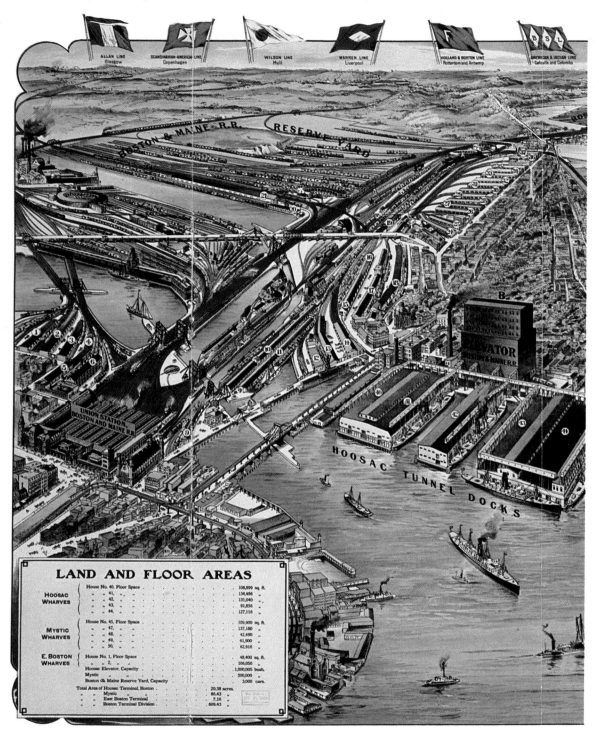

Portion of a 1905 bird's-eye-view map of Boston's railroad and shipyards, identifying individual buildings and facilities. Courtesy Library of Congress.

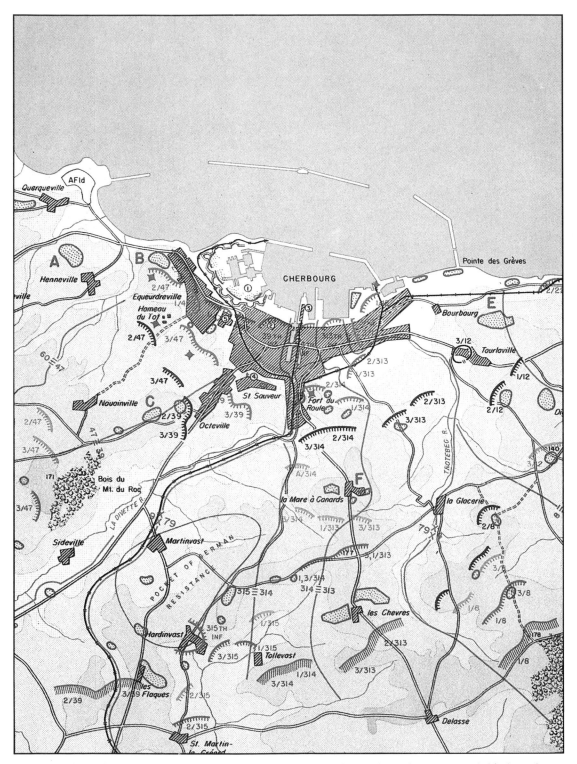

Portion of "The Final Drive On Cherbourg," a World War II military map showing the "night positions reached by forward elements" on June 21-25, 1944. Courtesy Government Printing Office.

(continued from page 124)

rock and surficial geologic maps illustrating structures and stratigraphy across the U.S.; and "Geologic Quadrangles" ($8.50), with four full-color geologic maps and cross sections. Most portfolios come with a guide.

W.H. Freeman and Company (41 Madison Ave., New York, NY 10010; 212-576-9400) publishes "The Bedrock Geology of the World" ($34.95; 48″ X 76″), a full-color map drawn at a scale of 1:23,230,300, showing the bedrock geology of continents as well as ocean basins.

Geological Map Service (P.O. Box 920, Sag Harbor, NY 11963; 516-725-0780) distributes a wide range of international geologic maps, including many hard-to-find foreign-produced maps, some translated into English. Offerings include: "Arctic Atlas" ($125); "Deep Geological Sections of Northern Asia" ($94); "Geological Map of Eurasia" in 12 sheets ($94); and maps of China and the Far East.

Geological Society of America (3300 Penrose Pl., P.O. Box 9140, Boulder, CO 80301; 303-447-2020; 800-472-1988) publishes several geologic maps in their Map and Chart series. Examples are "Geologic Map and Cross Section of the Eastern Ouachita Mountains, Arkansas" ($7.20 folded; $8.80 rolled), a color 46″ X 28″ map with an 8-page text; and "Map and Cross Section of the Sierra Nevada from Madera to the White Mountains, Central California" ($8 folded; $9.60 rolled), two full-color sheets with a 4-page text. A catalog is available.

Geoscience Resources (2990 Anthony Rd., P.O. Box 2096, Burlington, NC 27216; 919-227-8300; 800-742-2677) is an excellent source for geologic maps of all types—bedrock geology maps, bouguer gravity maps, quaternary geology maps, seismic maps, and tectonic maps of the U.S., states, and foreign countries. Selections include "Surficial Geology Map of Maine" ($8.95; 62-7105), "Magnetic Anomaly Map of South Australia" ($19.95; 64-1368), and "Seismotectonic Map of Iran/Afghanistan/Pakistan" ($15; 64-7820). A complete catalog is available.

Gulf Publishing Company (Book Div., P.O. Box 2608, Houston, TX 77252; 713-520-4444) publishes two geologic maps: "The Breakup and Dispersion of Pangea" ($20; 30″ X 40″) and "Sinai Area Geological and Geomorphological Map" ($95; in two 38″ X 20″ sheets), created from Landsat imagery.

Mountain Press Publishing Co. (2016 Strand Ave., P.O. Box 2399, Missoula, MT 59806; 406-728-1900) publishes "Roadside Geology" guides giving a road-by-road study of geologic features as revealed from a car window; sketch maps appear throughout. The guides ($9.95 to $14.95 each) are available for Alaska, Arizona, Colorado, Idaho, Montana, northern California, New York, Oregon, Pennsylvania, Vermont/New Hampshire, Utah, Washington, Wyoming, and "Yellowstone Country."

National Geographic Society (17th & M Sts. NW, Washington, DC 20036; 202-921-1200) sells "Earth's Dynamic Crust" ($4; 33″ X 42″) and "Earth's Dynamic Crust/Shaping of a Continent" ($4; 23″ X 18″), laminated maps illustrating world geology and tectonics.

Pt. Reyes National Seashore Association (Pt. Reyes National Seashore, Pt. Reyes, CA 94956; 415-663-1155) publishes a colorful 17″ X 28″ map, "Geology and Active Faults in the San Francisco Bay Area," showing the major fault lines as well as the four major rock types, color coded by relative stability in an earthquake. The poster is $3 plus $2 postage; California residents add 18 cents sales tax.

Unipub (4611-F Assembly Dr., Lanham, MD 20706; 301-459-7666; 800-274-4888; 800-233-0504 in Canada) distributes United Nations Educational, Scientific, and Cultural Organization (UNESCO) publications and maps. The selection varies from year to year; send for the latest catalog. Maps available for 1990 include: "International Hydrogeological Map of Europe," a 49-sheet trilingual map printed in 55 colors ($18 to $26 per sheet); "Metallogenic Map of South and East Asia" ($105), a 4-sheet English/French map co-published with the Commission for the Geological Map of the World; "International Geological Map of Africa," a five-sheet map of the continent ($20 per sheet); "International Tectonic Map of Africa" ($74), a 9-sheet map with explanatory brochure; and *Geological World Atlas*, 22 sheets and explanatory notes in a binder ($455).

Western GeoGraphics (Box 2204, Canon City, CO 81212; 719-275-8948) produces full-color state geologic highway maps for Colorado, Kansas, and Wyoming. The maps ($6 each, plus $1 shipping if folded, $2 if rolled) are drawn on a scale of 1:1,000,000; sheet size varies.

Williams & Heintz Map Corp. (8119 Central Ave., Capitol Heights, MD 20743; 301-336-1144) sells "geologic portfolios" designed for teaching geology: Portfolio No. 1, "Physical Geology" ($8.50), consists of eight full-color geologic maps and sections, with accompanying text; Portfolio No. 2, "Historical Geology" ($8.50), consists of thirteen full-color maps that cover the major geologic provinces of the United States; and Geologic Portfolio No. 3, "Physical Geology," consists of eight full-color bedrock and surficial geologic maps ($9.50). Another teaching package is Quadrangle Portfolio No. 1 ($5), containing four full-color geologic maps of four quadrangles with interesting geology, with accompanying text.

Williams & Heintz also produces and prints, but does not publish or sell, geologic maps. Examples include:

■ "Geology and Waste Disposal in Florida" ($1), available from **Florida Bureau of Geology**, Dept. of Natural Resources, 903 W. Tennessee St., Tallahassee, FL 32399.
■ "Geologic Map of the Northwest Cascades, Washington, 1987" ($14.50), available from **Geologic Society of America**, P.O. Box 9410, Boulder, CO 80301.
■ "Geologic Map Showing Magnetics, Gravity, and Rock Alterations, Bodie Mining District, California" ($18), available from **California Division of Mines and Geology**, P.O. Box 2980, Sacramento, CA 95812.
■ "Geologic Map of Alabama" ($17.50 flat; $19.50 rolled; $3 postage), available from **Geological Survey of Alabama**, Publication Sales, P.O. Box 1924, Tuscaloosa, AL 35486.

M.P. Weiss (Dept. of Geology, Northern Illinois Univ., DeKalb, IL 60115; 815-753-1943) publishes "Map of Modern Reefs and Sediments of Antigua, West Indies" ($14 rolled; $11 folded), a full-color 27″ X 40 1/2″ map with descriptions and illustrations of bottom sediments and communities on the reverse side, and a record of changes over time. Also available is a 116-page guide, "Antigua" ($11), describing the bedrock geology, reefs, and archaeology of the island; a tourist map of the island is included.

The guide and map are available together for $19.

■ **See also: "Emergency Information Maps," "Energy Maps," and "Natural Resource Maps."**

Land-Use Maps

In a world where every conceivable resource, from sunshine to soil to salt water, has a utilitarian purpose for people and nature, maps depicting land use are essential for keeping track of where and how our land is being used. Land-use maps can provide answers to such questions as what percentage of a region is used for industry, or how much of an agricultural region is cropland rather than pasture, and which land is prime for residential development.

Environmentalists, industrialists, foresters, farmers—anyone who needs to understand what land is being used and where it is being over- or underexploited—find land-use maps necessary tools of their trade. The development of virtually any transportation route, utility, energy, or urban-renewal project depends on the information found on such maps. The "big picture" about past, present, and future land use aids in making the best possible use of a region's soil, rock, sediment, vegetation, wildlife, and water.

Modern technology has led to the digitization of much of the Earth's surface, resulting in a new generation of extremely accurate land-use maps. The early stages of these digitized maps look a bit like a child's paint-by-numbers art kit. The field of the map is usually a basic light-blue topographic map, with major roads, rivers, forests, and parks outlined and named. The sections of the land given to specific uses are outlined with thin black lines (many of the outlined areas being so small that they resemble tiny amoebas on the map) and numbered according to use. These early versions, although not particularly pretty or simple to read, are quite adequate for most land use purposes.

The next step in the process produces more attractive and readable land-use maps. With computer assistance, or even diligent hand-coloring, these basic maps can be turned into vivid illustrations of land use. Splashes of red, orange, and green are applied according to the digitized information within each outline, and the once dull-looking map suddenly rivals the most lovely satellite image for beauty and vibrance.

The **U.S. Geological Survey** is presently involved in using such digital methods to map the land use and land cover for the entire United States. Land-cover maps illustrate, in less detail than topographic maps, the type of natural or man-made "cover" (buildings, vegetation, and water, for example) over the Earth's surface. Because land cover is often an indicator of past, present, or potential land use, these maps are helpful supplements to land-use maps.

The USGS project, begun in 1975, uses a two-level classification system to depict the various types of land use and land cover in America. The first classification level is divided into nine segments relating to different land characteristics: urban or built-up land, agricultural land, range land, forest land, water, wetland, barren land, tundra, and perennial snow and ice. These classifications are divided into subclassifications, comprising the second level of classification. The result is that USGS's system for describing the land use and cover of, say, barren lands, is broken into such subcategories as dry salt flats; beaches; strip mines, quarries, and gravel pits; and transitional areas.

All this is less complicated than it sounds, and the highly specific system of delineating land types and uses actually simplifies the process further by taking the guesswork out of map reading. At a glance, one can now see exactly what kind of farming or ranching is taking place on a given plot of land; earlier land-use maps were considerably more vague about depicting such specifics.

Aside from USGS maps, other sources of land-use maps are the state, county, or local departments involved with land use and

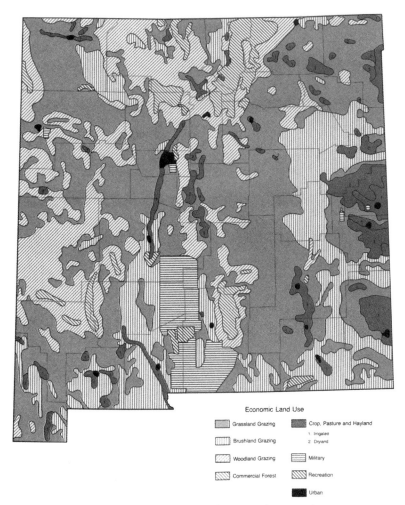

Economic Land Use

Grassland Grazing · Crop, Pasture and Hayland
1. Irrigated
2. Dryland

Brushland Grazing

Woodland Grazing · Military

Commercial Forest · Recreation

Urban

U.S. Agriculture Department's economic land-use map of New Mexico.

is the "Index of Land-Use and Land-Cover and Associated Maps."

GOVERNMENT SOURCES, CANADA

Canada Map Office (615 Booth St., Ottawa, Ontario K1A 0E9; 613-952-7000) distributes the "Northern Land Use Information" map series, produced by the Energy, Mines and Resources Canada office. The series, produced since 1971, presents information on wildlife, fish resources, Indian and Inuit land use, ecological land classification, and socio-economic and cultural data, both in map and text form. Maps are available for all areas in northern Canada; they are $3.50 Canadian each or can be purchased in sets. An index is available upon request.

development. Regional boards or bureaus of agriculture, natural resources, and land zoning are good starting points for finding specific kinds of land-use maps (see Appendix A for addresses of many of these agencies).

GOVERNMENT SOURCES, UNITED STATES

U.S. Geological Survey. USGS is the best source for maps created specifically for land-use and land-cover studies. USGS's maps are available as stable-base film positives, semi-stable diazo foils, paper diazos, or in digital format. Prices vary; a price list may be obtained free from the Earth Science Information Center (see Appendix B). Also free from USGS

COMMERCIAL SOURCES

Geoscience Resources (2990 Anthony Rd., P.O. Box 2096, Burlington, NC 27216; 919-227-8300; 800-742-2677) stocks a vast inventory of all types of maps. Some land-use maps are available, such as the "China—Land Use Map" ($10.95; 64-4093), showing major land-use classifications. A catalog is available upon request.

■ **See also: "Agriculture Maps," "Business Maps," "Geologic Maps," "Energy Maps," "Natural Resource Maps," "Recreation Maps," "River, Lake, and Waterway Maps," and "Utilities Maps.**

Natural Resource Maps

The modern world is paradoxically dependent on the natural world: The more technology we invent, the greater our need for resources to build or fuel this technology. From minerals to forests, natural gas to sunshine, all that the earth, sea, and air have to offer has been mapped, charted, and diagrammed.

The ecology movement of the 1960s and 1970s, with its philosophy of conservation rather than depletion, led the way to a new consciousness about the earth. "Use it or lose it" became "save it or else." Today, forests that are harvested are simultaneously replanted, "exotic" technologies such as solar and wind power are becoming commonplace, and the search for new sources of traditional energy has intensified. Maps often lead the way to finding and exploiting these resources wherever they lie—below the ground, beneath the ocean floor, or shining from above.

In 1871, when John Wesley Powell and his brother-in-law, Almon Harris Thompson, set out on a mapping survey of the American West, they were expected merely to fill in the details of an earlier, less-comprehensive map. But Powell had a new vision of the untamed land: He wanted to go beyond the preconceived notions of geologic and topographic mapping. Earlier maps had concentrated on geology for merely historical or industrial purposes. Topography, likewise, was treated with a narrow-minded eye toward the shape of the land and its suitability for agriculture and ranching. Powell believed that understanding the land and the way it was formed led to a deeper understanding of its usefulness for man. By surveying and mapping the land in relation to its ongoing development and specific uses, Powell predicted that future discovery and development of natural resources would be enhanced.

The Powell-Thompson maps are geologic landmarks, concentrating for the first time on mineral and soil development as well as studying the catalysts for the chosen path of a river or formation of a mountain. These studies shed new light on the search for and extraction of natural resources. One cartographic idea that Powell espoused but was never able to see carried out was the concept of "scientific classification" of the land. He believed that topographic maps were not specific enough in the nature of available resources; he wanted mappers to categorize the land in more specific terms of usefulness. Instead of "farmland," for example, Powell believed maps should read "irrigation land" or "pasturage land," while mining lands should be labeled with the type of mineral or coal beneath the surface rather than just the generic term "mining." Although Powell's idea did not catch on at the time, his concept of mapping land classification now plays an important role in the exploration and use of natural resources.

The mapping of natural resources in America is an ongoing process for both government and commercial cartographers. As older resources are depleted, new ones found, and different methods of retrieval created, there are maps made to reflect the changes. And because even greater care is taken to protect the environment and nurture its riches, there will always be work for mappers of natural resources.

Energy and mining companies, farmers, builders, foresters, conservationists, and anyone else interested in land use find natural resource maps invaluable. The diagramming of natural resources includes solar energy atlases, forest, oil field, and mineral, coal, and gas investigations maps. There are maps of past, present, and potential energy sources, as well as maps detailing lands where energy exploration is prohibited. These and other maps are available from the government and from selected commercial organizations.

Northern portion of the "California Water Map," published by the Water Education Foundation. ©1987, revised 1989.

GOVERNMENT SOURCES, UNITED STATES

Fish and Wildlife Service. In 1977, the Fish and Wildlife Service began the National Wetlands Inventory in an effort to classify and map America's remaining wetlands. As part of the National Wetlands Inventory Program, the USGS has been mapping the marshes, swamps, ponds and bogs of America, which currently represent about 5 percent of the total land surface of the lower 48 states. Of the 215 million acres of wetlands that originally existed in the conterminous U.S., the Fish and Wildlife Service estimates that there are only about 99 million acres left. More than 25,000 composite and overlay maps of the wetlands have been created to date and are used for a variety of

purposes, including town planning, zoning, flood-hazard planning, waste treatment, and water-quality planning. Information about NWI and the maps is available through Earth Science Information Center offices.

National Geophysical Data Center. NGDC produces maps depicting geothermal resources in the U.S., such as "Geothermal Energy in Alaska and Hawaii," "Geothermal Resources of Kansas," and "Geothermal Energy in the Western United States." Most maps are $5.

Tennessee Valley Authority. TVA has a mapping services branch that creates maps and charts for the local region, including:

■ "Geologic Map and Mineral Resources Summaries" ($2.50 each plus $1.50 postage). These geologic maps cover separate portions of Tennessee and come with a mineral resources summary in a 7 1/2-minute quadrangle format. A free listing of these maps is available from TVA.

■ "Mineral Resources of the Tennessee Valley Region" ($4 plus $2 for mailing tube and $1.50 postage). This 1970 map measures 35″ X 49″ and is printed in full color.

■ "Mineral Resources and Mineral Industries of Tennessee" ($2.50 plus $2 for mailing tube and $1.50 for postage). This 40″ X 66″ map illus-trates the state's rich lode of mineral resources.

Other government sources of natural resource maps include state, county, and local departments of natural resources, ecology, forestry, and mining (see Appendix A for addresses).

U.S. Geological Survey. USGS's National Atlas Program includes the following maps on natural resources:

■ "Major Forest Types" ($3.10; US00420-38077-AM-NA-07M-00), a 1967 map showing the distribution of eastern, western, Hawaiian, and Alaskan forest and nonforest lands.

■ "Networks of Ecological Research Atlas, 1983" ($3.10; US00725-38077-AO-NA-07M-00) shows the location of research areas designated by public and private agencies as secure sites for basic and applied studies of natural processes.

■ "Principle Lands Where Exploration and

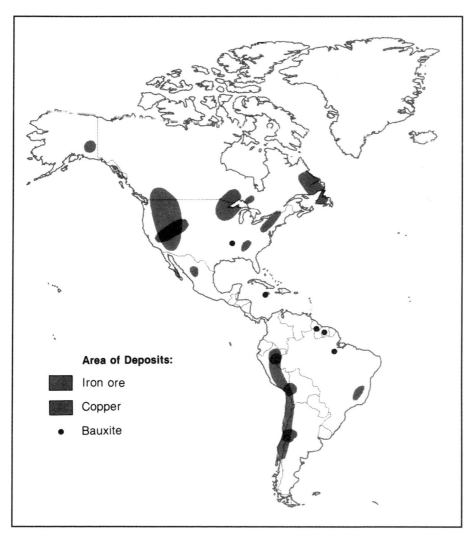

Portion of map showing mineral deposits of iron ore, copper, and bauxite, from the State Department's Atlas of United States Foreign Relations.

Area of Deposits:

▮ Iron ore

▮ Copper

• Bauxite

Development of Mineral Resources are Re-stricted" ($3.10; US00421-38077-AP-NA-07M-00). This 1981 map shows areas of 5,000 acres or more where mineral development is prohib-ited or severely, moderately, or slightly re-stricted. Alaska and Hawaii are included on the reverse.

USGS also produces coal, oil, and gas investi-gations maps (see "Energy Maps"), as well as a series of "Mineral Investigations" maps and studies. Write to the nearest USGS Earth Science Information Center for listings of available maps and surveys (see Appendix B).

GOVERNMENT SOURCES, CANADA

Canada Map Office (615 Booth St., Ottawa, Ontario K1A 0E9; 613-952-7000) distributes "Canada's Wetlands," a series of two maps of wetland areas in Canada: "Canada—Wetland Distribution" (MCR 4107E), showing the regional occurence of wetlands, which cover 14 percent of the country; and "Canada—Wetland Regions" (MCR 4108E), illustrating the types of wetlands found in Canada, and including descriptions of the ecology of major wetland types in each region. The maps are $5.50 Canadian each. It also distributes maps from the *National Atlas of Canada, 5th Edition,* including maps of the entire country depicting "Coal" (MCR 4053) and "Mineral Commodity Flows" (MCR 4081). The maps, $5.50 Canadian each, are available in either English or French.

Maps Alberta (Land Information Services Div., Main Fl., Brittania Bldg., 703 6th Ave. SW, Calgary, Alberta T2P 0T9; 403-297-7389) distributes a variety of natural resource maps of Alberta as part of the *Alberta Resource and Economic Atlas,* including: "Coal" (1984), "Move-ment of Coal Within Canada" (1983), "Forest Industry Development Area" (1984), "Potential Annual Forest Yield" (1984), "Metallic Minerals" (no date), "Minerals for Chemical and Metallur-gical Industries" (1982), "Minerals for Construc-tion Materials Industries" (1982), "Main Pipe-lines, Refineries and Gas Plants" (1984),

"Movement of Oil and Gas" (1983), and "Oil, Gas and Oil Sands" (1984).

Maps Alberta also produces "Aggregate Resource Maps," a series of 13 maps showing the distribution of major gravel deposits in the province, available in two scales, 1:250,000, and 1:50,000 ($3 Canadian each; 22 1/2" X 29").

Also available is an "Oil Sands Agreements Map" ($7.50 Canadian; 40" X 28"), showing oil sands leases and dispositions.

COMMERCIAL SOURCES

Several commercial map publishers and geologic organizations produce maps of natural resources:

Circum-Pacific Map Project (345 Middlefield Rd., MS 952, Menlo Park, CA 94025; 415-329-4002) is an international cooperative effort between the Circum-Pacific Council for Energy and Mineral Resources and the USGS to assemble and publish new geological and geophysical maps of the Pacific Ocean and surrounding areas. The maps are intended to show the relation of geology, tectonics, and crustal dynamics to known energy and mineral resources, and also to aid in exploration for new resources. A "Mineral Resources" series of working mineral resources maps is currently being created for the area ($14 each); a Mineral Resources map of the Northeast Quadrant has been completed to date. Circum-Pacific maps are available through the **American Associa-tion of Petroleum Geologists Bookstore** (P.O. Box 979, Tulsa, OK 74101; 918-584-2555).

George F. Cram Co. (P.O. Box 426, Indianapo-lis, IN 46206; 317-635-5564) publishes "Mineral Production of the United States," depicting locations of important minerals, with percent-ages of the total production indicated in figures and colors. The map is 52" X 40"; price ranges from $63 to $72, depending on mounting.

Geoscience Resources (2990 Anthony Rd., P.O. Box 2096, Burlington, NC 27216; 919-227-

8300; 800-742-2677) distributes natural resource maps for various countries, including the U.S. and Canada. Offerings include:

■ "China—Coal Map" ($29.95; 64-4040), a 34″ X 23 1/2″ generalized map showing newly discovered deposits, mines, and processing plants.
■ "Mozambique—Mineral Map" ($14.95; 65-02940; 29 1/2″ X 40″), illustrating the major ore deposits of metals, gemstones, gas, and uranium, comes with a 60-page booklet.
■ "Czechoslovakia—Mineral Deposits Map" ($18.95; 64-4640; 69″ X 38″), two sheets depicting mineral deposits on a geologic base map with Slavic legend.
■ "Wyoming—Mineral Deposits Map" ($9.95; 63-0241; 49 1/2″ X 29 1/2″), a three-color map illustrating 176 deposits and including a list of 92 references.
■ "Nevada—Gold Districts Map" ($5.45; 62-8044; 29″ X 33″), locating the major and minor gold districts of the state.
■ "Colorado—Geothermal Resources Map" ($3.95; 62-5785; 41 1/2″ X 53″) illustrates known geothermal areas, thermal areas and springs, and heat-flow values.

National Geographic Society (17th & M Sts. NW, Washington, DC 20036; 202-921-1200) offers "America's Federal Lands" ($4; 02005; 42 1/2″ X 29 1/2″). This 1982 map includes descriptive notes and a listing of natural resources by region.

Rand McNally (P.O. Box 7600, Chicago, IL 60680; 312-673-9100) has the "Our America" series of classroom maps, which includes

"Manufacturing and Minerals" ($59), showing the locations of manufacturing centers and mineral resources.

Water Education Foundation (717 K St., Ste. 517, Sacramento, CA 95814; 916-444-6240) produces the "California Water Map," a 28″ X 36″ colorful map featuring natural and man-made water routes in the state ($7.50).

Williams & Heintz Map Corporation (8119 Central Ave., Capitol Heights, MD 20743; 301-336-1144) produces but does not sell geologic and mineral maps (see also "Geologic Maps"), although some organizations offer its resources maps, including:

■ "Map of Alaska's Coal Resources" ($5), from **Division of Geological and Geophysical Surveys**, 794 University Ave., Ste. 200, Fairbanks, AK 99709.
■ "Mineral Resources of Guyana, 1987," from **Domingo Estay**, Rm. DC1-0830, United Nations, New York, NY 10017.
■ "Mineral Resources of West Virginia" ($9 folded; $10 rolled), from **West Virginia Geological Survey**, Publications Sales, P.O. Box 879, Morgantown, WV 26507.
■ "Availability of Federal Mineral Land Compared With Areas Favorable for Selected Locatable Minerals in Oregon," from **U.S. Bureau of Mines**, Branch of Production and Distribution, Cochrans Mill Rd., P.O. Box 1807, Pittsburgh, PA 15236.

■ **See also: "Agriculture Maps," "Energy Maps," "Geologic Maps," and "Recreation Maps."**

Topographic Maps

A topographic map is a line-and-symbol representation of natural and selected artificial features plotted to a definite scale. "Topos," as they are often called, show the shape and elevation of the terrain in precise detail by using contour lines. Topos show the location and shape of objects as big as mountains and as small as creeks and dirt roads.

The many uses of topographic maps make them the workhorses of cartography. They are of prime importance in planning airports, highways, dams, pipelines, and almost any type of building. They play an essential role in ecological studies and environmental control, geologic research, water-quality research, conservation, and reforestation. And, of course, they are widely used by hikers, hunters, bikers, and other outdoors enthusiasts. Topographic maps are also used as the basis for a wide range of other cartographic products,

from aeronautical charts to road maps.

Topo maps come in a variety of scales, usually stated as a ratio or fraction showing the measurement of the map in relation to the land it covers (see "Map Scale" for more on this). In a 1:24,000 (or 1/24,000) topo map, for example, one unit (an inch, centimeter, or whatever) on the map equals 24,000 of the same unit on the ground.

Map scale is the basic classification of topographic maps, and each scale series fills a range of map needs:

■ Large-scale maps, such as 1:24,000, are useful for highly developed or rural areas where detailed information is needed for engineering planning or similar purposes. Large-scale topo maps are the ones used most often by hikers and campers and for use in other recreational activities.

Portion of 1:24,000 topographic map of "Mt. Desert Island & Acadia National Park" in Maine, showing area around Northeast Harbor and Seal Harbor. Published by DeLorme Mapping Co.

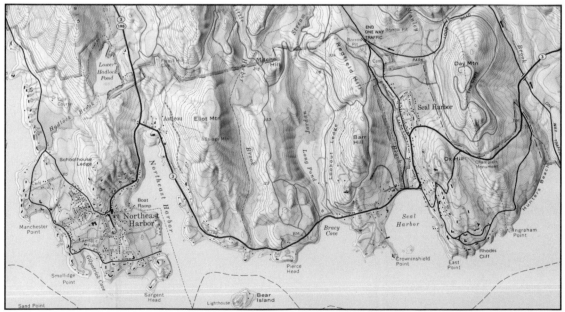

USGS 1:24,000-scale topographic map of "Washington West Quadrangle," showing portions of Washington, D.C., Maryland, and Virginia.

■ Intermediate-scale maps, ranging from 1:50,000 to 1:100,000, cover larger areas and are best suited for land management and planning.
■ Small-scale maps—1:250,000, 1:500,000, and 1:1,000,000—cover very large areas, on a single sheet, that are useful for comprehensive views of extensive projects or for regional planning.

See page 128 for examples of topo maps covering the same area at different scales.

For more than a century, the **U.S. Geological Survey** has been creating and revising topographic maps for the entire country at a variety of scales. There are about 60,000 USGS-produced topo maps, covering every square inch of U.S. territory. Each map covers a specific quadrangle (or "quad"), defined as a four-sided area bounded by parallels of latitude and meridians of longitude. Generally, adjacent maps of the same quadrangle can be combined to form a single large map.

USGS produces five series of topographic maps, each covering a different-size quad:

■ In the 7 1/2-minute series, each quad covers an area 7 1/2 minutes square (a minute is one-sixtieth of a degree in latitude and longitude). In this series, the scale is 1:24,000 (1:25,000 for Alaska and 1:20,000 for Puerto Rico); one inch represents about 2,000 feet, and each quad covers about 49 to 71 square miles.
■ In the 15-minute series, each quad covers an area 15 minutes square. In this series, the scale is 1:62,500 (1:63,360 for Alaska); one inch represents about one mile, and each quad covers about 197 to 282 square miles.
■ In the intermediate-scale quadrangle series, each quad covers an area 30 minutes by one degree. In this series, the scale is 1:100,000; one inch represents about 1 1/2 miles, and each quad covers about 1,145 to 2,167 square miles.
■ In the U.S. 1:250,000 series, each quad covers an area one degree by two degrees. In this series, as the name indicates, the scale is

1:250,000; one inch represents about four miles, and each quad covers about 4,580 to 8,669 square miles. Maps of Alaska and Hawaii vary from these standards.
■ In the International Map of the World series, each quad covers an area four by six degrees. In this series, the scale is 1:1,000,000; one inch represents about sixteen miles, and each quad covers about 73,734 to 102,759 square miles.

For comparison purposes, the area covered by one 1 X 2 degree map requires four 30 X 60 minute maps, 32 15-minute maps, 64 7 1/2 X 15-minute maps, and 128 7 1/2-minute maps.

When it comes to topos, the basic topographic map is just the tip of the cartographic iceberg. USGS also produces other special-purpose map data. These include:

■ *Orthophotomaps*, produced for selected topographic quadrangles, show land features primarily by color-enhanced photographic images that have been processed to show detail in true position. Orthophotomaps may or may not include contours. Because imagery naturally depicts an area in a more true-to-life manner than on a conventional topographic map, the orthophotomap provides an excellent portrayal of extensive areas of sand, marsh, or flat agricultural areas.
■ *Orthophotoquads* are a basic type of photo-image map prepared in a quadrangle map format. They can be produced quickly because they are printed in shades of gray without image enhancement or cartographic symbols. Orthophotoquads are valuable as map substitutes in unmapped areas and as complements to existing line maps.
■ A number of county maps at scales of 1:50,000 or 1:100,000 have been prepared cooperatively with some states. The maps are multicolored, show political boundaries, a complete road network, and a variety of topographical and cultural features. A series of 1:100,000-scale quad maps provides much of the detail shown on larger-scale maps yet covers enough geographic area to be useful as base maps for county-wide and regional study.

■ State maps at scales of 1:500,000 and 1:1,000,000 are available for all states except Alaska and Hawaii, which are covered by state maps at other scales. There also are shaded-relief editions for many states.

■ The National Park Series, at various scales, covers many parks, monuments, and historic sites. Many of these maps are available with shaded-relief overprinting in which the topography is made to appear three-dimensional by the use of shadow effects.

■ U.S. maps are available in sizes and scales ranging from letter size (1:16,500,000) to a two-sheet wall map (1:2,500,000). The complete series includes two maps, one at 1:6,000,000 and one at 1:10,000,000, that show all 50 states in their correct position and scale.

■ USGS also is engaged in a mapping program for Antarctica. The Antarctic maps are published at several scales, primarily 1:250,000, with shaded relief.

Although most topographic maps are produced by the USGS's National Mapping Program, other federal agencies—including the **Defense Mapping Agency**, **National Ocean Survey**, **Tennessee Valley Authority**, and the **National Park Service**—also prepare topographic maps as part of their regular activities, although such maps have been incorporated into the topographic map series published by USGS. They also may be obtained from the producing agencies (see Appendix B).

Colors, lines, and symbols. More than on most other map types, symbols are the graphic language of topographic maps. Color plays a key role, too:

■ symbols for water features are shown in *blue*;

■ manmade objects like roads, railroads, buildings, transmission lines, and political boundaries, are shown in *black*;

■ *green* distinguishes wooded areas from clearings;

■ *red* represents or emphasizes the more important roads, route numbers, fence lines, land grants, and the lines of townships, ranges, and sections of states subdivided by public-land surveys;

■ heavily built-up areas larger than three-quarters of a square mile are given a *pink* tint;

■ features added from aerial photographs during map revision are *purple*;

■ the contour lines that show the shape and elevation of the land surface are *brown*.

Color is just the beginning. The type of line on a topographic map also provides valuable information. For example, one type of black line designates national boundaries, another shows state boundaries, and still others indicate counties, parishes, civil townships, precincts, towns, barrios, incorporated cities, villages, hamlets, reservations, small parks, cemeteries, airports, and land grants. Additional black lines denote certain types of roads, trails, railroads, telephone lines, bridges, and tunnels. Similarly, there are a dozen or so different forms of blue lines that designate various types of rivers, streams, aqueducts, and canals.

Symbols are another key feature. Some topo map symbols are "pictographs," resembling the objects they represent — a pick and ax designates a mine or quarry, for example; an exposed boat wreck appears as a partially submerged vessel.

All told, there are about 140 different topographic map lines and symbols. A complete directory of colors, lines, and symbols is contained in a free brochure, *Topographic Map Symbols*, available by mail from USGS distribution centers. You may pick up the brochure in person from the distribution centers, the 14 sales counters, or the many commercial dealers that sell USGS topographic maps.

GOVERNMENT SOURCES, UNITED STATES

U.S. Geological Survey. The USGS distributes several indexes to its topographic maps, including "Index to Topographic and Other Map Coverage and Catalog of Topographic and Other Published Maps," available for each state, "Status and Progress of Topographic Mapping, U.S.," "Index to USGS/DMA 1:50,000-

Scale, 15-Minute Mapping," "Index to Interme-diate-Scale Mapping/Index to County Mapping," and "Index to USGS Topographic Map Coverage of National Park System." The USGS also distributes the pamphlet, *Topographic Map Symbols*.

GOVERNMENT SOURCES, CANADA

Canada Map Office (615 Booth St., Ottawa, Ontario K1A 0E9; 613-952-7000) distributes topographic maps covering Canada and its territories, as part of the "National Topographic System" series. Maps are available in various scales; indexes are available.

Manitoba Natural Resources (Surveys and Mapping Branch, 1007 Century St., Winnipeg, Manitoba R3H 0W4; 204-945-6666) distributes National Topographic System maps of Manitoba, in 1:50,000- and 1:250,000-scale. A pamphlet is available.

Maps Alberta (Land Information Services Division, Main Fl., Brittania Bldg., 703 6th Ave. SW, Calgary, Alberta T2P 0T9; 403-297-7389) distributes various 1:250,000- and 1:100,000-scale topographic maps as part of the "Provincial Access Series," as well as maps from the National Topographic series. A catalog is available.

Québec Ministère de l'Energie et des Ressources (Photocartothèque Québecoise, 1995 Blvd. Charest Ouest, Sainte-Foy, Québec G1N 4H9; 418-643-7704) distributes topographic maps for areas in Quebec drawn on 1:20,000- and 1:12,000-scales. An index of the maps, "Repertoire des Cartes, Plans, et Photographies Aeriennes," is available.

Saskatchewan Property Management Corp. (Central Survey and Mapping Agency, Distribution Center, 2045 Broad St., 1st Fl., Regina, Saskatchewan S4P 3V7; 306-787-6911) distributes topographic maps of areas in Saskatchewan, available in the following scales: 1:25,000, 1:50,000, and 1:125,000.

COMMERCIAL SOURCES

DeLorme Mapping Co. (P.O. Box 298, Freeport, ME 04032; 207-865-4171) produces highly detailed topographic maps of recreation areas in New England, such as "Allagash & St. John Map and Guide" ($4.95), "Mount Desert Island/Acadia National Park" ($6.95), and "Trail Map & Guide to the White Mountain National Forest" ($6.95). DeLorme also publishes a 4' X 6' "Maine Wall Map," a laminated topographic and political map ($95).

Geoscience Resources (2990 Anthony Rd., P.O. Box 2096, Burlington, NC 27216; 919-227-8300; 800-742-2677) distributes USGS topographic maps, Canadian National Park topographic maps, and topographic maps of many foreign countries. A sample of foreign titles available includes:
■ "Benin—Topographic Map" ($9.95; 64-2050).
■ "Cameroon—Topographic Quadrangle Maps" ($7.95 each; 64-2565). Seven of these bilingual maps are currently available.
■ "Norway—Topographic Quadrangle Map" ($7.95 each; set of 17 sheets, $99.95), a series offering complete coverage of Norway.
■ "Hungary—Topographic County Maps," by Cartographia ($6.95 each; set of 19 sheets, $76), 19 sheets offering complete coverage of Hungary, county by county.

Green Trails Inc. (P.O. Box 1932, Bothell, WA 98041; 206-485-9144) publishes topographic maps for Washington state's Olympic Peninsula, the North Cascades, the Central Cascades, the South Cascades, and Oregon. The four- and five-color maps are numbered with a system compatible with the Uniform Map System, used by search and rescue groups. Green Trails maps are available in a variety of retail outlets in the Pacific Northwest; a list of dealers and further information is available.

Raven Images (34 N. Central Ave., Medford, OR 97501; 503-773-1436; 800-237-0798) produces colorful shaded topographic maps of Arizona, California, Colorado, Hawaii, Montana, Nevada, New Mexico, Oregon, Utah,

Washington, and Wyoming. The maps are $20 each, $45 laminated.

Trails Illustrated (P.O. Box 3610, Evergreen, CO 80439; 303-670-3457) publishes more than 50 topographic maps of recreational areas in Colorado as well as selected national parks. The waterproof, tearproof maps are updated annually and are available in retail stores or can be ordered directly. Trails Illustrated produces the following series:

■ "Colorado Series" ($7 each), includes maps for Indian Peaks, Tarryall Mountains, Poudre River, Steamboat Springs, Flat Tops, Maroon Bells, Kebler Pass West, and Weminuche/Chicago Basin.
■ "National Park and Recreation Areas Series" ($7 each) includes Rocky Mountain National Park, Sequoia/Kings Canyon National Park, Grand Canyon National Park, Glacier/Waterton National Park, and Dinosaur National Monument.
■ "Quad Series" includes maps for Triple Divide Peak, California ($4), and Longs/McHenrys Peak, Colorado ($7).
■ "Cross Country Series" ($5 each), maps for Frisco-Breckenridge, Nederland-Georgetown, Vail-Leadville, and Aspen-Carbondale.
■ "Alpine Ski Series" consists of the "Vail Ski Trail Map" ($4) and the "Vail Wall Map" ($8).

University of Hawaii Press (Univ. of Hawaii at Manoa, 2840 Kalawalu St., Honolulu, HI 96822; 808-948-8255) distributes the "Reference Maps of the Islands of Hawai`i" series by James A. Bier. The full-color, shaded relief, topographic maps available are: "Hawai'i" ($2.95; 25" X 31"); "Kaua'i" ($2.25; 22" X 15 1/2"); "Maui" ($2.95; 24" X 18 1/2"); "Moloka'i-Lana'i" ($2.95; 20" X 17"); and "O'ahu" ($2.95; 29" X 25"). Also available is James Bier's topographic "Reference Map of Tutuila, Manu'a, Upolu, and Savai'i" ($2.95; 32" X 16 1/2").

Warren Communications (P.O. Box 8635, San Diego, CA 92102; 619-531-0765) distributes topographic maps produced by the Mexican government. Warren sells the maps only on a wholesale basis; however, they can be purchased from **The Map Center**, 2611 University Ave., San Diego, CA 92104; 619-291-3830.

Wilderness Press (2440 Bancroft Way, Berkeley, CA 94704; 415-843-8080) publishes a series of topographic maps to popular hiking areas in California. The maps contain updated changes in trails, roads, creeks, lakes, meadows, contour lines, and other features. Most maps are at a scale of 1:62,500 and measure 18" X 21"; all have a handy map-grid-and-index system. An added feature is the Polyart 2 plastic stock on which the maps are printed, which is tear-, water-, and grease-resistant, and easy to write on. Titles available include: "Devils Postpile," "Hetch Hetchy Reservoir," "Lassen Volcanic National Park and Vicinity," "Mt. Goddard," "Tuolumne Meadows," and "Yosemite National Park and Vicinity."

BOOKS ON READING TOPOGRAPHIC MAPS

Here are four titles on using maps and compasses especially geared to reading topographic maps. All include sample maps and explanations of map symbols.

■ *The Basic Elements of Map and Compass*, by Cliff Jacobson (ICS Books, One Tower Plaza, 107 E. 89th Ave., Merrillville, IN 46410; 219-769-0585). 1988. $4.95 paper.
■ *Be Expert With Map & Compass—The Orienteering Handbook*, by Bjorn Kjellstrom (Charles Scribner's Sons, 866 Third Ave., New York, NY 10022; 212-702-2000). 1976. $4.95 paper.
■ *Land Navigation Handbook—The Sierra Club Guide to Map and Compass*, by W.S. Kals (Sierra Club Books, dist. by Random House, 201 E. 50th St., New York, NY 10022; 301-848-1900; 800-726-0600). 1983. $9.95.
■ *Outward Bound Map and Compass Handbook*, by Glenn Randall (Lyons & Burford, 31 W. 21st St., New York, NY 10010; 212-620-9580). 1989. $8.95.

■ **See also: "Tourism Maps and Guides."**

Wildlife Maps

Ah, wilderness! While we toil and play daily in our concrete and glass jungles, wildlife, by and large, has been relegated to a relative handful of sanctuaries, refuges, and other protective enclaves. Thanks largely to the federal government and an array of environmental groups, though, there is plenty of wildlife left, although the number of endangered species increases daily. Campers, bikers, hikers, botanists, entomologists, ornithologists, and zoologists are among those who use wildlife maps to keep track of Mother Nature.

Uncle Sam, through the **Fish and Wildlife Service** (part of the Department of the Interior), is landlord to more than 90 million acres of wildlife habitat—home to more than 1,200 species of mammals, birds, reptiles, amphibians, and fish—the majority of which is in national wildlife refuges. These are lands where the only shooting permitted is done with a camera, where winged, finned, and furred creatures reign, and where humans are mere guests. The Fish and Wildlife Service administers these lands and provides maps, many of them free, to the public for each of its holdings. The **National Park Service** and the **U.S. Forest Service** also provide maps for the lands, rich with flora and fauna, that they administer.

State, county, and city governments are also good sources of wildlife maps, especially in regions where there is an abundance of locally protected forest and game land. Write to the appropriate conservation board or fish-and-game bureau to learn what maps are available (see Appendix A).

As for foreign wildlife, the travel and tourism bureaus for some countries known for their unusual wildlife (including Australia and many African nations) provide wilderness maps to visitors, as do privately run game parks both in the United States and abroad.

Other sources for wildlife maps include many of the organizations that promote conservation and protection of the environment, and scientific associations that study various species (although these are often technical and available only to members). Zoological parks and museums can also help locate wildlife maps. If there's nothing in their bookstores and gift shops, check with a curator or resident expert.

See "Recreation Maps" for sources of hunting and fishing maps.

GOVERNMENT SOURCES, UNITED STATES

Defense Mapping Agency. DMA produces a "Whale Chart" ($2.50) that shows whale species and their locations along the North American coastline. The chart can be obtained directly from the DMA or through the Government Printing Office.

Fish and Wildlife Service. In 1977, the Fish and Wildlife Service began the National Wetlands Inventory in an effort to classify and map America's remaining wetlands. As part of the National Wetlands Inventory Program, the USGS has been mapping the marshes, swamps, ponds and bogs of America, which currently represent about 5 percent of the total land surface of the lower 48 states. Of the 215 million acres of wetlands that originally existed in the conterminous U.S., the Fish and Wildlife Service estimates that there are only about 99 million acres left. More than 25,000 composite and overlay maps of the wetlands have been created to date, and are used for a variety of purposes, including town planning, zoning, flood-hazard planning, waste treatment, and water-quality planning. Information about NWI and the maps are available through Earth Science Information Center offices. The Fish and Wildlife Service also has free maps and brochures of wildlife refuges and fish hatcheries under federal protection. They are available at the locations themselves or by writing to the

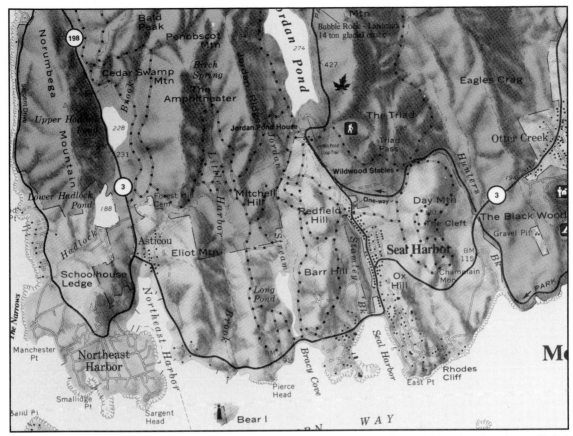

Portion of "Wildlife Map of Acadia National Park" in Maine, showing area around Northeast Harbor and Seal Harbor. Published by American Nature Maps.

service's Division of Realty at the Washington, D.C., headquarters. The maps, which measure 8 1/2" X 11", are black-and-white diagrams showing boundaries, wilderness areas, refuge headquarters, and driving directions from surrounding areas. Also available are three U.S. maps showing the land under the Fish and Wildlife Service's jurisdiction. The free maps, which are available on 8 1/2" X 11" or 17" X 22" sheets, are titled "National Wildlife Refuge System," "National Fish and Wildlife Management Areas," and "National Fish Hatcheries and Fishery Assistance Stations." The U.S. maps, which are black-and-white with red lines denoting regional boundaries, are updated annually.

Government Printing Office. GPO offers a

small assortment of wildlife maps and atlases, including:

■ "National Wildlife Refuges: A Visitor's Guide" ($1; S/N 024-010-00680-3), a 1988 map of the U.S. showing locations of all national wildlife refuges, including names, addresses, and descriptions of available activities.
■ *Gulf of Mexico Coastal and Ocean Zones Strategic Assessment: Data Atlas* ($138; S/N 003-017-00523-2), a 1985 atlas that includes 71 maps covering 73 species of living marine resources, as well as maps of physical environments, biotic environments, economic activities, environmental quality, and jurisdictions.

Tennessee Valley Authority. TVA has created a series of topographic maps of wildlife and game refuge areas in the Tennessee Valley region. The maps illustrate management boundaries, interior trails and roads, checking stations, and private lands within the TVA boundary. Available maps are:

- "Catoosa Wildlife Management Area" ($2).
- "Chuck Swan Wildlife Management Area" ($1.50).
- "Prentiss Cooper Wildlife Management Area" ($1.50).
- "Tellico Wildlife Management Area" ($2).
- "Ocoee Wildlife Management Area" ($2).
- "AEDC Wildlife Management Area" ($1.50).
- "Fall Creek Wildlife Management Area" ($1.50).

Include 50 cents postage for each map ordered.

U.S. Geological Survey. USGS produces a series of topographic maps of the National Park System that includes maps of wildlife preserves and refuges under the care of the federal government. The "Index to USGS Topographic Map Coverage of the National Park System," a fold-out map pinpointing park and wildlife areas covered in the topographic program, is available free from the USGS (see "Recreation Maps" and "Topographic Maps" for more information). The maps in the National Park Series cost $4 each and are drawn at various scales; some parks require more than one sheet. Also available from USGS is a series of "Coastal Ecological Inventory Maps" ($4 each) that vividly illustrate the ecological setting, including flora and fauna, of the North American coastline. Maps are available for each major ecological coastal area.

GOVERNMENT SOURCES, CANADA

Canada Map Office (615 Booth St., Ottawa, Ontario K1A 0E9; 613-952-7000) distributes "Canada's Wetlands," a series of two maps of wetland areas in Canada: "Canada—Wetland Distribution" (MCR 4107E), showing the regional occurence of the wetlands, which

U.S. Fish and Wildlife Service's "National Wildlife Refuges" map.

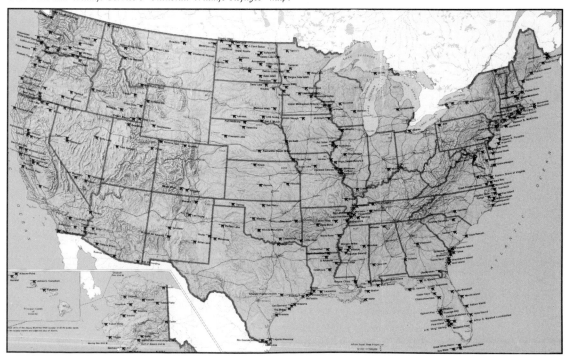

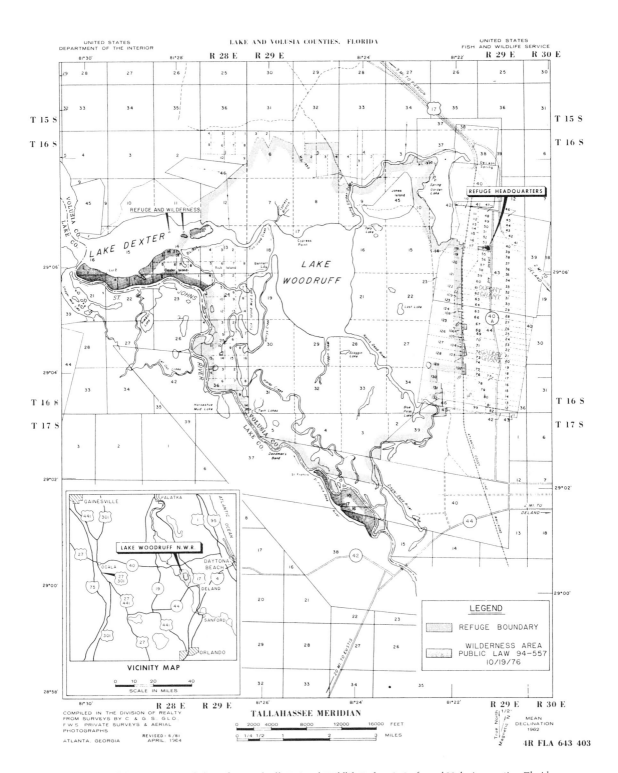

U.S. Fish and Wildlife Service map of the Lake Woodruff National Wildlife Refuge in Lake and Volusia counties, Florida.

cover 14 percent of the country; and "Canada—Wetland Regions" (MCR 4108E), illustrating the types of wetlands found in Canada, and including descriptions of the ecology of major wetland types in each region. The maps are $5.50 Canadian each.

Maps Alberta (Land Information Services Div., Main Fl., Brittania Bldg., 703 6th Ave. SW, Calgary, Alberta T2P 0T9; 403-297-7389) distributes Canada Land Inventory/Alberta Land Inventory maps, a survey of land capability, designed for resource planning relating to agriculture, forestry, recreation, and wildlife. Maps depicting waterfowl wildlife as well as those showing wildlife ungulates (hoofed animals) are available in five-color paper or diazo for $1.50 to $3.50 Canadian each. An index of the maps is available upon request.

Saskatchewan Property Management Corp. (Central Survey and Mapping Agency, Distribution Center, 2045 Broad St., 1st Fl., Regina, Saskatchewan S4P 3V7; 306-787-6911) distributes "Habitat Inventory Maps" of the province. These show the location of critical habitat for important wildlife species, as well as public wildlife areas and "Wildlife Lands." Maps in this series include: "Coloured Land Systems Map" ($4 Canadian); "Blueprint Land Use Map" ($1.50 Canadian); "Blueprint Critical Wildlife Habitat Map" ($4 Canadian); and an in-depth report on the wildlife resources of each map ($3 Canadian each). The entire package can be purchased for $18 Canadian.

Also available are Department of the Environment, Lands Directorate maps depicting land capability for both waterfowl wildlife and ungulate (hoofed) wildlife. Maps are $1 each; an index of maps published in Saskatchewan is available upon request.

COMMERCIAL SOURCES

American Nature Maps (11620 Ivystone Ct., Ste. 201, Reston, VA 22091; 703-476-0378) produces wildlife maps for national parks. The maps also depict tidal pools, fall-color areas, ranger stations, campgrounds, and trails. Titles include "Wildlife Map of the Great Smoky Mountains National Park" ($4.95) and "Wildlife Map of Acadia National Park" ($5.95).

Hammond Inc. (515 Valley St., Maplewood, NJ 07040; 201-763-6000) distributes a "Dinosaur Fossil Find Map" ($2.95; 0144-7), a 38" X 25", full-color map depicting regions where there have been significant findings of dinosaur fossils. Bartholomew maps, distributed in the U.S. by Hammond, has a Johnston and Bacon pictorial map series that includes a map titled "Loch Ness Monster" ($9.50; 6381-7). The map depicts Scotland's Loch Ness and shows where the monster has been sighted.

National Geographic Society (17th & M Sts. NW, Washington, DC 20036; 202-921-1200) has several wildlife maps and charts, including:

■ "Alaska's Ice Age Mammals" ($3; 02252; 31 1/2" X 23"), with illustrations.
■ "Bird Migration in the Americas" ($3; 02810; 23" X 36"), with illustrations of birds and migratory routes.
■ "The Great Whales" ($3; 02820; 30" X 20 1/2"), with illustrations and notes.
■ "Pinnipeds of the World" ($4; 20016; 21" X 30"), showing the locations of a class of aquatic mammals, including seals and walruses.

The World Conservation Union (Avenue Mont-Blanc, CH-1196, Gland, Switzerland; 022-647-181) produces authoritative maps on the distribution of species and protected areas in all parts of the world, at the national system level and at the level of the particular site. They also have detailed maps of certain coastal and marine ecosystems, primarily in the Caribbean Sea, the Mediterranean Sea, and the enclosed seas of the Middle East. Write for more information.

■ **See also: "Bicycle Route Maps" and "Recreation Maps."**

Maps ◆ Travel Maps ◆ Bicycle Route Maps ◆ Mass Transit Maps ◆ Railroad Maps
Maps ◆ Tourism Maps ◆ World Status Map ◆ County Maps ◆ Foreign Country Map
and Maps ◆ State Maps ◆ United States Maps ◆ Urban Maps and City Plans ◆ Wo
orld Game Map ◆ Boundary Maps ◆ Congressional District Maps ◆ Land Owners
al Maps ◆ Scientific Maps ◆ Agriculture Maps ◆ Geologic Maps ◆ Land Use Maps
rce Maps ◆ Topographic Maps ◆ Wildlife Maps ◆ Antique Maps ◆ Researching C
ic Site Maps ◆ History Maps ◆ Military Maps ◆ Treaure Maps ◆ Business Maps
Emergency Information Maps ◆ Energy Maps ◆ Utilities Maps ◆ Water Maps ◆ N
Ocean Maps ◆ River, Lake, and Waterway Maps ◆ Tide and Current Maps ◆ Sky
tical Charts ◆ Star Charts ◆ Star Magnitudes ◆ Weather Maps ◆ How to Read a W
mages as Maps ◆ Aerial Photographs ◆ Space Imagery ◆ How to Buy an Atlas ◆
Geography Education Mat   ◆ Map Organizations ◆ Map
◆ Map Stuff ◆ The Turnabout Ma   ◆ Map Projections ◆
t Map Skills ◆ Copying Maps ◆ Travel Maps ◆ Bicycle Route Maps ◆ Mass Transit
Maps ◆ Recreation   County Maps ◆
Maps ◆ Indian Land Maps ◆ State Maps ◆ United States Maps ◆ Urban Maps and
World Maps ◆ The World Game Map ◆ Boundary Maps ◆ Congressional District M
nership Maps ◆ Political Maps ◆ Scientific Maps ◆ Agriculture Maps ◆ Geologic
e Maps ◆ Natural Resource Maps ◆ Topographic Maps ◆ Wildlife Maps ◆ Antique
ing Old Maps ◆ Historic Site Maps ◆ History Maps ◆ Military Maps ◆ Treaure Ma
Maps ◆ Census Maps ◆ Emergency Information Maps ◆ Energy Maps ◆ Utilities
aps ◆ Nautical Charts ◆ Ocean Maps ◆ River, Lake, and Waterway Maps ◆ Tide an
s ◆ Sky Maps ◆ Aeronautical Charts ◆ Star Charts ◆ Star Magnitudes ◆ Weather
Read a Weather Map ◆ Images as Maps ◆ Aerial Photographs ◆ Space Imagery
tlas ◆ Atlases ◆ Globes ◆ Geography Education Materials ◆ Map Accessories ◆
tions ◆ Map Software ◆ Map Stuff ◆ The Turnabout Map ◆ How to Choose a Map
ns ◆ Learning About Map Skills ◆ Copying Maps ◆ Travel Maps ◆ Bicycle Route
nsit Maps ◆ Railroad Maps ◆ Recreation Maps ◆ Tourism Maps ◆ World Status M
Maps ◆ Foreign Country Maps ◆ Indian Land Maps ◆ State Maps ◆ United States
aps and City Plans ◆ World Maps ◆ The World Game Map ◆ Boundary Maps ◆ C
strict Maps ◆ Land Ownership Maps ◆ Political Maps ◆ Scientific Maps ◆ Agricul
Geologic Maps ◆ Land Use Maps ◆ Natural Resource Maps ◆ Topographic Maps
◆ Antique Maps ◆ Researching Old Maps ◆ Historic Site Maps ◆ History Maps
Treaure Maps ◆ Business Maps ◆ Census Maps ◆ Emergency Information Maps
Utilities Maps ◆ Water Maps ◆ Nautical Charts ◆ Ocean Maps ◆ River, Lake, and
Tide and Current Maps ◆ Sky Maps ◆ Aeronautical Charts ◆ Star Charts ◆ Star M
er Maps ◆ How to Read a Weather Map ◆ Images as Maps ◆ Aerial Photographs
◆ How to Buy an Atlas ◆ Atlases ◆ Globes ◆ Geography Education Materials ◆
ies ◆ Map Organizations ◆ Map Software ◆ Map Stuff ◆ The Turnabout Map ◆
a Map ◆ Map Projections ◆ Learning About Map Skills ◆ Copying Maps ◆ Travel
oute Maps ◆ Mass Transit Maps ◆ Railroad Maps ◆ Recreation Maps ◆ Tourism M
atus Map ◆ County Maps ◆ Foreign Country Maps ◆ Indian Land Maps ◆ State M

HISTORY
THROUGH MAPS

Antique Maps and Reproductions

Old maps are beautiful and thought-provoking windows into our past, rich with the history of generations that have been outlived by the pictures they drew of their world. Short of stumbling across finds at auctions or on a store's dusty shelves, the best places to find antique maps are through the numerous collections of map libraries (see Appendix E), historical societies, or museums. There are also a number of publishers that create reproductions of historically significant—or simply artistically beautiful—maps and charts, as well as dealers with the resources to locate both originals and reproductions of these maps for a fee.

GOVERNMENT SOURCES

Two government agencies in Washington, D.C., represent the lion's share of existing antique maps:

■ The Geography and Map Division of the **Library of Congress** holds one of the world's largest collections of antique maps and atlases, boasting more than 4 million maps and 52,000 atlases, plus hundreds of antique globes.
■ The **National Archives** has approximately 2 million maps and charts produced by the federal government between 1750 and 1950.

Both the Library of Congress and the Archives have reproduction services for their collections (see "Uncle Sam's Treasure Chests," page 166). These collections, as well as other government sources, are described in more detail in the *Scholar's Guide to Washington, D.C. for Cartography and Remote Sensing Imagery* ($15), available from

Right: *Portion of 1755 map of Virginia and Maryland.*
Reprinted with permission of Historic Urban Plans, Ithaca, N.Y.

the **Smithsonian Institution Press**, 955 L'Enfant Plaza, Rm. 2100, Washington, DC 20560; 202-357-2888; 800-678-2675.

Federal government agencies that have older maps on file for study and reproduction include the USGS (dating to 1879) and the National Ocean Service (various marine and navigation charts from the 18th century on; Civil War maps detailing certain marches and campaigns).

The Library of Congress also sells reproductions of several antique maps in its collection. These may be ordered from the Library of Congress, Information Office, Box A, Washington, DC 20540. Include $2 postage for orders under $50, $3.50 for orders over $50. Titles include:

■ "Map of Manhattan" (17 3/4″ X 26 1/2″; $15), drawn in 1639, the earliest known map of the island and its environs.
■ "Map of the World" (34″ X 22 1/2″; $20), drawn in 1565 by Paolo Forlani of Verona, an excellent example of the maps printed from copper plates in Italy in the mid-sixteenth century.
■ "Map of the World" (19 3/4″ X 13 5/8″; $10), drawn in 1544 by Battista Agnese, showing Magellan's route around the world.
■ "Map of the North Pacific" (31 1/2″ X 20 3/4″; $20), drawn around 1630 by Joao Teixeira Albernaz I, the most notable Portuguese cartographer of his day.
■ "A Chart of the Gulf Stream" (17 1/4″ X 11″; $10), commissioned by Benjamin Franklin in 1786, the first map to show the Gulf Stream as a continuous feature flowing from Florida to Newfoundland.
■ "Chart of the Mediterranean Sea and Western Europe" (27″ X 39″; $20), a 1559 drawing by Mateus Prunes, a leading member of one of the well-known families of sixteenth-century Majorcan cartographers.

COMMERCIAL SOURCES, REPRODUCTIONS

Several map publishers carry one or more antique map reproductions. Examples include:

American Geographical Society (156 Fifth Ave., Ste. 600, New York, NY 10010; 212-242-0214) distributes "World Map" ($4; 14″ X 14″), reproduced from the small "Under Heaven Map Book," published in Korea in the late 17th or early 18th century.

Class Publications Inc. (237 Hamilton St., Hartford, CT 06106; 203-951-9200) distributes "Old World Map," a 27″ X 39″ reproduction of a 1636 map of the world ($5, plus $3 shipping; C3030).

Historic Urban Plans (P.O. Box 276, Ithaca, NY 14850; 607-273-4695) sells reproductions of old city plans, both U.S. and foreign. Prices range from $5 for maps in the "Historic America Maps and Urban Views" series to $7.50 to $15 for maps in the "Historic City Plans and Views" series.

International Map Services Ltd. (P.O. Box 2187, Grand Cayman, Cayman Islands, British West Indies; 809-94-94700) reproduces a map showing North America and the Caribbean in 1776. The map, originally printed in London in 1777 in two sheets, has been reproduced in a single sheet. The hand coloring is based upon that used in a similar map in the collection of the British Museum.

COMMERCIAL SOURCES, ANTIQUE MAPS

Two excellent sources for information about antique maps and map dealers are: *The Map Collector*, a British quarterly published in March, June, September, and December. The journal features articles on types of maps ("Understanding Engraved Maps" is one example), events in cartographic history ("The Geographic and Cartographic Work of the American Military Mission to Egypt, 1870-1878"), and specific maps, such as Johann Gabriel Doppelmayr's 1720 map of the world, as well as news of events in the antique map world, and, of course, advertisements from antique map dealers. Subscription is £24 annually for UK residents, £27 for airmail delivery to other

Researching Old Maps

Because there are many different kinds of old maps, stored in many different collections, they are difficult to research. However, with a little imagination and a lot of perseverance, you can probably find just the map you want. There are many sources for you to investigate, ranging from historical societies to the cartographic offices of your state or local government to the **National Archives** and the **Library of Congress**.

The best place to begin a search for an old map is your local library. Two reference books—*The Directory of Historical Societies in the* *United States and Canada* (**American Association for State and Local History**, 172 Second Ave. N, Ste. 102, Nashville, TN 37201; 615-255-2971) and *Map Collections in the United States and Canada*: A *Directory* (**Geography and Map Division of the Special Library Association**, 1700 18th St. NW, Wasington, DC 20009; 202-234-4700)—are excellent sources for finding local antique map collections; they may be found on many libraries' reference shelves. These books may lead you to organizations that have their own collections, or who can suggest other places to look.

countries. And **Imago Mundi** (c/o Secty./Treas., G.R.P Lawrence, Geography Dept., King's College, Strand, London WC2R 2LS), an international journal with articles on early maps and the history of cartography. For subscription details, write to the address above.

One event of note to antique map buffs is the **Antique Map & Print Fair** (26 Kings Rd., Cheltenham GL52 6BG, England; Tel: 0242-514287; FAX: 0242-513890), called "the only monthly antique map & print fair in the world." It is held at the Bonnington Hotel, Southampton Row, London WC1 4BH; admission is free.

ANTIQUE MAP DEALERS

Following is a select list of antique dealers and their specialties. A more complete list can be found in the *International Directory of Map Dealers*, available from **Map Collector Publications Ltd.**, 48 High St., Tring, Herts HP23 5BH, England; 044-282-4977.

Alaska
The Observatory
202 Katlian Street
Sitka, AK 99835
907-747-3033
(rare and common books, maps, prints—Alaska and other Polar regions)

California
Argonaut Book Shop
792 Sutter St.
San Francisco, CA 94109
415-474-9067
(rare books, maps, prints)

Holy Land Treasures
1200 Edgehill Dr.
Burlingame, CA 94010
415-343-9578
(rare and antique maps of the Holy Land)

John Howell—Books
434 Post St.
San Francisco, CA 94102
415-781-7795
(rare and early maps and atlases)

Manning's Fine Books
1255 Post St., Ste. 609
San Francisco, CA 94109
415-673-1900
(Americana, including maps and atlases)

John Scopazzi
278 Post St., Ste. 305
San Francisco, CA 94108
415-362-5708
(antique maps)

Colorado
The Chinook Bookshop Inc.
210 N. Tejon St.
Colorado Springs, CO 80903
719-635-1195

Connecticut
Cedric L. Robinson, Bookseller
597 Palisado Ave.
Windsor, CT 06095
203-688-2582
(American atlases, maps, and travelers' guides prior to 1900)

Florida
Capt. Kit L. Kapp
Box 64
Osprey, FL 34229
(antiquarian maps, the Americas and rest of world; catalog, $3)

Mickler's Floridiana
PO Box 38, Lake Dr.
Chuluota, FL 32766
305-365-3636
(books, atlases, maps on Florida; search service for out-of-print maps)

Illinois
J.T. Monckton Ltd.
Speculum Orbis Press
730 N. Franklin
Chicago, IL 60610
(early maps, prints and books; cartographic facsimiles; scholarly monographs; catalogs upon request)

Kenneth Nebenzahl, Inc.
333 N. Michigan Ave.
Chicago, IL 60601
312-641-2711
(15th- to 19th-century atlases, maps & globes)

George Ritzlin
P.O. Box 6060
Evanston, IL 60204
312-328-1966
(antiquarian maps and atlases; cartographic history)

Harry L. Stern, Ltd.
620 N. Michigan Ave.
Chicago, IL 60611
312-787-4433
(all types of rare and out-of-print maps and atlases)

Maryland
Old World Mail Auctions
5614 Northfield Rd.
Bethesda, MD 20817
301-657-9074
(16th- to 19th-century maps; free catalog)

Massachusetts
David C. Jolly Publishers
P.O. Box 931
Brookline, MA 02146
617-232-6222
(mailing lists of antique maps dealers; also publishes *Price Record and Handbook* and *Maps of America in Periodicals Before* 1800)

Michigan
Clifton F. Ferguson
4999 Meandering Creek Dr.
Belmont, MI 49306
616-874-9297
(antique maps, atlases and sea charts)

Missouri
Phyllis Y. Brown
736 Demun
St. Louis, MO 63105
314 725 1023
(17th-, 18th- and 19th-century engraving, woodcuts and lithographs—antique prints, maps, books)

Elizabeth F. Dunlap, Books & Maps
6063 Westminster Pl.
St. Louis, MO 63112
314-863-5068
(specializes in maps of North America prior to 1885)

New Hampshire
G.B. Manasek Inc.
35 S. Main St., Box 961
Hanover, NH 03755-0961
603-643-2227
(Japan, Mideast, medieval maps)

New Jersey
Grace Galleries
75 Grand Ave.
Englewood, NJ 07631
201-567-6169
(antiquarian maps and sea charts from 17th century to early 20th century)

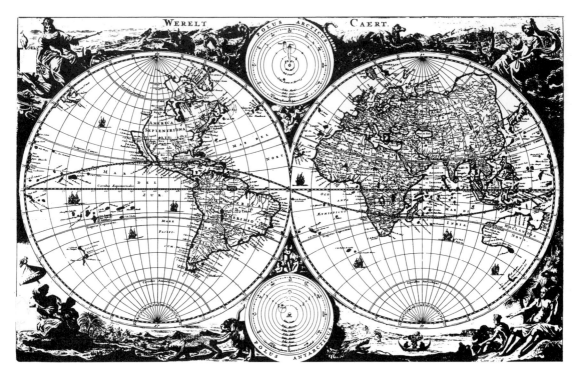

A *Dutch "Bible-map" from 1690, probably created by Daniel Stoopendaal, a pre-eminent Dutch map maker.*

New Mexico
Richard Fitch, Old maps & prints & books
2324 Calle Halcon
Santa Fe, NM 87501
505-982-2939
(specializes in North American antique maps and
prints)

New York
W. Graham Arader III
29 E. 72nd St.
New York, NY 10021
212-628-3668
(antique maps and books; offices located throughout
U.S.)

Richard B. Arkway Inc.
538 Madison Ave.
New York, NY 10022
212-751-8135; 800-453-0045
(antique maps)

JoAnn and Richard Casten
4 Dodge Ln.
Old Field, NY 11733
(antique maps, atlases and books)

Graphic Arts Unlimited Inc.
225 Fifth Ave.
New York, NY 10010
212-255-4805

High Ridge Books Inc.
Box 286
Rye, NY 10580
914-967-3332
(19th-century maps of America)

Martayan Lan
10 W. 66 St.
New York, NY 10023
212-595-1776
(maps of the world and the Americas)

Thomas and Ahngsana Suarez
RD 2, Box 297
Yorktown Heights, NY 10598
914-248-6650
(maps, atlases and prints relating to exploration)

With Pipe and Book
117 Main St.
Lake Placid, NY 12946
518-523-9096
(books and maps of the Adirondacks)

Pennsylvania

The Cartophile
934 Bridle Ln.
West Chester, PA 19382
215-692-7697
(U.S. explorations, battle maps, land surveys,
gazeteers, pocket maps, geographies)

The Lamp
William G. Mayer, Jr.
204 Auburn St.
Pittsburgh, PA 15206
412-661-6600
(19th-century geographical atlases)

William Perry Jr.
Mill Creek Farm
Rushland, PA 18956
215-598-3662
(American historical maps and prints)

The Philadelphia Print Shop Ltd.
8441 Germantown Ave.
Philadelphia, PA 19118
215-242-4750
(prints, maps and rare books; send $4 for most
recent catalog, or $18 for next six)

Tennessee

Murray Hudson Books & Maps
109 S. Church St., Box 163
Halls, TN 38040
901-836-9057
(antiquarian maps and books with maps; specialty:
U.S. South and West prior to 1900)

Vermont

Tuttle Antiquarian Books Inc.
P.O. Box 541
28 S. Main St.
Rutland, VT 05701
802-773-8930
(maps and atlases of America)

Virginia:

Cartographic Arts
P.O. Box 2202
Petersburg, VA 23804
804-861-6770
(antique and rare maps and atlases)

James Barnes Gallery
222 S. Washington St.
Alexandria, VA 22314
703-548-8008
(16th- to 19th-century rare prints, maps and documents)

Paul Roberts Stoney
P.O. Box 'F'
Williamsburg, VA 23187
804-220-3346
(print and mapseller; specializing in colonial America;
catalogs periodically)

Washington

Shorey's Bookstore
110 Union St.
P.O. Box 21626
Seattle, WA 98111
206-624-0221
(Western Americana, Alaska, and the Arctic)

Wisconsin

Sadlon's Ltd.
109 N. Broadway
De Pere, WI 54115
414-336-6665
(antique maps)

FOREIGN COUNTRIES

Australia

Antique Print Room
130 King William Rd.
Goodwood 5034, SA
08-272-3506
(Australian and world maps)

Bibliophile
24 Glenmore Rd.
Paddington, Sydney
New South Wales 2021
02-331-3411

Read's Rare Book Shop
62 Charlotte St.
Brisbane 4000
07-229-3278
(Maps of Australia to 1850)

Spencer Scott Sandilands
546 High St.
East Prahran, 3181
Victoria
03 529 8011

Belgium

Librarie van Loock
51, rue Saint Jean
1000 Bruxelles
02-512-7465

Reproduction or Original?
by David C. Jolly

How can you tell an original antique map from a reproduction?

An **original** generally refers to a copy printed more or less at the time the map or view first appeared. In some cases, maps were printed for a century or more from the same copperplate or woodblock, occasionally with updating to include newer information. In such cases, as long as the plate or block was being employed commercially, impressions from it were considered to be "originals." Sometimes the block or plate survived to a later time, and was used to print restrikes. These are often identified by special watermarks or stamps. In any event, restrikes of old maps are almost unheard of on the commercial market.

Reproductions are a different matter. They can be defined as impressions made by some process, nowadays usually photographic, based on an original impression. Reproductions are not necessarily printed recently. Some 19th-century reproductions exist, many of excellent quality, but most reproductions encountered will have been done in the last few decades. A small number of reproductions can be distinguished only by experts. Most, however, require only a little knowledge. Some of the factors to be considered in detecting reproductions include:

■ **Size.** Reproductions not intended to deceive are often produced slightly larger or smaller than the original. Of course, one must know the size of the original. My book, *Antique Maps, Sea Charts, City Views, Celestial Charts, and Battle Plans, Price Record and Handbook*, gives dimensions that can be helpful in this regard, but remember that some dimensions are rounded off by dealers and that paper can expand by several percent as the humidity rises.

■ **Coloring.** Colored reproductions often employ halftone colors. These consist of patterns of small dots, geometrically arranged, which can be seen quite readily with a magnifying glass. A few reproductions, however, are colored by hand, just like the originals.

■ **Printing Quality.** Sometimes reproductions have a slightly blurred appearance. The black lines do not have the fine, dense quality of a true engraving. This can show up especially in cross-hatched areas, where the lines may fuse together.

■ **Plate Mark.** When an engraved map is printed, the impression of the metal plate crushes the paper, resulting in a depressed area. The depressed area   (continued)

Canada
Butler Galleries
341 W. Pender St.
Vancouver, B.C. V6B 1T3
604-681-6537
(Pacific Northwest exploration)

D&E Lake Ltd.
239 King St. E.
Toronto, Ontario M5A 1J9
416-863-9930
(antique maps, specializing in Americana/Canadiana; catalog available)

North By West
1016 Fort St.
Victoria, B.C. V8V 3K4
604-383-3442
(antique maps)

Ptolémée Plus
CP344, Succ. Cartierville
Montreal, Quebec H4K 2J6
(North America)

is usually rectangular and extends slightly beyond the printed area. This can often be seen, or felt, as a slight step or ridge. In a few cases, where the paper is thin, or where the map has been trimmed to the border, the plate mark may not be visible. On a very few reproductions, plate marks have been added to enhance realism. While visible on steel and copper engravings, a plate mark is normally not found on woodcuts, because much less pressure is used in printing them.

■ **Legends.** On most reproductions, there is a legend, usually in fine print, saying something like "Copyright 1968" or "From an original in the Library of Congress." It may seem unnecessary to even mention this, but people sometimes overlook the obvious. The legend can be hard to find. I have seen tiny legends embedded in the borders of woodcut maps. If outside the border, the legend may have been trimmed. Someone may also have tried to erase it, though this should leave a thin spot or scuff marks.

■ **Paper.** Probably the best method of distinguishing is to study the paper. Chain marks, rectangular grids of lines on handmade rag paper that can be seen when paper is held up to the light, are visible on paper made before about 1800. Some reproductions are printed on modern paper having chain marks, but the markings tend to be more regular on the modern paper.

Watermarks can also be an important clue, but some expertise may be needed in interpreting them. Many of the folio-size maps have watermarks, but smaller maps often do not. The paper on originals tends to have an aged appearance, perhaps browned, or even brittle, and sometimes having spotting or "foxing" (small, usually brown, spots on the paper caused by mold, often resulting from storage in damp conditions). Some originals show signs of use, such as stains, soiling, wear, and tears. Originals often show slight offsetting, either of color, or printer's ink, depending on how they were originally bound or folded. If colored, the pigments sometimes oxidize the paper, which can be seen by looking for browning on the verso (other side) corresponding to the colors on the map.

The above advice can help, but there is no substitute for long-term experience. Before investing money in an antique map, it is best to consult an experienced dealer. Some may charge a small amount for authenticating an item. Local libraries, art galleries, or museums may also be able to help.

Excerpted from Antique Maps, Sea Charts, City Views, Celestial Charts and Battle Plans, Price Record and Handbook, *David C. Jolly, ed., an annual. Available for $34.50 from David C. Jolly, Publishers, P.O. Box 931, Brookline, MA 02146; 617-232-6222.*

England
Avril Noble
2 Southampton St., Covent Garden
London WC2E 7HA
01-240-1970
(16th- to 19th-century antique maps and engravings)

The British Library
Marketing and Publishing
41 Russell Sq.
London WC1B 3DG
01-323-7535
(publishes reproductions)

Ivan R. Deverall
Duval House
The Glen, Cambridge Way
Uckfield, Sussex TN22 2AB
0825-2474
(professional coloring of maps and prints)

Susanna Fisher
Spencer, Upham
Southampton
Durley 291
(16th- to 19th-century sea charts)

The Goad Map Shop
Salisbury Sq.
Old Hatfield, Hertfordshire
AL9 5BE
07072-71171
(Fire Insurance Plans, 1878-1970)

Mrs. D. M. Green
7 Tower Grove
Weybridge, Surrey KT13 9LX
0932 241105
(antique county maps, town plans, & road maps)

Intercol London
1A Camden Walk
London N1 8DY
Cheltenham (0242) 522669
(coloring, cleaning, original style 16th- to 19th-
century atlases, maps, prints; world maps)

J.A.L. Franks Ltd.
7 New Oxford St.
London WC1A 1BA
01 405-0274
(early maps of most types and areas)

Jonathan Potter Ltd.
21 Grosvenor St.
Mayfair, London W1X 9FE
01 491-3520
(rare, interesting, and decorative old maps)

Leycester Map Galleries Ltd.
Well House
Arnesby, Leicester
LE8 3WJ
053 758 462
(16th- to 19th-century antique maps)

Harry Margary
Lympne Castle
Kent
0303 67571
(antique maps)

The Map House of London
54 Beauchamp Pl.
Knightsbridge, London
SW3 1NY
01-589 4325
(antique maps)

Olwen Caradoc Evans
Bodfor, The Esplanade
Penmaenmawr
Gweynedd LL34 6LY
0492 623955
(collotype prints; charts and maps of United
Kingdom)

Paul Orssich
117 Munster Rd.
London SW6 6DH
01-736 3869
(rare books and maps; lists on request)

Prinny's Antiques Gallery
3 Meeting House Ln.
Brighton BN1 1HB
0273 776943
(antique maps)

Warwich Leadlay
5 Nelson Rd.
Greenwich, London
SE10 9JB
01-858 0317
(antique maps worldwide; specialize in maps of
southeast London and northwest Kent)

Waterloo Fine Arts Ltd.
Paul & Mona Nicholas
40, Bloomsbury Way
London WC1
0734 713745
(antique maps)

Finland
Atikki-Kirja
Kelevankatu 25
SF-00100 Helsinki
90-611775
(northern regions, Russia, Scandinavia)

France
Librairie Ancienne—Curiosités
3, rue de l'Université
75005 Paris
42-60-75-94
(rare and antique maps)

Librairie Dudragne
86, rue de Maubeurge
75010 Paris
48-78-50-95
(France, Africa, Middle East, America)

Greece
K.E.B.E.
Odos Sina 44
10672 Athens
21-361-5548
(Cyprus, Greece, Malta, and the Near East)

Ireland
Neptune Gallery
41 S. William St.
Dublin 2
01-715021

Italy
Libreria Antiquaria Soave
Via Po 48
10123 Torino
011-878957
(Italian topographic, rare and antique maps)

The Netherlands
C. Broekema, F.R.G.S.
PO Box 5880
1007 AW Amsterdam
020-662 9510
(old maps, atlases, travel books)

Cartographica Neerlandica
Lester Pearsonweg 6
3731 CD de BILT
030-202396

Scotland
Billson of St. Andrews
15 Greyfriars Garden
St. Andrews
Fife KY 16 9DE
St. Andrews 0334 75063
(antiquarian Scottish atlases, maps, county maps,
town plans, road maps, and sea charts)

The Carson Clark Gallery
Scotia Maps-Mapsellers
173 Canongate
The Royal Mile
Edinburgh EH8 8BN
031-556-4710
(antique maps and charts of all parts of the world)

Switzerland
Antik-Pfister Antique Maps, Prints and Books
Zahringerplatz 14
Box 784
CH-8025 Zurich
01-47-62-32
(antique Swiss maps and prints of all parts of the
world)

West Germany
Kiepert KG
Hardenbergstrasse 4-5
1000 Berlin 12
(collection of maps of the urban development of
Berlin from 1650 to present)

Kunstantiquariat
Monika Schmidt
Turkenstrasse 48
8000 München 40 RFA
089-284233
(15th- to 18th-century maps, Japanese woodcuts)

History Maps

Maps of history are modern diagrams that illustrate a region's trends or events over decades or centuries. Instructors use history maps to teach everything from archaeology and anthropology to military history and literature. Scholars use them to track trends in a society or culture. As study tools, they provide valuable overviews of the evolution of a society's traditions, explorations, and advances.

There is a wide range of history maps available. There are maps that trace the routes of famous explorers, and maps of the history of medicine, religion, theater, poetry, art, and technology. There are maps of the most significant developments in the history of civilization. Many history maps are poster-size and intended for both decoration and study. Others are accompanied by texts that provide both background and analysis of the events illustrated on the maps.

The best overall sources for history maps are the larger commercial map publishers, especially those with an educational market. Historical maps, as well as atlases of American and world history, military history, and social history, are produced in great numbers by these publishers. Many smaller companies often publish history maps related to the other products in which they specialize. The federal government produces little in the way of history maps, but some state and local governments produce maps related to local and regional development.

GOVERNMENT SOURCES, UNITED STATES

U.S. Geological Survey. USGS has published a *National Atlas* map, "Territorial Growth," that shows the territory of the United States at various years between 1775 and 1920, including a final map of the U.S. and territories as of March 1986 ($3.10; US00664-38077-AV-NA-34M-00). USGS Base Map 8-A shows the routes used by principal explorers between 1501 and 1844 ($2.40; 05365). USGS also produced a special-edition map sheet to commemorate the bicentennial of the signing of the U.S. Constitution, "Maps of an Emerging Nation— The United States of America, 1775-1987." One side of the map sheet shows the first map of the U.S. produced by an American in 1774; the reverse side has a series of 14 maps showing the expansion of the U.S. between 1775 and 1987, along with depictions of flags appropriate to the years of the maps ($2.50; 38077-HI-UG-05M-18).

Some local or state agencies produce or distribute maps describing local history. Write to the local historical societies for further information, or contact state mapping agencies where applicable (see Appendix A).

GOVERNMENT SOURCES, CANADA

Canada Map Office (615 Booth St., Ottawa, Ontario K1A 0E9; 613-952-7000) distributes various historical maps, including:

■ "Historic Maps of Canada," a special series produced by the Department of Energy, Mines and Resources of Canada. The maps trace the expanding knowledge of the New World as well as changing cartographic techniques. The series includes: "America Septentionalis, 1639" (MCR 2300), "Amerique Septentrionale, 1695" (MCR 2302), "An Accurate Map of Canada, 1761" (MCR 2303), "British North America, 1834" (MCR 2304), and "Le Canada, ou Nouvelle France, 1656" (MCR 2305). The five maps can be purchased separately ($3.50 Canadian each) or as a set ($12.25 Canadian; MCR 2301).

Also available as part of the *National Atlas of Canada* series is "Canada Then and Now," a set

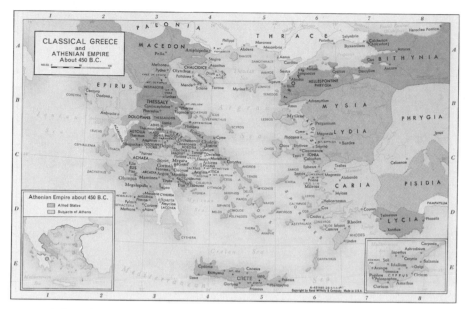

Map of Classical Greece and the Athenian Empire. Reprinted with permission from Rand McNally & Company's Historical Atlas of the World. ©1965.

814 A.D." and "Colonial Possessions of the World Powers, 1914." Both series include maps of major wars, explorations, and expansions. Each series costs $403, with mounting extra.

Hammond Inc. (515 Valley St., Maplewood, NJ 07040; 201-763-6000) publishes several atlases of world history in cooperation with Times Books of London, as well as a small series of its own, including Bible atlases and atlases of American and world history. Hammond is also the American distributor for Bartholomew Maps, which has a series of historical maps of British possessions, including England, Ireland, London, Scotland, and Wales. Maps in this series illustrate castles, battles, historical characters, and the coats of arms of cities and towns. Bartholomew also produces several pictorial maps on historical subjects, including "Cathedrals and Abbeys." Atlases are $85 for hardcover editions. Bartholomew historical maps are $9.50; pictorial maps, $8.50.

Modern School (524 E. Jackson St., Goshen, IL 46526; 219-533-3111; 800-431-5929) produces 13 history maps for classroom use, including "Early Exploration," "Territorial Expansion to 1848—The Mexican War," and "World War II: Europe." The series is available in three different mountings, ranging from $244 to $336.

National Geographic Society (17th & M Sts. NW, Washington, DC 20036; 202-921-1200) produces a large selection of history maps. Titles include "Historical Europe" ($4; 02021),

containing three maps that portray the early geography of Canada and the evolution of provincial and territorial boundaries. Maps in the series include "Canada—Confederation," "Canada—Territorial Evolution," and "Canada," a 1982 map of the country. The maps are $5.50 Canadian each.

Maps Alberta (Land Information Services Div., Main Fl., Brittania Bldg., 703 6th Ave. SW, Calgary, Alberta T2P 0T9; 403- 297-7389) also distributes maps from the *National Atlas* series.

COMMERCIAL SOURCES

George F. Cram Co. (P.O. Box 426, Indianapolis, IN 46206; 317-635-5564) publishes two series of 52″ X 40″ maps, one of American history, the other of world history. The American History Series contains 33 titles, ranging from "Voyages and Discoveries to 1610" to "Transportation, River and Canal Period, 1816-1840." The World History Series, with 43 titles, includes "Europe at the Death of Charlemagne,

Uncle Sam's Treasure Chests

The world of maps is so vast that you practically need a map to find one. Two of the best resources for historical map research are the map collections of the Library of Congress and the National Archives, both in or around Washington, D.C. Funded and operated by the federal government, both collections contain millions of maps, charts, atlases, and globes, as well as a myriad of valuable research tools for the study of cartographic information.

Library of Congress, Geography and Map Division (Washington, DC 20541; 202-707-6277). When the Library of Congress was established in 1800, some of its earliest acquisitions were maps and atlases. In 1897, the Geography and Map Division was given its own room; it has since moved to specially designed quarters on Capitol Hill. The collection, the largest in the world, consists of more than 4 million maps, 52,000 atlases, 2,000 plastic-relief models, 300 globes and 8,000 reference works. Among its most prized possessions are three sailing atlases and 19

sailing charts dating back to the 14th through 17th centuries. There are also numerous early American maps and charts, some predating the Revolutionary War, and the Sanborn Map Co.'s 19th-century fire insurance maps, which detail the growth of the 12,000 American cities and towns. The rest of the world is well represented, too, including rare Asian maps in the Hummel and Warner collections.

Although there is no single catalog of all the division's holdings, there are specialized card and book catalogs, as well as computerized magnetic tapes listing newer maps. The library publishes selected lists on specialized topics, including annotated lists of Civil War maps, railroad maps, land-ownership maps, and Indian land maps. The Geography and Map Division's List of Publications is free upon request. The Geography and Map Division's collections are meant for research purposes only, and lending privileges are restricted to members of Congress, federal agencies, and authorized libraries. A reference service is available to the public, and staff can handle phone and mail requests. *(continued)*

"Shakespeare's Britain" ($3; 02535), "Colonization and Trade in the New World" ($3; 02880), and "Classical Lands of the Mediterranean" ($3; 02271). NGS also publishes several historical charts, including "Alaska's Ice Age Mammals" and "History Salvaged From the Sea." National Geographic maps are full-color with illustrations and are usually suitable for framing. A list of publications is available upon request.

Nystrom (3333 Elston Ave., Chicago, IL 60618; 312-463-1144; 800-621-8086) specializes in learning materials for schools. Its series of American and world history maps are 50″ X 38″. Among the 33 maps in the American history series are "Indian Tribes and Settlements in the New World, 1500-1750" and

"Transportation Unites the Nation." Titles in the series of 32 world history maps include "Christian Europe and the Crusades" and "U.S.S.R. Territorial Expansion." Prices of single maps range from $43 to $66, depending on type of mounting. Set prices range from $309 to $778.

Rand McNally (P.O. Box 7600, Chicago, IL 60680; 312-673-9100) produces three sets of American history and one set of world history maps intended for school use. It also distributes the Breasted-Huth-Harding series of world history maps. The three U.S. series are "American History Maps" (for intermediate grades), "American Studies Series" (all levels), and the "Our America Series," which includes time lines and text on each map. The world history series

The Library's Photoduplication Service can make reproductions of maps and atlases for a fee, except where copyright or other restrictions apply.

The Library's Photoduplication Service handles all reproduction orders. Prices range from $5 for a 2" X 2" color slide (minimum order of three) and $10 for a black-and-white photodirect paper print, to literally hundreds of dollars for exhibition-quality prints, depending on size and reproduction quality desired. The minimum reproduction time is six weeks.

National Archives, Cartographic and Architectural Branch (841 S. Pickett St., Alexandria, VA 22304; 703-756-6700). The Archives' Cartographic and Architectural Branch holds one of the largest collections of American maps in the world. Established as the repository for maps commissioned or created by the government, its holdings date from 1750 to 1950. The Branch has more than 2 million maps and charts 800,000 in manuscript form, as well as 500,000 architectural drawings and engineering plans and 8 million aerial photos.

The maps of the Lewis and Clark and other explorations of North America are among the treasures in the collection. Virtually every aspect of American society and expansion can be found in the files of the branch, located in a suburb of Washington, D.C. A number of special catalogs and reference information papers are available through the Archives, their subjects running the gamut from 18th-century Indian lands to 20th-century transportation growth. The free pamphlet *Cartographic and Architectural Branch, National Archives and Records Administration*, available directly from the NARA, lists many of the categories and reference materials in the collection. Reproduction services of varying degrees of quality and price are available; orders usually require four to six weeks for completion.

The Archives staff can produce black-and-white photocopies of maps while you wait for $1.60 per linear foot of copy. If you need a better-quality reproduction, the Archives Reproduction Service can make many types of copies, but not while you wait. Prices range from $2.45 for a 2" X 2" color slide, or $10.85 for a 4" X 5" color negative, to as much as $192.80 for a 30" X 40" color print. Reproduction time can take up to eight weeks.

covers the development of civilization in most areas of the globe. The 62 maps in the Breasted-Huth-Harding series include "Barbarian Migration" and "Air Age."

OTHER SOURCES

Following is a sample of the history maps produced by smaller map companies or organizations involved in studying or promoting specific subjects:

Celestial Arts (P.O. Box 7327, Berkeley, CA 94707; 415-524-8755; 800-841-2665) publishes a "wall visual" poster ($9.95) called "The Aztec Cosmos," a stylized ancient historic map demonstrating the Aztec culture's artistic,

mathematical, and stone-carving accomplishments. The map includes an educational 32-page booklet describing Aztec symbols and religion.

Facts On File Inc. (460 Park Ave. S., New York, NY 10016; 800-322-8755; 800-443-8323 in Canada) produces *Historical Maps on File* ($145). The more than 300 copyright-free maps included in this loose-leaf collection are simple black-and-white diagrams of major historical trends or events. Titles include "Greece During the Persian Wars," "Expedition of Coronado, 1540-1542," and "Adams-Onis Treaty, 1819." These are useful for students or researchers who need to copy maps for reports and presentations.

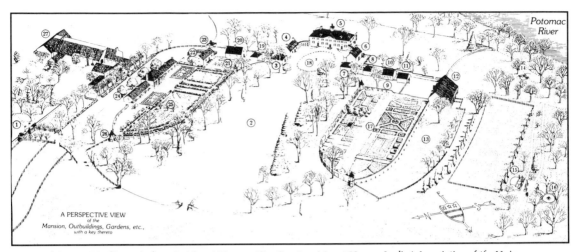

A map of George Washington's home in Mount Vernon, Va. Courtesy Mount Vernon Ladies' Association of the Union.

Geological Society of America (3300 Penrose Pl., P.O. Box 9140, Boulder, CO 80301; 303-447-2020) publishes a wall-size geologic "time scale" in black and white ($8.50; 30" X 36"), which traces the geologic development of the North American continent.

Williams & Heintz Map Corp. (8119 Central Ave., Capitol Heights, MD 20743; 301-336-1144) primarily a producer of technical maps for various customers, also publishes a series of geologic maps. Its "Historical Geology" portfolio ($8.50) has 13 maps that trace the history of major U.S. geologic provinces.

HISTORIC SITE MAPS

Civil War battlefields, presidents' homes, and monuments to long-gone explorers dot the American landscape, as do countless other historic sites of major and minor importance to our national heritage. Some sites are operated under the auspices of the federal government or a state or local historic society. Others are privately run. For virtually all, there is a map available to guide one to and through the places where history was made.

GOVERNMENT SOURCES

National Park Service. The National Park Service is the guardian of national historic sites. The "House Where Lincoln Died," Bunker Hill, and Mesa Verde are three of the hundreds of historic sites operated and preserved under the Park Service's care. Maps are usually available at these sites or through the agency's Office of Public Inquiries.

U.S. Geological Survey. USGS publishes National Park maps, which include some maps of national historic landmarks, such as "Custer Battlefield National Monument" in Montana, "Home of Franklin D. Roosevelt" in New York, and "Colonial National Historical Park" covering Williamsburg and Yorktown in Virginia ($4). USGS also pinpoints certain historic landmarks on its maps of America. There must be a substantial object (a house, a battlefield, or a grave, for example) for the site to be included on a USGS map; small monuments or signs do not merit note. Maps depicting national historic sites and areas are listed by state, in an "Index to USGS Topographic Map Coverage of National Park System," available from Earth Science Information Centers and the USGS Distribution Centers in Denver, Colorado, and Fairbanks, Alaska.

Government Printing Office. GPO has maps and guidebooks to several popular historic sites. Prices range from $2.25 for a simple guidebook to $10 or more for comprehensive texts, complete with maps, of various sites. Examples of maps and guides available from GPO include: "Campaign for Petersburg: Petersburg National Battlefield, Virginia" ($3.75; S/N 024-005-00979-5), "Fort Vancouver National Historic Site, Washington" ($7; S/N 024-005-00816-1), and "Guide to United States Army Museums and Historic Sites" ($5.50; S/N 008-020-00561-4).

State historic societies (see Appendix A) and building preservation societies sometimes publish maps or guides to historic sites preserved by states or local jurisdictions. State tourist agencies (Appendix A) and local chambers of commerce often publish walking-tour maps or guidebooks to area attractions.

■ **See also: "Recreation Maps," "Selected Atlases," and "Tourism Maps and Guides."**

Military Maps

Among the precious few consolations of war are the resulting technological advances that often have applications in nonmilitary life. Over the years, the need for accurate military maps has led, time after time, to the creation of newer and more sophisticated mapping techniques.

In 1777, the Military Cartographic Headquarters was established at Ringwood, New Jersey, to create maps for General George Washington's campaign against the British. The attacks and marches planned with these maps led the Americans to an unprecedented victory against Mother England, and the legacy of American military mapping began.

In the decades that followed, the young nation found itself embroiled in a series of conflicts with other nations, and finally, with itself. Cartographic agencies were created to keep pace with the changing needs of the military, and these agencies in turn gave birth to numerous advances in surveying and mapping techniques.

The War of 1812 pointed out the need for nautical military maps to accommodate an expanding Navy, which resulted in the establishment of the Navy Depot of Charts and Instruments in 1830. The Civil War led to the development of aerial-reconnaissance mapping, with surveyors rising above the fray in hot-air balloons to diagram enemy territory. In 1907, aerial reconnaissance reached new heights with the establishment of the Aeronautical Division of the Army Signal Corps.

The Spanish-American War gave U.S. military cartographers their first real chance at mapping foreign soil, an exercise that paid off during the trench and field battles of the first World War. The use of aerial photography for military purposes was also perfected during World War I. The fine art of cartographic cooperation was honed during World War II, as the federal government's civilian map agencies joined with the Army Map Service and the Navy Hydrographic Office to help battle the Nazi menace.

GOVERNMENT SOURCES, UNITED STATES

The mapping agencies of the Defense Department were combined in 1972 to form the Defense Mapping Agency. Today, DMA's Topographic, Hydrographic, and Aerospace Centers produce many of the maps used by the American military. Aside from updating and producing a steady stream of military maps for everyday use, the DMA is currently creating a digitized map of the world to help guide missiles and other military systems. The DMA provides a free 104-page catalog, "Digitizing the Future," which explains all DMA digitizing activities and lists products available (see "Mapping the Future" for more on these efforts). Many of DMA's regular maps are available to the public, although they are among the most expensive in the federal government's map collection.

Besides DMA, there are several other government sources of military maps:

■ The **National Oceanic and Atmospheric Administration** distributes military aeronautical and nautical charts and maps.
■ The **Central Intelligence Agency** creates maps of foreign countries (see "Foreign Country Maps"), which are distributed by the National Technical Information Service.
■ The **Defense Department**, **State Department**, and other agencies produce maps sometimes used for military purposes; they are distributed by the GPO.
■ The collections of the **Library of Congress** and the **National Archives** are filled with military maps available for study. The Archives contains maps created by government mapping agencies, while the Library of Congress's collection includes both domestic and foreign military maps. For a fee, the reproductions

MAP NO. 2

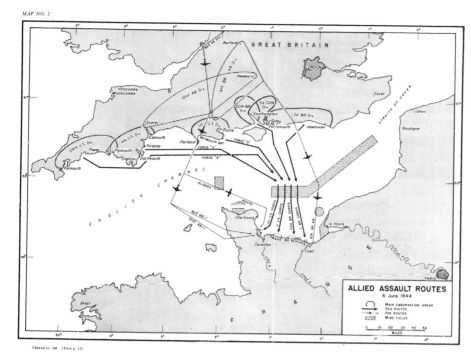

The routes of the Allied assault on D-Day, from a set of maps contained in the GPO book Utah Beach to Cherbourg.

services at both collections make copies of their maps (see "Uncle Sam's Treasure Chests").
■ Maps of historic battlegrounds that are preserved by the federal government can be obtained from the **National Park Service**, while state and local bureaus of tourism or historical preservation usually hand out free maps of the memorials or battlegrounds under their auspices.

If you're searching for a military map, here are some strategic places to look:

Government Printing Office. GPO has several publications that contain military maps, including:

■ *South to the Naktong, North to the Yalu,* June-November 1950 ($25.50; S/N 008-029-00079-2).
■ *Military Advisors in Korea:* KMAG in *Peace and War* ($12; S/N 008-029-00002-4).
■ *Soviet 1945 Strategic Offensive in Manchuria* ($8.50; S/N 008-020-00984-9).

■ *United States Army in the Occupation of Germany,* 1944-1946 ($16; S/N 008-029-00090-3).
■ *Time Runs Out in CBI* ($15; S/N 008-209-00014-8).
■ *Ardennes, Battle of the Bulge* ($21; S/N 008-029-00069-5).
■ *Omaha Beachhead* ($8.50; S/N 008-029-00128-4).
■ *Saint Lo, July 7-19, 1944* ($8.50; S/N 008-029-00127-6).
■ *Lorraine Campaign* ($29; S/N 008-029-00019-9).
■ *Mediterranean Theater of Operations: Cassino to the Alps* ($19; S/N 008-029-00095-4).
■ *Global Logistics and Strategy,* 1940-1943 ($19; S/N 008-029-00056-3).
■ *War in the Pacific: Campaign in the Marianas* ($19; S/N 008-029-00040-7).
■ *Guadalcanal, The First Offensive* ($20; S/N 008-029-00067-9).

Tennessee Valley Authority. TVA produces several inexpensive blueline lithoprint reproductions of original Civil War maps depicting campaigns fought in the Tennessee Valley region. When ordering, include 50 cents postage for each map. Available maps include:

■ "Map of the approaches and defense of Knoxville during the Civil War" ($1.50; 455 K 90), a blueline print measuring 25" X 31", surveyed in 1863.
■ "Map of the Army movements around Chattanooga made to accompany the report of Major General Grant in January, 1864" ($1; G MD 453 G 754), a litho measuring 18" X 24".
■ "Battlefield of Chickamauga, Georgia, April,

1864" ($1; G MD 453 G 754-1), a blueline print measuring 18" X 22".
■ "Map showing the operations of the National Forces under the command of General W.T. Sherman during the campaign of Atlanta, September 1864" ($1; G MD 453 G 754-2), a blueline print measuring 16" X 21".
■ "Chattanooga and its approaches" ($1.50; G MS 453 K 754-3), a blueline print measuring 22" X 28" that shows the Union and Rebel works during the battles of November 1863.
■ "Maneuver Ground, Chickamauga Park and Vicinity" ($1; G CF 8141 G), a blueline print measuring 18" X 24" of a topographic map made in April 1910.
■ "Map of Battlefields of Chattanooga" ($1; 5-438), a blueline print measuring 18" X 24" dated 1901, showing movement against Orchard Knob.
■ "The First Epoch of the Atlanta Campaign, from Chattanooga to Oostanaula River" ($1.50; 453 K 754-6), a blueline print measuring 22" X 35". This map shows the positions held by Generals Sherman and Johnston.
■ "Military Map Showing the Marches of U.S. Forces Under General Sherman, 1863-1865" ($1.50; 453 M 754-8), measuring 23" X 38".

U.S. Geological Survey. USGS's topographic mapping program, which covers the National Park System, has mapped many historic forts, battlegrounds, and military memorials and parks. A free map pinpointing topographic maps, "Index to USGS Map Coverage of the National Park System," is available from USGS. The maps in the National Park Series cost $4 each and are drawn at various scales, some parks requiring more than one sheet (see "Recreation Maps" and "Topographic Maps" for more information).

National Archives and **Library of Congress.** Both agencies contain thousands of military maps providing worldwide coverage.

National Oceanic and Atmospheric Administration. For reasons no one can explain, the NOAA maintains an extraordinary collection of

Civil War charts and diagrams drawn by the cartographers of the Confederate Army. A catalog of NOAA's Civil War collection is available from GPO.
The collections of the Library of Congress, National Archives, and NOAA are available to the public for study only. Reproduction facilities are available at each collection, although reproductions can take up to eight weeks. See "Uncle Sam's Treasure Chests" for more information on the collections of the Library of Congress and National Archives.

GOVERNMENT SOURCES, CANADA
Canada Map Office (615 Booth St., Ottawa, Ontario K1A 0E9; 613-952-7000) distributes, as part of the *National Atlas of Canada, 5th Edition*, the map "Canada-The Northwest Campaign, 1885" ($5.50 Canadian; MCR 4106).

COMMERCIAL SOURCES
Older maps of military campaigns or battles can be found through antique map and military paraphernalia dealers; many print and art dealers carry antique military map reproductions. Reproductions can also be obtained from bookstores at many museums and libraries.
    Other sources for military maps include those commercial map companies that produce classroom maps tracing military campaigns or battles. Some companies also publish reproductions of military maps for decorative purposes, as well as up-to-date, utilitarian maps pinpointing current military bases.
    Another good source for historical military maps are local or regional military clubs for devotees of a certain era or war. Pennsylvania and West Virginia have several Civil War organizations, for example, while the New England area is rife with Revolutionary War buffs. The museums and gift shops of military battlefields and parks are also good hunting grounds for commercial military maps.
    Several educational map publishers produce maps depicting military history. Here's a sampling:

George F. Cram Co. (P.O. Box 426, Indianapolis, IN 46206; 317-635-5564) publishes several large, full-color maps of military history for classroom use. Maps in the American History Series include "The Revolution in the Middle and Northern Colonies," "The Revolution in the South and West," "The Mexican War and Compromise of 1850," "The Civil War, 1861-1865," and "The World War, 1914-1918." Maps in Cram's World History Series include "Caesar's Conquest of Gaul, 58-50 BC," "Norman Conquest of England," "Mohammedan Conquests at Their Height, 750 A.D.," "Europe in 1648," "The Nations at War in 1918," "The World War, 1914-1918," "World War II, European Theater," and "World War II, Pacific Theater." The maps measure 52″ X 40″ and range in price from $63 to $72.

Europe Map Service/OTD Ltd. (1 Pinewood Rd., RD 7, Hopewell Junction, NY 12533; 914-221-0208) distributes reproductions of the official German Reich administrative World War II maps showing borders of the Third Reich from 1937 to May 1945. The series of 1:25,000-scale maps consists of 2,557 sectional maps showing topographic features, buildings, railways, and roads in exceptional detail. The maps, some in color, some in black and white, measure approximately 23″ X 24″ and cost from $9 to $12 each.

National Geographic Society (17th & M Sts. NW, Washington, DC 20036; 202-921-1200) has "Battlefields of the Civil War" ($4; 20019), a full-color illustrated map that measures 30″ X 23″.

Nystrom (3333 Elston Ave., Chicago, IL 60618; 312-463-1144; 800-621-8086) sells two richly colored series of history teaching maps similar to Cram's. The Nystrom maps ($43 to $66, depending on mounting) measure 50″ X 38″. Titles relating to military history in the American series include "The Revolutionary War," "The War of 1812," "The War Between the States, 1861-1865," and "Korean War; Vietnam

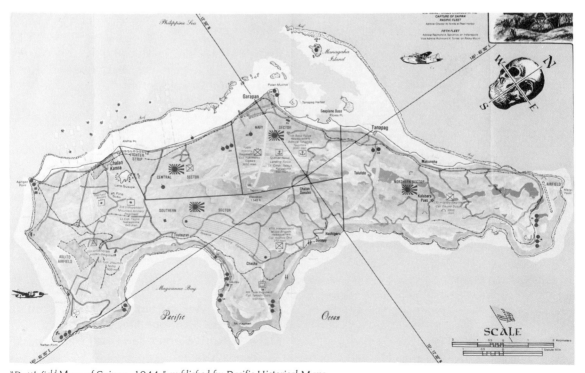

"Battlefield Map of Saipan—1944," published by Pacific Historical Maps.

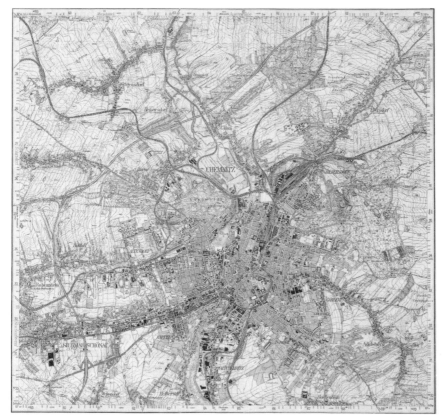

1937 *topographic map of Chemnitz, Germany, from Europe*
Map Service/OTD Ltd. The area, now known as Karl Marx
Stadt in East Germany, was devastated during World War II.

War." Maps in the Nystrom's World History
series include "Mongol-Turkish Conquests,"
"Napoleonic Empire, 1812," "World War II in
Europe and Northern Africa," and "Russian and
Japanese Expansion in the Far East."

Pacific Historical Maps (P.O. Box 201, CHRB,
Saipan, Mariana Islands 96950) publishes
historical maps and books of World War II
Pacific battlefields, which are distributed by
other companies and agencies. Offerings
include:

■ "Battlefield Map of Saipan—1944" ($3.50), a
four-color map illustrating the U.S. route of
advance and Japanese defensive positions,
Japanese sectors, and railways on the island,
with copious text in the margins providing a

historical summary
of events, along
with information on
everything from
"Coastal Defenses"
to "Japanese Strat-
egy." The reverse
side of the map
features a 1944
aerial photograph
of the island, with
various war photos
on the margins.
The map is avail-
able from **Mari-
anas Visitors Bu-
reau**, Box 861,
Saipan MP 96950.
■ "Dive Map of
Ghost Fleet of Truk
Lagoon" ($3.50), a
combination map
and pictorial de-
picting the 1944
U.S. aircraft carrier
attack on Japanese
fleets. Shipwrecks
from this battle are located and described. The
map is available from the **Department of Resources and Development**, P.O. Box 280,
Moen, Truk, F.S.M. 96942.
■ "Battlefield Map of Peleliu" ($4.50), describ-
ing the American attack of 1944 and the
Japanese defense. The map is available from
Western Caroline Trading Co., Box 280,
Koror, Palau 96940.
■ "Ghost Fleet of Truk Lagoon" ($11.50), a
book describing the U.S. naval attack on the
Japanese fleet, illustrated with maps and
photos, available from **Pictorial Histories
Publishing**, 713 S. Third W., Missoula, MT
59801.

■ **See also: "Aerial Photographs," "Aeronau-
tical Charts," "Antique Maps," "History
Maps," "Native American Land Maps," and
"Selected Atlases."**

UTILITY AND
SERVICE MAPS

Business Maps

For the traveling salesperson, the road map is as vital a tool as the telephone and the motel, but it isn't the only kind of map available to people in sales and other aspects of business. Maps are being used increasingly in businesses large and small, as computer and other data are combined with territorial maps to create visual representations for a variety of business applications, primarily in the areas of sales and marketing.

Business maps come in many shapes and forms. Among them:

■ demographic maps ranking areas by population, income, age, race, and other factors;
■ thematic maps showing manufacturing and trade areas;
■ simple, uncluttered black-and-white or color maps of countries, states, counties, cities, and major transportation routes;
■ zip code maps with population data;
■ maps showing the major television and radio markets;
■ travelers' maps and atlases showing efficient routes, convention centers, hotels, and other amenities; and
■ various marketing and sales atlases combining many of the above elements.

Most of these are manufactured by a few large companies that either specialize in, or have departments devoted to, business maps. The federal government's contribution to the business map field comes in the form of census maps (see "Census Maps"), many of which are also available, often with enhancements, from commercial publishers. The cartographic offices of foreign countries (see Appendix C) can often locate the necessary business maps for their regions.

One of the fastest-growing areas of business maps is "desktop mapping"—the creation of maps on desktop and other computers. A growing number of businesses, as well as

government agencies and educational institutions, are finding myriad uses for these technologies, particularly in the areas of marketing and sales. For more on desktop mapping, see "Map Software."

GOVERNMENT SOURCES

Several Census Bureau maps are available through the Government Printing Office, including:

■ "Metropolitan Statistical Areas (CMSAs, PMSAs and MSAs)" ($5.50; S/N 003-024-06904-1). This 1988 map shows the locations of three standardized census regions—consolidated metropolitan statistical areas (CMSAs), primary metropolitan areas (PMAs), and metropolitan statistical areas (MSAs)—for the United States and Puerto Rico. The map (16″ X 34″) is printed on both sides; the continental U.S., Alaska, Hawaii, and Puerto Rico are on the front and a large map of New England is on the back. The map is updated annually.
■ "Population Distribution in the United States: 1980" ($2.25; S/N 003-024-06445-6), a 20″ X 30″ map illustrating the densely populated areas of the United States with spots of white on a black background. This map, known also as the "Nighttime View," was revised in 1987 (see "Census Maps" for additional titles).

COMMERCIAL SOURCES

Many commercial business maps were made with utility in mind: Some atlases can slide into glove compartments; larger maps are designed to be mounted and displayed for presentation. Graphics tend to be minimal, as are details about tourist attractions or other extraneous features found on most general-use maps.

American Map Corp. (46-35 54th Rd., Maspeth, NY 11378; 718-784-0055; 800-432-6277) produces a variety of business-related maps. Its "Cleartype" series is a collection of black-and-white outline maps in varying sizes showing place-names and borders for U.S. cities, counties, and states, as well as maps of Canada and the world. Cleartype maps can be drawn or written on and are available with a variety of mounting styles. The selection includes:

■ "U.S. State Outline" shows the continental United States, Alaska, and Hawaii with each state outlined and identified. There are six sizes available, from 11″ X 8 1/2″ (55 cents) to 74″ X 50″ (24 cents).
■ "Individual States/County Outline" shows all counties within a single state. Maps are available for all 50 states, in 8 1/2″ X 11″ (55 cents).
■ "Individual States/County and Town Outline" are the same as above, but show both towns and counties.
■ "World Outline" shows outlines and names

of continents and countries on a map drawn on a Mercator projection. The three sizes and prices available range from 11″ X 8 1/2″ (55 cents) to 50″ X 38″ ($22).
■ "Continents and Oceania Outline Maps" show outlines of countries, individual continents, and Oceania (17″ X 22″; $2.75).

Other Cleartype maps include sets of principal U.S., Canadian, and world cities, and major U.S. metropolitan areas.
    Cleartype thematic maps available include:

■ "U.S. MSA Markets" ($15), a Metropolitan Statistical Area map measuring 44″ X 30″.
■ "U.S. ADI Markets" ($15), Areas of Dominant Influence, industry jargon for major media markets, measuring 44″ X 64″.

"Colorprint" maps are similar to Cleartype maps in detail but are produced in full color.

New York City and surrounding areas are illustrated in this portion of a "Principal City Map." Reprinted with permission of the American Map Corporation.

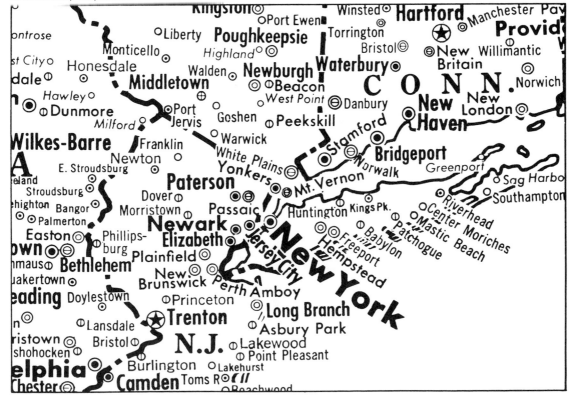

This simple, concise U.S. map is intended for use in sales planning, marketing, and other business applications. ©1982 *by Rand McNally & Company. Reprinted with permission.*

Available sales maps include:

■ "World Map Murals" ($25 to $150; 64″ X 42″ to 110″ X 68″).
■ "Laminated Sales Maps," county maps available showing: towns of 2,500+ population ($40), towns of 5,000+ population ($28), and towns of 10,000+ population ($18); and a U.S. map showing principal cities of 20,000+ population ($5), 10,000+ population ($6), and 5,000+ population ($6).

Cismap Inc. (P.O. Box 770369, Houston, TX 77215; 713-783-5728) distributes *Gulf Coast Industrial Handbook* ($100), a directory/atlas showing over 200 processing facilities, featuring full-color maps. Information on the facilities is cross-referenced to maps, making it easy to locate key businesses.

Hammond Inc. (515 Valley St., Maplewood, NJ 07040; 201-763-6000) produces a small series of maps for business use, as well as a

comprehensive *Sales Planning Atlas*. Available business maps include:

■ "City/State Sales Planning Maps": black-and-white U.S. maps showing state boundaries and major cities/trade centers. Three sizes: 11″ X 17″ ($.95); 19″ X 25″ ($3.50); 25″ X 38″ ($4.90).
■ "State/County Outline Maps": single state maps showing county outlines, major cities and towns, state capitals, and county seats. An index, zip code directory, and census information are printed on the back of each map. The 8 1/2″ X 11″ maps are sold in packages of 10 for $4.
■ "County Outline Map": A 25″ X 38″ wall map of the U.S. with counties outlined within each state ($3.95).
■ *Sales Planning Atlas* ($14.95): This 128-page atlas includes U.S. state maps with county outlines and place designations for places with more than 1,000 people, census statistics, U.S. and world time zones, highway mileage and air distances, city weather information, and other

pertinent sales and travel planning features. A "U.S. County Wall Map" is included in an inside pocket.

M.A.P.S. Midwestern (City Park Rd., Oelwein, IA 50662; 319-283-3912) produces county/township maps depicting ownership and residency, targeted toward commercial use ($15). Maps are available for rural areas in Illinois, eastern Iowa, Minnesota, Missouri, South Dakota, and Wisconsin. Also available for these areas are books of residency, showing who lives where ($15 each).

The National Survey (Chester, VT 05143; 802-875-2121) produces "Township Outline Maps" as well as "School Outline Maps" for the New England states. Map sizes vary from 8 1/2" X 11" to 22" X 34"; prices range from 50 cents to $4 each.

Phoenix Mapping Service (2626 W. Indian School Rd., Phoenix, AZ 85017; 602-279-2323) publishes various maps of the Phoenix and Tucson metropolitan areas. Titles include black-and-white, color, and zip code wall maps of both metropolitan areas (60" X 42"; $36.95). It also produces arterial street maps for both areas, as well as a "Proposed Freeway and Expressway Routes" map detailing existing, plannned, or under-construction freeways and expressway routes (80 cents each; 8 1/2" X 11").

Pierson Graphics Corp. (899 Broadway, Denver, CO 80203; 303-623-4299) produces maps for Colorado metropolitan areas. Titles range from "Greater Denver Wall Map" (36" X 46"; $45), including block numbers, zip codes, municipal and county boundaries, and township ranges, to the "Close Up Series," maps of the Denver regional and metropolitan areas, as well as Boulder/Longmont, with prices from $7.95 to $29.95.

Pittmon Map Co. (732 SE Hawthorne Blvd., Portland, OR 97214; 503-232-1161; 800-547-3576; 800-452-3228 in Ore.) sells "Outline Maps," used to mark sales areas. Maps are

available for most western states; sizes range from 8 1/2" X 11" to 22" X 34" ($1.25 to $3.00). Pittmon also distributes zip code and freeway maps for the Portland metropolitan area.

Rand McNally (P.O. Box 7600, Chicago, IL 60680; 312-673-9100) has a line of office and business maps that includes outline, thematic, and marketing maps, and several business-oriented atlases.

Rand McNally outline maps include:

■ "MarketMaps—U.S. Titles," a series of black-and-white outline maps in various sizes and concentrations: "U.S. State Outline," with state outlines only (five sizes, from 8 1/2" X 11"; 50 cents, to 46" X 66"; $12); "U.S. County/State Outline," with state and county outlines only (four sizes from 22" X 19" to 66" X 46"); "U.S. City/State Outline," cities designated within state outlines (three sizes, from 8 1/2" X 11" to 21" X 28"); and "U.S. City/County/State Outline," cities shown within counties (two sizes, 19" X 22" and 24" X 28"). There are also single state MarketMaps in both county and city/county outline formats in three sizes: 8 1/2" X 11", 17" X 22", and 21" X 28".
■ "Color MarketMaps II": This is a color version of the black-and-white MarketMaps series. Color MarketMaps include "U.S. City/County/State Outlines" in three sizes: 24" X 28", 33" X 42", and 46" X 66". There is also a 21" X 28" "U.S. City/County/State Outline."
■ "Sectional Sales Control Maps": These show sections of the U.S. in black-and-white detail, featuring state and county names, county seats, and cities with populations of 1,000 or more, as well as insets of major metropolitan areas. Four of the five maps ("North Central," "Northeast," "Southeast," and "Southwest"), are available in 41" X 45 1/2"; the Western Sales Control Map is available only in 41" X 45 1/2".
■ "Zip Code Map of the U.S.": A four-color map with state and county names, showing names of zip code sectional areas (first three zip digit) and other zoned cities, with large-scale insets of 13 major metropolitan areas (25" X 32"; $19.95).

Rand McNally business atlases include:

■ *Zip Code Atlas + Market Planner* has full-color, 11″ X 15″ state, city, and metro area maps. Five-digit zip code boundary maps are printed on acetate overlays for the maps. The atlas comes in a portfolio with a 10-ring binder and Velcro closure ($295).

■ *Commercial Atlas + Marketing Guide* ($265), the oldest continuously published atlas in the country, revised annually.

■ *The Business Travelers Road Atlas* ($9.95), a 160-page wire-bound atlas that contains the standard *Rand McNally Road Atlas* maps, plus maps of major airports and reference data on area codes, state gas taxes, and driving time. Also included is a state-by-state list of convention facilities, toll-free hotel and car-rental telephone numbers, and expense-account information.

■ *Handy Railroad Atlas of the United States* ($9.95), a 64-page paperback that includes one- and two-color maps of the American and Canadian rail systems. U.S. railroad distance tables are also included.

■ *Metro Area Planning Atlas* ($29.95), a 190-page wire-bound atlas containing full-page black-and-white maps suitable for coloring or shading to illustrate business plans. Fully indexed, with demographic information.

■ *3-Digit Zip Code Planning Atlas* ($29.95), a 216-page paperback with color maps of every state to illustrate three-digit zip code areas, plus county boundaries and seats, major cities, and population centers. Demographic information about population size and distribution is included, as are zip code maps for major cities.

■ *City/County Planning Atlas* ($29.95), a 300-page atlas with population, sales, and income data for all U.S. counties and full-page maps of each

An American Map Corporation zip code map of Virginia. ©1986 American Map Corporation. Reprinted with permission.

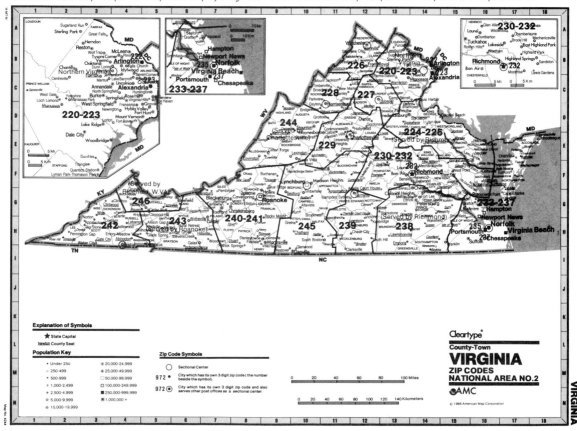

state. State indexes list populations, zip codes and FIPS codes (see "Census Maps" for more on FIPS codes).

■ *Business Planning Atlas* ($16.95) has black-and-white maps for all states and an index of counties and places with populations greater than 1,000.

Thomas Bros. Maps (17731 Cowan Dr., Irvine CA 92714; 714-863-1984; 800-432-8430 in Calif.) offers business maps for California. Products include:

■ "Arterial Wall Maps" of the Bay Area and Los Angeles, showing only freeways, major highways, and major streets. Prices range from $15 for the "Reduced California Wall Map" (25" X 38") to $119.95 for the "Southern California Freeway and Artery Map" (69" X 52").
■ "Thomas Zip Code Guides" ($18.95), for Los Angeles-Orange counties, San Bernardino-Riverside counties, San Diego counties, Santa Barbara-Ventura counties, and Santa Clara County; also available for the Portland Metropolitan Area ($24.95).
■ "Thomas Census Tract Guides" ($25.95-$49.95), for the following counties: Los Angeles-Orange, San Bernardino-Riverside, San Diego, Santa Barbara-Ventura, Alameda-Contra Costa, San Francisco-Marin, Santa Clara-San Mateo, and Sacramento; also available for King-Pierce-Snohomish counties in Washington ($59.95).

T.R. Map Co. (16403 W. 126th Ter., Olathe, KS 66062; 913-782-6168) publishes full-color "Metro Area Office Wall Maps" (38" X 50"; $74.50), which include house numbering systems, railroads, water features, city limits, subdivision names, public buildings, shopping centers, and various other landmarks. Maps are available for metro areas in Arkansas, Iowa, Kansas, Missouri, Nebraska, and Oklahoma. Also available is a one-color map of suburban Johnson County, Kansas (52" X 63"; $49).

Visual Planning Division, MagnaPlan Corp. (6805 Blvd. Decarie, Montreal, Canada H3W 3E4; 514-739-3116) distributes business maps from various sources. The company claims to be able to fulfill virtually any request for business maps and map accessories.

Western Economic Research Co. Inc. (8155 Van Nuys Blvd., Ste. 100, Panorama City, CA 91402; 818-787-6277) produces a full line of business maps for most metropolitan areas. Map categories include "5-Digit Zip Code Outline Maps," "Census Tract Outline Maps," "Updated 5-Digit Zip Code Annual Demographic Estimates," "1986 Business Reports," "Major Street and Highway—Freeway Community Maps," and "1980 Census Demographic Maps." It also offers "Annual Demographic Estimates" for all zip codes in the U.S., either printed or on diskette. Western Economic Research offers extensive specialty business maps for the Los Angeles County area, including a "Hispanic Growth Map" (11" X 17"), a "Hi-Rise Office Building Map," and "Distribution of Manufacturing Employment Map."

■ **See also: "Census Maps," "Energy Maps," "Land-Ownership Maps," "Map Software," "Natural Resource Maps," "Selected Atlases," "Urban Maps and City Plans," and "Utilities Maps."**

Census Maps

Every 10 years, the Bureau of the Census compiles and tabulates millions of statistics about Americans. Between these decennial censuses, it gathers and disseminates information every five years in its economic, agriculture, and government censuses. Summary reports that present the data collected in the decennial, economic, agriculture, and government censuses usually include appropriate maps. A variety of other maps can be purchased separately from the publications and computer tape files, either from the **Government Printing Office** or directly from the **Census Bureau**. The census data and maps are useful to businesses looking for prospective markets; to developers in search of sites for new homes, offices, or factories; to analysts concerned with demographic data and population trends; and to anyone else with an interest in the numbers, distribution, incomes, life styles, and so on, of Americans.

For the 1990 census, the Census Bureau has prepared an automated cartographic data base for the United States and its possessions. This data base, called TIGER (for Topologically Integrated Geographic Encoding and Referencing), will allow the Bureau to produce several types of maps on large-format (36″-wide) electrostatic plotters. The 1990 census maps may be page-size or full-size (approximately 36″ X 42″) map sheets. For the 1990 census, as it did for the 1980 census, the Bureau plans to produce three categories of map types: 1990 census maps, summary reference outline maps, and thematic or statistical maps. The majority of map types will be of publication quality. Most maps will be black and white, but the Bureau will present a few publication maps in color.

Maps will be included in various printed 1990 report series: the "1990 Census of Population and Housing" (1990 CPH series), the "1990 Census of Population" (1990 CP series), and the "1990 Census of Housing" (1990 CH

series). Similar to the 1980 census reports, plans call for the national summary report in each series to include a few thematic maps, as well as the pertinent summary-reference outline maps depicting the areas for which the report provides data.

The 1990 map products, with the notable exception of the 1990 census map series, are similar to the 1980 census products. The 1990 census map series replaces five map series that comprise the "1980 census maps." The 1980 census maps include the "Metropolitan Map Series," the "Vicinity Map Series," the "County Map Series," the "Place Map Series," and the "Place-and-Vicinity Map Series"; these maps are still available for purchase, primarily from the Census Bureau. Three map products are new for 1990: the "Voting District Outline Map Series," the "American Indian/Alaska Native Area Outline Map Series," and the "Urbanized Area/Highway Map Series."

The 1990 census maps are detailed summary-reference outline maps. The 1990 census map series is county-based, consisting of one or more—usually more—full-size map sheets for each county or statistically equivalent area, supplemented by an index map. The map sheets for each county are at one of several standard scales. Like the 1980 census maps, the 1990 census maps portray most of the elements of census geography, from census block to international boundary. Most of the various geographic entities recognized for the census are identified by boundary, name, and, in most cases, Federal Information Processing Standards (FIPS) code. The maps also show a wide range of linear base features—primarily roads, streams, and railroads—along with landmark features such as schools, military installations, and major parks.

Whereas the 1990 census map series portrays most geographic areas for which the Census Bureau tabulates decennial census data, the other summary-reference outline map series focus on specific levels of information and depict

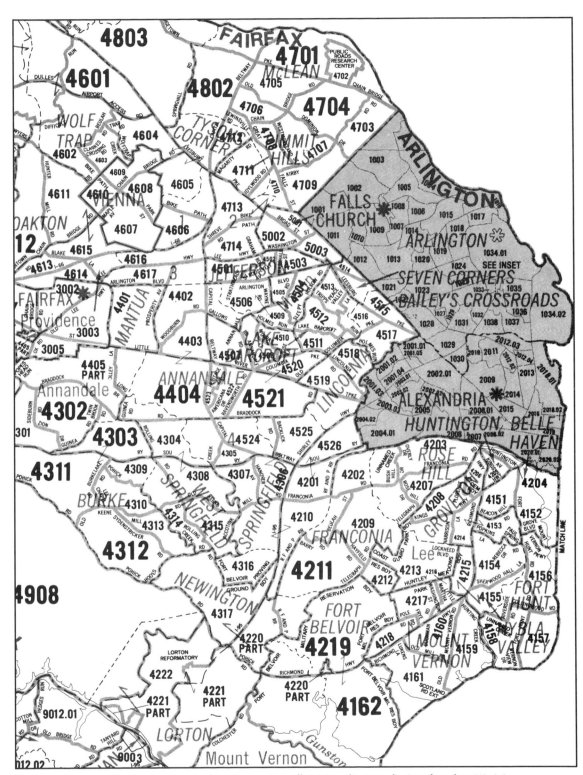

Portion of a Standard Metropolitan Statistical Area census tract, illustrating the census districts of northern Virginia.

only selected geographic areas. Geographic areas are identified by boundary symbology, type style, name, and, usually, code. Most of the summary-reference outline map series identify linear base features that coincide with the displayed geographic area boundaries.

Summary-reference outline maps include the following:

■ "Regions and Divisions of the United States." This is a single-page-size map of the conterminous U.S., with Alaska and Hawaii shown as

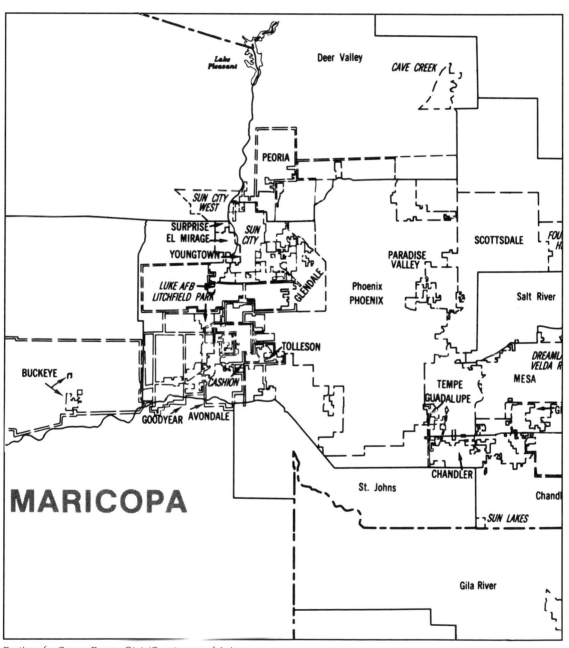

Portion of a Census Bureau State/County map of Arizona.

insets. The map shows the names and boundaries of the four census regions and nine divisions, as well as the states.

■ "Metropolitan Statistical Areas of the United States, Urbanized Areas of the United States, and American Indian/Alaska Native Areas of the United States." These are individual, two-page maps of the conterminous U.S., with Alaska and Hawaii shown as insets. Each map displays state and county boundaries, as well as the outlines and names of geographic areas specified in the map title.

■ "County Subdivision Maps." This is a state-based map series. The maps for each state or statistically equivalent area will be produced at one of several standard scales as a full-size map sheet. The full-size versions of the county subdivision maps, in sections, will be included in the 1990 CPH, 1990 CP, and 1990 CH series of printed reports for each state. The county subdivision map series portrays all appropriate international, state, American Indian/Alaska Native area (AI/ANA), county, county subdivision, and place-boundaries names.

■ "State/County Outline Maps and State/MSA Outline Maps." These state-based map series are issued at variable scales in a page-size format. For each state or statistical area, the maps portray the county areas. In addition, maps show the three types of metropolitan areas defined by the Office of Management and Budget: metropolitan statistical areas (MSAs), consolidated MSAs (CMSAs), and primary MSAs (PMSAs); they also show the capital and selected large cities in each state.

■ "Voting District (VTD) Outline Maps." This is a county-based map series, produced at one of several standard scales as one or more full-size map sheets. The maps show international, state, county, AI/ANA, county subdivision, place, and VTD boundaries and names. The maps also identify selected linear base features that coincide with VTD boundaries.

■ "Census Tract/Block Numbering Area (CT/BNA) Outline Maps." This is a county-based map series, produced at one of several standard scales as one or more full-size map sheets. The maps show international, state, AI/

ANA, county, county subdivision, place, and CT/BNA boundaries and names. The maps also identify selected linear base features that coincide with CT/BNA boundaries.

■ "AI/ANA Outline Maps and Urbanized Area (UA) Outline Maps." The AI/ANA Outline Maps are an AI/ANA-based map series, and the UA Outline Maps are a UA-based map series. Scale is variable within a page-size format. A large AI/ANA or UA may be shown as a multipage map, whereas several small AI/ANAs or UAs will be grouped on a single page. Geographic-area information shown on both maps includes international, state, AI/ANA, county, county subdivision, and place boundaries and names. Additionally, the UA Outline Maps depict the boundary and shade coverage of each UA.

■ "UA/Highway Maps." This map series shows the extent of each urbanized area in relation to major roads and water bodies, as well as international, state, AI/ANA, county, county subdivision, and place boundaries; it also displays and identifies other linear features that coincide with the UA boundaries.

Thematic, or statistical, maps depict a wide variety of topics. A thematic map presents the distribution or structure of a specific set of data. The GE-50 series is a wall-size (approximately 35″ X 46″) map series, generally at a scale of 1:5,000,000, that depicts data by county or displays names and boundaries of specific geographic entities for the entire United States. Subject titles in this series include "United States Base Map," "Metropolitan Statistical Areas" (for the most recent year for which changes were announced), and "1980 Population Distribution" (both "day-time" and "night-time" versions). Most of the GE-50 series maps are four-color maps.

The Census Bureau also provides page-size thematic maps for inclusion in the national summary reports for the several censuses. The maps in the Census Bureau's 1980 *Number of Inhabitants, U.S. Summary* report are either state- or county-based maps. In addition to a population map of the United States, 1980 census map titles include "Major Acquisitions of

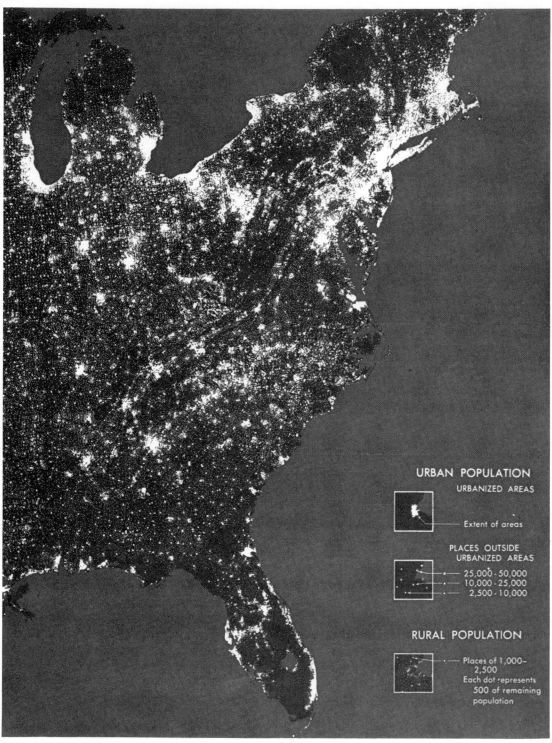

Eastern portion of the Census Bureau's "Nighttime View" map, illustrating population distribution by depicting how the country might appear from above at night.

Territory of the United States and Dates of Admission of States," "Metropolitan Areas of the United States," "Urbanized Areas of the United States," "Regions and Census Divisions of the United States," and "Center of Population for the United States: 1790 to 1980." There are also maps of population density and "percentage urban" by county, and maps that trace changes in total, urban, rural, metropolitan, and nonmetropolitan populations, by state, from 1960 to 1970 and/or 1970 to 1980.

The Census Bureau has published graphic summaries portraying data for the 1977 Economic Censuses and for each Census of Agriculture from 1967 through 1982; a similar atlas for the 1987 Census of Agriculture is in preparation. The maps are for the United States, showing state outlines, with the data displayed by dots or in color shading.

The Census Bureau published a Congressional District (CD) Atlas for the 100th Congress (whose districts are in effect for the 101st Congress and are expected to be in effect for the 102nd Congress), and plans to produce one for the 103rd Congress to show the new congressional districts resulting from the use of the 1990 census data; replacement pages will be made available as required to reflect any revisions for the 104th and subsequent Congresses. The atlas contains one page-size map for each state, and page-size insets for counties located in two or more CDs. The state maps include international, state, CD, and county boundaries and names; selected incorporated places are shown as point symbols. The inset maps display and identify appropriate geographic entities as well as the features that coincide with CD boundaries.

■ See also: "Business Maps," "Energy Maps," "Native American Land Maps," "Political Maps," "United States Maps," and "Urban Maps and City Plans."

Emergency Information Maps

The adage "forewarned is forearmed" is especially true when it applies to rising flood waters or tremors signaling a major quake. Maps of potential or past disaster can often avert serious injury and property loss. But mapping for emergencies can be a tricky thing. Not knowing what emergency will occur, or when and where it might strike, makes it virtually impossible to plan for every event. Still, areas that are flood- or earthquake-prone, seaside towns frequently hit by hurricanes or tidal waves, and any place with a history of disaster will probably have maps for evacuation and insurance purposes.

Emergency maps have been around for as long as there have been emergencies. Over the centuries, there have been "plague maps" showing the spread of disease, flood and fire maps showing damage to life and property,

and countless other diagrams, charts, and maps attempting to illustrate and evaluate the havoc nature, man, or both can have on the environment. There have been preventive maps as well: Countless lives and valuable property have been saved by use of maps of flood plains, faults, unstable areas prone to mudslides and avalanches, and tornado "alleys" that seem to attract nearby twisters.

Among the most common emergency information maps are flood maps. While flooding damage takes its toll in billions of dollars and hundreds of lives annually, flood maps have been highly successful in helping potential victims prepare for the onslaught of mud and water. In some regions, simply having a map of flood plain locations—and avoiding building on them—can make the difference between financial ruin and survival.

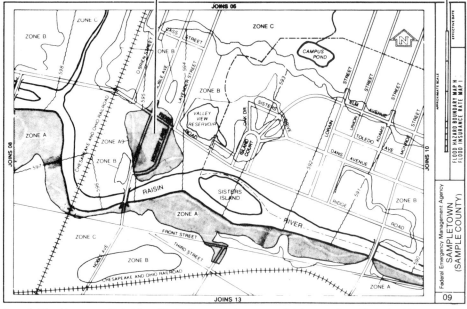

Sample Flood Insurance Rate Map, illustrating property areas and risk zones, from the Federal Emergency Management Agency's Flood Insurance Program.

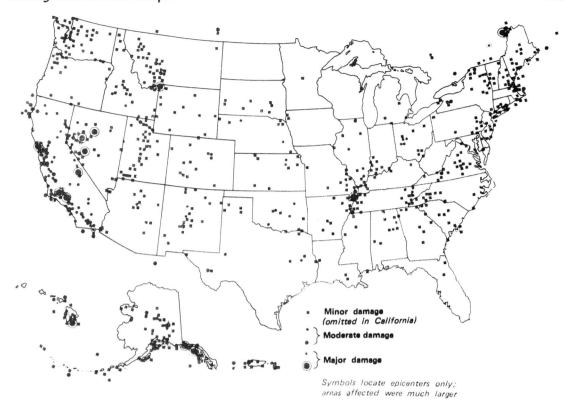

Minor damage
(omitted in California)

} Moderate damage

} Major damage

Symbols locate epicenters only;
areas affected were much larger

National Oceanic and Atmospheric Administration map pinpointing the epicenters of damaging earthquakes in the United States.

Seismicity maps of earthquake regions are also frequently compiled. Their usefulness for pinpointing regions where earthquakes have occurred or may strike make them valuable tools for planning and constructing new roads, industries, or communities. Seismicity maps are also helpful in the study of faults, shifts, and movement in the Earth's crust.

Other emergency maps available include civil evacuation maps, maps of dangerous tides or currents, maps of active volcano regions, and maps predicting or tracing the paths of such natural disasters as avalanches and hurricanes. But emergency information maps are useful in another way: They have historical value as well.

The fire-insurance maps the **Sanborn Co.** began creating in 1867 were certainly utilitarian at the time. Now, however, they are among the **Library of Congress's** most prized collections. Because of the regularity with which Sanborn updated its maps, historians use the company's works to study the growth of hundreds of American towns and cities. The maps are so detailed that over a single decade one can trace the advances made in building materials or the changing demographics of a neighborhood. Even the earliest Sanborn map in the Library's collection, an 1867 fire atlas of Boston, categorizes buildings, materials, construction, and contents by level of fire hazard.

Certain emergency maps, especially those for civil evacuation, are created by local, county, or state officials rather than by the federal government. Uncle Sam does, however, produce numerous seismicity, flood plain, storm-tracking, and tidal-warning maps through its agencies, including the **U.S. Geologic Survey**, the **National Geophysical Data Center**, and the **Federal Emergency Management Administration**.

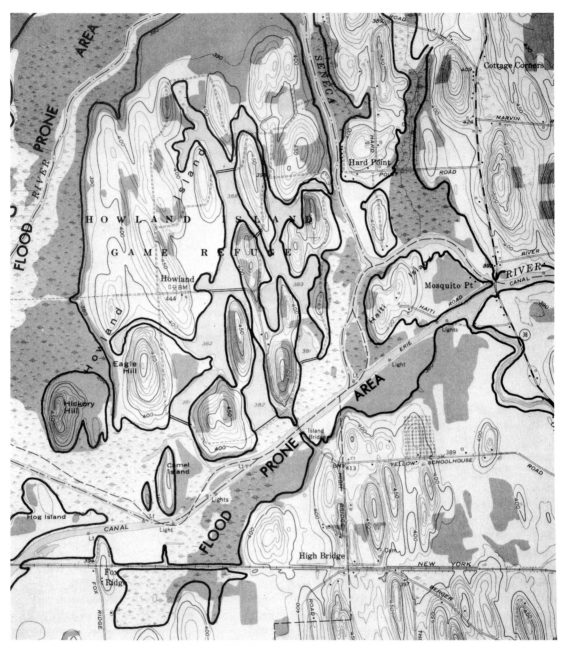

Portion of a Montezuma, N.Y., quadrangle map of flood-prone areas, drawn at a scale of 1:24,000.

GOVERNMENT SOURCES, UNITED STATES

Federal Emergency Management Agency.

FEMA runs the National Flood Insurance Program, which creates and distributes "Flood Hazard Boundary Maps" and "Flood Insurance Rate Maps." The maps in both series are created in two forms: the "flat-map format" contains a map index, a legend on the cover sheet, and 11" X 17" map pages covering the community; the "multiple-fold format" (also known as the "Z-Fold" format) resembles a standard folding road map and includes a map index only if more than one map page is required for the community.

Another FEMA series is "Flood Hazard Boundary Maps," which show hazard areas where flooding may occur within communities. "Special Flood Hazard Areas" are depicted with shading against a base grid of the community. A free pamphlet, "How to Read Flood Hazard Boundary Maps," is available upon request from FEMA. "Flood Insurance Rate Maps" (or FIRMs) replace the Flood Hazard Boundary Maps, illustrating base flood elevations and varying degrees of flood hazard zones. FIRMs aid in establishing flood insurance rates for specific properties within a community. A free pamphlet, "How to Read a Flood Insurance Rate Map," is available from FEMA.

The best way to obtain these maps for your area is to call FEMA toll-free (800-333-1363). To obtain the pamphlets only, call 800-638-6620.

National Geophysical Data Center. The NGDC, part of the **National Oceanic and Atmospheric Administration**, produces numerous seismicity and tsunami maps, useful for seeing where earthquakes and tsunamis have occurred historically. Examples of maps available include:

■ "Significant Earthquakes Map, 1900-79" ($5 rolled, $10 folded; 41" X 54"), a world map produced in 1980.
■ "Seismicity Map of the Middle East, 1900-83" ($5 folded, $10 rolled; 41" X 54"), produced in 1985.
■ "Seismicity Map of Middle America, 1900-79" ($5 folded, $10 rolled; 35" X 41"), created in 1982.
■ "Regional Seismicity Maps," a series of 13 maps produced by the National Earthquake Information Center, showing data collected from 1962 to 1969 on seismically active areas of the Earth ($5 each; approximately 20" X 29" each).
■ "Tsunamis in Pacific Basin, 1900-83" ($5 folded, $10 rolled; 43" X 60"), produced in 1984.
■ "Tsunamis in Peru-Chile, 1562-1985" ($5), created in 1985.

COMMERCIAL SOURCES

Williams & Heintz Map Corp. (8119 Central Ave., Capitol Heights, MD 20743; 301-337-1144) produces, but does not sell, many kinds of geologic maps. The maps are custom-made for a number of companies, state geologic offices, and other groups needing specialized work. Recently, they have produced "Flooding, Tempe Quadrangle, Maricopa County, Arizona" ($2), available from the **Arizona Bureau of Geology and Mineral Technology**, 845 N. Park Ave., Tucson, AZ 85719; 602-626-2733.

■ **See also: "Geologic Maps" and "Weather Maps."**

Energy Maps

From lightbulbs to limousines, we are as dependent on electricity, gas, and oil as our primitive ancestors were on clubs, stones, and knives. The search for new energy sources has become a serious endeavor for scientists and engineers—and mappers. Without the means to pinpoint these resources, the wealth of the Earth's natural energy would remain hidden.

There are energy maps for every type of energy source. Sunshine maps show the locations on Earth where sunlight is most plentiful; coal- and oil-investigation maps point out where these resources can be found. There are maps of hydroelectric and nuclear power plant sites, gas pipelines, and oil refineries, as well as maps of the sediment on the ocean floor where retrievable hydrocarbons can be found.

The federal government produces many of these maps, reports, and charts used by those in both the public and private sectors. Other sources of energy maps are public utilities, mining or oil companies, and geologic map companies or organizations.

GOVERNMENT SOURCES, UNITED STATES

The U.S. government is a primary source for energy maps because of its subsidized utilities projects (such as the Tennessee Valley Authority) and its awareness of the strategic importance for energy independence. Some of the agencies producing energy-related maps include:

Federal Energy Regulatory Commission (FERC). FERC produces a map illustrating interstate natural gas pipelines in the U.S. "Major Natural Gas Pipelines" ($3.25; 13″ X 18″) shows major existing pipelines, as well as those proposed or under construction; locations of natural gas fields; and imports and exports of gas from Canada and Mexico. It is available from the Government Printing Office

and the FERC Public Reference Branch.

Central Intelligence Agency. The CIA has several international energy maps available through the National Technical Information Service, including:

■ "China: Fuels, Power, Minerals and Metals" ($7.50; PB 83-928207).
■ "Eastern Europe: Major Power Facilities" ($11.50; PB 82-927905).
■ "Middle East Area Oilfields and Facilities" ($11.95; PB 88-928368).
■ "Soviet Union Electric Power," including its petroleum-refining and chemical industry ($10; PB-88-928311).
■ "Soviet Union, East & South," includes electric and other power facilities, minerals, and petroleum-refining sites ($11.50; PB 82-928113).

National Climatic Data Center. A good source for historical maps of solar energy is the NCDC (Federal Bldg., Asheville, NC 22810; 704-259-0682), which produces A *History of Sunshine Data in the United States,* 1891-1980 ($2, available in fiche only) as part of its "Historical Climatology Series." The study gives digitized, summarized monthly and annual totals, when available, of "duration of sunshine" from 239 observation sites between 1891 and 1980. An accompanying map shows the sunshine station network as it existed in 1891, 1900, 1920, 1940, 1960, and 1980.

U.S. Geological Survey. USGS's National Atlas Program has two sunshine map sets:

■ "Monthly Sunshine." This 1965-edition sheet map contains theoretical maximum and mean actual hours of yearly sunshine for selected locations, as well as 12 monthly maps of mean actual sunshine ($3.10; US00478 38077-AI-NA-17M-00).

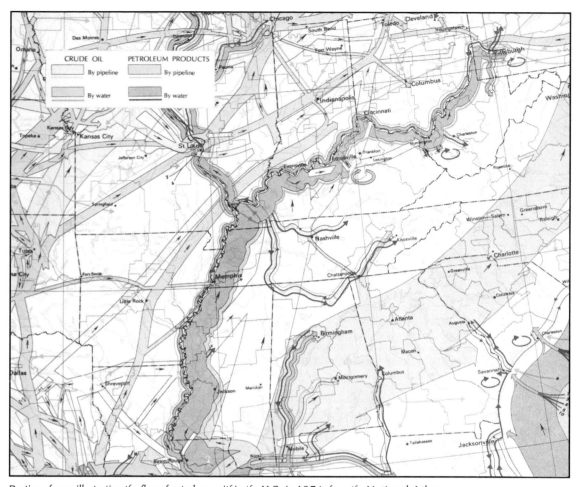

Portion of map illustrating the flow of petroleum within the U.S. in 1974, from the National Atlas.

■ "Annual Sunshine, Evaporation, and Solar Radiation." This 1969 sheet contains three maps of mean actual sunshine, annual pan evaporation, and May-October evaporation, and five maps of annual solar radiation and mean solar radiation for January, April, July, and October ($3.10; US00565 38077-AJ-NA-17M-00).

Also available from the U.S. Geological Survey are a number of gas-, oil-, and coal-investigation maps, many of which are produced by the Bureau of Mines. (See also "Geologic Maps.") For listings of these maps, write or call the nearest USGS Earth Science Information Center (see Appendix B).

Tennessee Valley Authority. TVA produces two maps illustrating the transmission lines and private utilities of TVA's roughly 40,910-square-mile territory, covering parts of Tennessee, Alabama, Georgia, Kentucky, Mississippi, North Carolina, and Virginia. The "L" series ($5 plus $2 for mailing tube and postage), is a 31″ X 40″ map showing TVA and private utilities transmission lines, dams, generating plants, highways, railroads, county and state boundaries, cities and towns, and rivers and lakes. The "S" series map ($1 plus $2 for mailing tube and postage) measures 17″ X 22″ and shows transmissions lines, service areas of TVA electric power, cities and towns, rivers and lakes, and political boundaries.

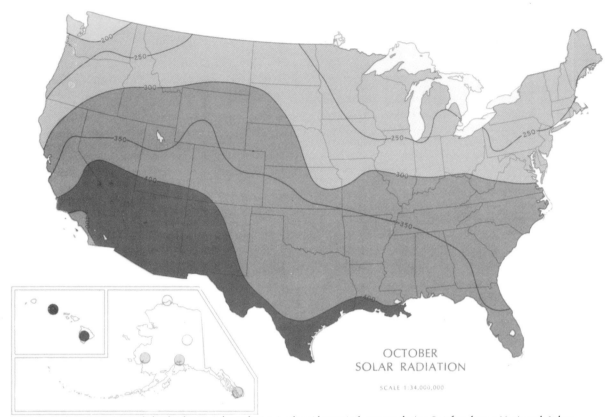

OCTOBER
SOLAR RADIATION

SCALE 1:34,000,000

Inset map depicting nationwide levels of mean solar radiation in the mid-twentieth century during October, from a National Atlas *sheet of solar radiation maps.*

GOVERNMENT SOURCES, CANADA

Canada Map Office (615 Booth St., Ottawa, Ontario K1A 0E9; 613-952-7000) distributes maps from *National Atlas of Canada, 5th Edition* series, including "Oil Pipelines" (MCR 4048), "Natural Gas Pipelines" (MCR 4049), and "Electricity" (MCR 4144). Maps from the 4th Edition (1973) are also available, and include: "Fossil Fuels and Pipelines, Eastern Canada" (MCR 1199), "Fossil Fuels and Pipelines, Western Canada" (MCR 1200), "Electricity, Eastern Canada" (MCR 1227), and "Electricity, Western Canada" (MCR 1228). The maps cost $2 Canadian each, plus shipping. All maps are available with either English or French text.

Maps Alberta (Land Information Services Div., Main Fl., Brittania Bldg., 703 6th Ave. SW, Calgary, Alberta T2P 0T9; 403-297-7389) produces various energy maps as part of its *Alberta Resource and Economic Atlas*, including:
■ "Main Pipelines, Refineries and Gas Plants, 1984" (75 cents Canadian; D44).
■ "Movement of Oil and Gas, 1983" (75 cents Canadian; D45).

Maps Alberta also distributes maps from the *National Atlas of Canada, 5th Edition*, including:
■ "Electricity Generation and Transmission" ($5.50 Canadian; MCR 4069).
■ "Energy" ($5.50 Canadian; MCR 4002).
■ "Mineral Commodity Flows" ($5.50 Canadian; MCR 4081).

COMMERCIAL SOURCES

Some of the best sources for energy maps are geologic map companies or associations. Here are several that produce high-quality maps:

The American Association of Petroleum Geologists (AAPG Bookstore, P.O. Box 979, Tulsa, OK 74101; 918-584-2555) is a membership organization for petroleum geologists. Although members receive a discount on AAPG products, the publications are also available to nonmembers. Products include:

■ Geothermal Gradient Maps ($10 each; $250 per set; $3 and up for shipping): This portfolio of 39 computer-contoured and labeled maps covers all of North America.

■ "Sedimentary Provinces of the World" ($14 plus $3.50 shipping): This 40" X 54" map uses the latest data to classify, inventory, and rate those sedimentary deposits that may contain recoverable hydrocarbons. A 36-page booklet included with the map lists each province and shows potential hydrocarbon productivity.

■ "Geological Provinces: Contiguous 48" ($6 plus $1.75 shipping): This code map by the AAPG Committee of Statistics of Drilling was prepared for reporting well information from a standardized index map. The map lists codes for the 48 contiguous states.

■ "Geologic Provinces: Alaska" ($6 plus $1.75 shipping): Same as above, for Alaska.

■ AAPG also sells maps made by the Circum-Pacific Map Project, a series of which show mineral resources in the Pacific Ocean region (see "Geologic Maps" for more information on this project).

Canadian Society of Petroleum Geologists (206 7th Ave. SW, #505, Calgary, Alberta T2P 0W7; 403-264-5610) produces a variety of specialized maps of sedimentary basins in Canada. A listing of maps is available.

Geoscience Resources (2990 Anthony Rd., P.O. Box 2096, Burlington, NC 27216; 919-227-8300; 800-742-2677) distributes a vast array of energy maps of U.S. states and foreign countries, including:

■ "Illinois—Oil & Gas Industry Map" ($5.95; 62-6580), a four-color map illustrating oil and gas fields, pipelines, refineries, pumping stations

and underground storage fields.

■ "Oklahoma—Oil & Gas Fields Map" ($8.95; 62-8880) outlines 3,083 active and 35 abandoned fields.

■ "Finland—Energy Production and Supply" ($24.95; 64-5583). This Finnish/English map depicts energy data for nuclear, steam, gas-turbine, hydroelectric, and byproducts of heat from industry.

■ "Venezuela—Petroleum Resources Map" ($14.95; 65-5180). This black-and-white, Spanish-text map shows the petroleum-producing regions, producing and nonproducing wells, oil and gas pipelines, and refineries.

Gulf Publishing Co. (Book Div., P.O. Box 2608, Houston, TX 77252; 713-520-4444) publishes an 80-page USSR *Energy Atlas* ($35), with full-color maps and photos along with a 22" X 36" fold-out color reference map.

MAPSearch Services (9800 Richmond Ave., Ste. 375, Houston, TX 77042) publishes pipeline and facilities maps for the liquid petroleum gas, petrochemical and refining industries. Their maps include: *Gulf Coast Olefins Atlas* ($400), a wire-bound book describing the ethylene, propylene, butadiene, and butylene systems from Corpus Christi, Texas, through southern Louisiana; *Refined Products Atlas of the United States & Southern Canada* ($300), a 64-map atlas depicting the integrated refined products pipelines system in the area; and the "LPG Systems Map of the United States & Canada" ($400), a multi-color conference-room style 5' X 3 1/2' map showing the LPG system.

PennWell Books (P.O. Box 21288, Tulsa, OK 74121; 918-831-9421; 800-627-3212) produces several detailed maps depicting energy resources in the U.S. Among the maps currently available:

■ "Oklahoma Pipeline Map Series" ($130), a two-map series of full-color maps: "Natural Gas Pipelines of Oklahoma" ($75; 40" X 58") portrays pipeline operators, pipe diameters, compressor stations, gas-storage fields, gas-processing plants, and petrochemical plants using natural gas; and "Petroleum Liquids Pipelines of Oklahoma" ($75; 40" X 58") shows pipeline operators, pipe diameters, pump stations, interconnections, terminals, refineries, and petrochemical plants.

■ "Products Pipelines of the United States & Canada" ($90; 40" X 57") uses separate colors to show liquid petroleum gas and natural gas liquid pipelines, CO_2 pipelines, and refined-product pipelines, along with insets of several major metropolitan areas.

■ "World Sedimentary Basins and Related Features" ($40; 44" X 63"), a six-color shaded-relief world map depicting major petroleum and geologic basins, shelves, platforms, thrust and fold belts, and grabens and rises.

■ "Polar Oil & Gas Map" ($75; 40" X 58"), showing offshore lease ownership; bathymetry; pump and compressor stations; oil, gas, and petrochemical pipelines with diameters; refineries; petrochemical plants; and gas processing plants.

Petroleum Information (PI MAPS) (P.O. Box 2612, Denver, CO 80201; 303-740-7100; 800-645-6277) publishes a variety of maps useful to the petroleum industry. It offers base maps, which depict township, range, section, latitude, longitude, major water features, roads, towns, and posted well spots; lease maps, which show current lease status, historical and current well completions, surface ownership, township, range, section, highways, lakes, rivers and creeks; display maps, full-color maps illustrat-

ing oil and gas fields, major geologic features, and townships; and photogeologic-geomorphic maps, interpretations of Landsat imagery depicting geologic outcrop patterns, strike and dip of beds, anticlinal and synclinal axes, fractures, faults, and alignments, as well as geomorphic features. Data are also available in digital form. Free catalogs are issued for the following regions of the U.S.: Northeast and Midwest, Mid-Continent, West Coast, Rocky Mountain, and Southern.

Williams & Heintz Map Corp. (8119 Central Ave., Capitol Heights, MD 20743; 301-337-1144) produces, but does not sell, many kinds of geologic maps custom-made for companies, state geologic offices, and other groups needing specialized work. Williams & Heintz energy maps available from these sources include:

■ "New York State Oil and Gas Fields" ($10), drawn on a 1:250,000-scale. Available from **Department of Environmental Conservation**, Div. of Mineral Resources, Rm. 202, 50 Wolf Rd., Albany, NY 12233.

■ "Map of Alaska's Coal Resources" ($5), produced in 1986 at a scale of 1:2,500,000. Available from **Division of Geological and Geophysical Surveys**, 794 University Ave., Ste. 200, Fairbanks, AK 99709.

■ **See also: "Geology Maps," "Land-Use Maps," "Natural Resource Maps," and "Utilities Maps."**

Utilities Maps

Utilities maps generally aren't suitable for framing or gift-giving (unless the recipient is a lineman for the county); they are typically black and white, with perhaps a color or two to delineate utilities lines. Still, they are necessary tools for developers and construction workers as well as for utilities companies themselves. Even the weekend gardener may have need of these maps to avoid that awful moment when hoe meets underground cable. As important as it is to know where utility lines are, it some-times helps to know where they aren't.

The typical utility map is a straightforward representation of the lines, pipes, cables, and wires carrying one or more commodities, and pinpoints the stations or sources generating them. Utilities maps run the gamut from simple diagrams of phone lines, gas pipes, and water mains to complex depictions of sewer systems, television cables, hydroelectric plants, and any other utility that serves a given area.

The best sources for local utility maps are the utilities companies themselves. Sewer- and water-system maps are created and distributed by local and county departments of public works. Several branches of the federal govern-ment produce or distribute national and regional maps of power plants and telephone systems. Federal producers of utilities maps include the **Tennessee Valley Authority**, which has a number of power maps for its region, the **Bureau of the Census**, with a statistical map on home-heating fuel use, and the **Central Intelligence Agency**, which produces several maps detailing energy sources in the Soviet Union and the Middle East. The **Government Printing Office** also distributes a number of utilities maps from various federal agencies. Cartographic depart-ments of foreign countries can assist in finding utilities maps for those lands, while the local tourism, sanitation, or utilities commissions of specific provinces may also be of help in locating maps for their regions.

GOVERNMENT SOURCES, UNITED STATES

U.S. Geological Survey. USGS publishes the following energy maps, all produced in 1974: "Natural Gas Movements by Pipelines" ($3.10; US 00422), "Electric Power Transmission" ($3.10; US 00430), and "Total Interstate Energy Movement" ($3.10; US 00437), as part of the National Atlas series of maps.

GOVERNMENT SOURCES, CANADA

Canada Map Office (615 Booth St., Ottawa, Ontario K1A 0E9; 613-952-7000) distributes maps from *The National Atlas of Canada, 5th Edition*, which include ($5.50 Canadian each): "Air Transportation" (MCR 4102), "Electricity Generation and Transmission" (MCR 4069), and "Telecommunications Systems" (MCR 4112).

Maps Alberta (Land Information Services Div., Main Fl., Brittania Bldg., 703 6th Ave. SW, Calgary, Alberta T2P 0T9; 403-297-7389) also distributes the above maps from *The National Atlas of Canada, 5th Edition*.

COMMERCIAL SOURCES

ArcTrek Publishing (1231 Airport Rd., Ste. 177, Allentown, PA 181003) sells *Highway Radio: A Guide To Tuning in on America's Highways* ($4.95 plus $2.50 postage), listing more than 9,000 AM and FM stations by city and format, and including state maps illustrating broadcast boundaries.

Communications Publishing Service (Box 500, Mercer Island, WA 98040; 206-232-3464) distributes the "Motorola Cellular Coverage Map" ($9), a 25″ X 36″ fold-out road map of the U.S. and Canada that shows where cellular tele-phone service is available; on the reverse are facts about cellular telephone service. Also available are the following directories: "The Cellular Telephone Directory, 2nd Edition"

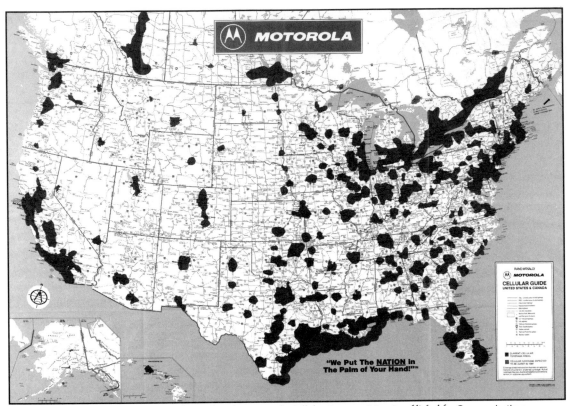

America's cellular-telephone regions are depicted in the "Motorola Cellular Coverage Map," published by Communications Publishing Service.

($14), a 448-page guide including maps of the coverage areas of all U.S. and Canadian cellular-telephone companies. The guide tells how to place and receive calls as a visitor in each city, lists rates charged, and includes a listing of cellular-telephone companies, with addresses and telephone numbers. "The Mobile Telephone Directory, 4th Edition" ($10) includes maps showing where conventional radio-telephone service is available in the U.S. and Canada, information on how to place calls using the noncellular UHF/VHF mobile-telephone service, and lists companies.

■ **See also: "Energy Maps" and "Natural Resource Maps."**

WATER MAPS

Nautical Charts and Maps

Rivers, lakes, oceans, and other bodies of water are portrayed on both maps and charts, the distinction between which even some veteran old salts don't really understand. *Charts* differ from *maps* primarily in the amount of navigation information shown. Charts are much more detailed, showing a main channel sailing line, for example, as well as safety harbors, the general shape and elevation of river or lake bottoms, hazard areas, and other key symbols that enable a boat pilot to wend safely through a potentially treacherous body of water. Maps, in contrast, are much more superficial, primarily showing landmarks. They are limited more to the requirements for small-craft navigation and for general recreation guidance.

The **National Ocean Service** (NOS), part of the National Oceanic and Atmospheric Administration (itself an agency of the U.S. Department of Commerce), produces several types of maps and charts:

■ **Coast charts** (scales from 1:50,000 to 1:150,000), the most widely used nautical charts, are intended for coast-wide navigation inside offshore reefs and shoals, entering bays and harbors of considerable size, and navigating certain inland waterways.

■ **General charts** (scales from 1:150,000 to 1:600,000) are designed for use when a vessel's course is well offshore, and when its position can be fixed by landmarks, lights, buoys, and characteristic soundings.

■ **Sailing charts** (scales smaller than 1:600,000) are plotting charts used for offshore sailing between distant coastal ports and for approaching the coast from the open water.

■ **Harbor charts** (scales larger than 1:50,000)

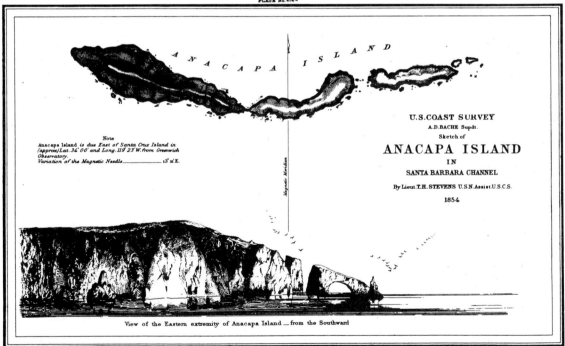

An 1854 U.S. Coast Survey sketch of Anacapa Island, off the southern California coast, engraved by a young James Whistler during his short career as an illustrator with the Survey.

are for navigation and anchorage in harbors and smaller waterways.

■ **Small-craft charts** (scales from 1:10,000 to 1:80,000) include specific information pertinent to small-craft operators. They show a great variety of information, such as tide and current data, marina and anchorage facilities, and courses. These charts are published both as folded sheet maps and in book form, for handling convenience on small boats.

■ **Canoe charts**, a series of charts of the Minnesota-Ontario border lakes, are designed to suit the needs of small, shallow-draft vessels.

■ **Coast pilots** and **Great Lakes pilots**, published in nine volumes, provide detailed navigation information that cannot be shown conveniently on charts, such as radio service, weather service, port data, sailing directions, and natural features.

■ **Special maps and data**, byproducts of the nautical charting program, are generally for non-navigational use. They include topographic surveys and planimetric shoreline maps, aerial photographs, hydrographic smooth sheets, graphic depth records, descriptive reports of surveys, and sedimentology sample data (see "Ocean Maps" for details of such map products).

GOVERNMENT SOURCES, UNITED STATES

Defense Mapping Agency. DMA, through its Hydrographic Center, publishes a variety of maps covering foreign regions, most priced around $11.50. A complete list of nautical charts available from DMA is contained in DMA's 10-volume *Catalog of Maps, Charts, and Related Products, Part 2—Hydrographic Products* ($2.50 each), available from DMA's Combat Support Center. The catalog includes DMA's series of "world charts," showing major oceans and shipping regions, and its several hundred "coastal, harbor, and approach charts," covering smaller regions throughout North America. The DMA catalog is color-coded, indicating which charts are available from DMA, NOS, and the Government Printing Office, and which must be obtained from other nations' hydro-graphic offices or sales agents. The catalog contains a complete name-and-address listing of DMA nautical chart agents throughout the U.S. and in 40 countries, as well as order forms for ordering charts available directly from DMA.

National Ocean Service. NOS publishes free chart catalogs—actually large, folded maps—that list nearly a thousand available charts. There are four catalogs: "Nautical Chart Catalog 1" covers the Atlantic and Gulf Coasts, including Puerto Rico and the Virgin Islands; "Nautical Chart Catalog 2" covers the Pacific Coast, including Hawaii, Guam, and Samoa; "Nautical Chart Catalog 3" covers Alaska, including the Aleutian Islands; and "Great Lakes" covers those lakes and adjacent waterways. The catalogs are useful for another reason: Each contains a listing of the hundreds of authorized NOS nautical chart dealers throughout the U.S. and Canada.

NOS also provides information and indexes on the following lake and river charts: "Lakes" (Cayuga, Champlain, Great Lakes, Mead, Minnesota-Ontario border lakes, Okeechobee, Oneida, Pend Oreille, Franklin D. Roosevelt, Seneca, Tahoe); and "Rivers" (Columbia, Connecticut, Delaware, Hudson, James, Kennebec, Neuse, New, New York State Barge Canal, Pamlico Sound, Penobscot, Potomac, Rappahannock, Savannah, St. Johns, St. Lawrence to Cornwall, York, and others).

A separate free publication, *Dates of Latest Editions of Nautical Charts*, is issued quarterly by NOS to aid mariners in obtaining up-to-date charts. *Notice to Mariners*, also available from NOS, is a pamphlet issued weekly by the Defense Mapping Agency in cooperation with NOS, the U.S. Coast Guard, and the U.S. Army Corps of Engineers to keep mariners advised of new publications and information on marine safety.

Tennessee Valley Authority. Since its inception in 1933, TVA has operated a Maps and Surveys Branch that produces a great variety of nautical maps and charts. Among them are navigation charts and maps, published for the TVA main-river reservoirs, and recreation maps for each of the TVA lakes (see also "Recreation Maps"). TVA publishes an *Index to Navigation Charts and Maps of TVA Reservoirs* and an *Index to Recreation Maps-Tennessee Valley Lakes*. A price catalog and indexes may be obtained free from the TVA Mapping Services Branch.

U.S. Army Corps of Engineers. The Corps of Engineers publishes nautical charts of selected rivers showing water depths and other navigation data, and indexes showing water areas and the number of charts required to cover them. Covered rivers include the Allegheny, Atchafalaya, Big Sandy, Big Sunflower, Calcasieu, Cumberland, Illinois, Kanawha, Mississippi, Missouri, Monongahela, Ohio, Tennessee, and the Gulf Intracoastal Waterway. Maps and indexes may be obtained from any of the nine Corps of Engineers district offices.

GOVERNMENT SOURCES, CANADA

Canadian Hydrographic Service (Dept. of Fisheries and Oceans, Ottawa, Ontario K1A 0E6; 613-998-4931) produces a wide range of publications, including nautical charts, "Sailing Directories and Small Craft Guides," and "Tide and Current Tables" for areas in Canada. A complete publications list is available.

Maps Alberta (Land Information Services Div., Main Fl., Brittania Bldg., 703 6th Ave. SW, Calgary, Alberta T2P 0T9; 403-297-7389) produces nautical charts for many areas in Alberta and neighboring provinces. The charts ($10 Canadian each) show bearings, soundings, heights, lake and river resolutions, monthly water levels, buoy markers, radar reflectors, mileage, beacon range, port and starboard targets, rocks, rapids and shoals. The charts are $9 Canadian each; a catalog containing chart information is available upon request. Also available is "Symbols and Abbreviations Used

on Canadian Nautical Charts" ($3 Canadian). Maps Alberta also sells hydrographic maps depicting water depths of lakes throughout the province ($3 Canadian each).

Saskatchewan Property Management Corp. (Central Survey and Mapping Agency, Distribution Center, 2045 Broad St., 1st Fl., Regina, Saskatchewan S4P 3V7; 306-787-6911) produces hydrographic charts for Lake Athabasca, Lac La Ronge, Poplar Point to Stony Rapid, and Slave River to Mackenzie River ($10 Canadian each). Bathymetric maps are also available for more than 525 lakes in the province. Map sizes and prices vary; a list is available upon request.

COMMERCIAL SOURCES

While the federal government is the primary producer of up-to-date nautical maps and charts, there are some commercial sources as well. At least one source distributes renderings of sailing days long past.

Better Boating Association Inc. (Box 407, Needham, MA 02192; 617-449-3314) produces "Chart Kits," books containing full-color reproductions of government charts. Titles available include: *The Bahamas; Canadian Border to Block Island, Rhode Island; Cape Cod Canal to Cape Elizabeth; Cape Elizabeth to Eastport, Maine; Cape Sable—Clearwater; Chesapeake and Delaware Bays; Florida East Coast; Florida Keys; Florida West Coast; The Intracoastal Waterway—Norfolk, Virginia to Jacksonville, Florida; Jacksonville to Miami; Lake Michigan; Long Island Sound; Naragansett Bay to Nantucket; New Orleans to Panama City, Florida; New Orleans to Texas Border; Southern California;* and *Texas.* Prices range from $39.95 to $84.95; sizes range from 12" X 17" to 17" X 22".

Bluewater Books & Charts (1481 SE 17th St., Ft. Lauderdale, FL 33316; 305-763-6533; 800-942-2583) distributes Better Boating Association Chart Kits, cruising guides, waterway guides, and various other publications, such as *Greek Waters Pilot* ($63), offering navigational advice and sketch charts for the Greek Islands.

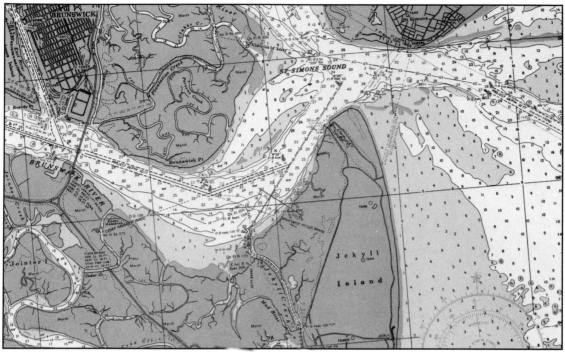

Portion of the southern Georgia coast, from the Better Boating Association Inc.'s Chart Kit for Region 6: "Norfolk, VA to Jacksonville, FL—The Intercoastal Waterway."

Chartifacts (P.O. Box 8954, Richmond, VA 23225; 804-272-7120) offers originals and reproductions of antique coastal charts and surveys. The company specializes in American coast and seaport charts from the earliest USGS surveys, and its knowledgeable cataloging includes detailed histories and descriptions of each map and chart it carries. In general, prices of original maps range from about $25 (for an 1853, 1:40,000-scale reconnaissance map of Sabine Pass between Louisiana and Texas Points) to about $125 (for a rare 1859 Progress Sketch of Florida from Biscayne Bay to Dry Tortugas). Reproductions are generally less than $20. Descriptive lists of originals and illustrated catalog pages of reproductions are available from Chartifacts. You are asked to specify your area of interest and include a stamped, self-addressed envelope.

Clarkson Map Co. (P.O. Box 218, Kaukauna, WI 54130; 414-766-3000) distributes nautical charts ($13 each) of Lake Michigan, Lake Superior, and "Canadian Boundary Waters." Also available is a handy 36-page booklet, *Chart Symbols and Abbreviations* ($2.50).

Gulf Publishing Co. (P.O. Box 2608, Houston, TX 77252; 713-520-4444) publishes ChartCrafter's "Mariner's Atlas" series of books packed with full-color government nautical charts and "chartlets," supplements of nautical data for harbors, inlets, rivers, and other tricky waters. The 10″ X 14″ paperbacks are available for the following areas: *Florida Gulf Coast/The Florida Keys; Lake Michigan; Long Island Sound/South Shore; New England; Southeast Florida/The Florida Keys; Southern California;* and *Texas Gulf Coast.*

Richardsons' Marine Publishing (P.O. Box 23, Streamwood, IL 60103; 312-741-4239) distributes nautical charts and guides, including:

■ *Waterway Guide Chartbooks,* 11″ X 13″ books containing accurate, full-color reproductions of current government charts. Chart Books titles include: *Cape Cod to Maine, New York Waters, Chesapeake and Delaware Bays/New York to Norfolk, The Intracoastal Waterway/Norfolk to Jacksonville, Florida's East Coast/Jacksonville to Miami, Florida's West Coast/Tampa to Pensacola.* Prices range from $21.95 to $49.50. Also available is a *Boater's Photographic Chartbook,* a set of three books for southern Florida waters, combining color aerial photos with NOS charts ($60).

■ "Chart Kits" published by the Better Boating Association.

■ *Richardson's Chartbook and Cruising Guides,* 12" X 18" books combining reproduced government charts with information on marinas, towns, cities, navigation and cruising information, harbor photos, and radio and cruising logs; available for the Great Lakes ($51.95).

Richardson's also sells various cruising guide-books; a catalog of charts, chart kits, and cruising guides is available upon request.

Waterway Guides (6255 Barfield Rd., Atlanta, GA 30328; 404-256-9800; 800-233-3359) sells a variety of nautical charts for pleasure boaters, including "Waterway Charts" ($45.50 to $49.50 each) for seven ocean areas such as "Newport to Canada," "Norfolk to Jacksonville," and "Lower Florida and Keys." Also available are Waterway Guides ($23.95 each) containing useful information on locations of bridges, docks, and the like. Guides are available for the following regions: Northern U.S., Southern U.S., Mid-Atlantic, and Great Lakes.

■ **See also: "Ocean Maps," "Recreation Maps," "River, Lake, and Waterway Maps," and "Tide and Current Maps."**

Ocean Maps

When it comes to mapping, there are essentially three "oceans." One is the geographic ocean: the ocean as it appears from space. Another is the geologic ocean: the nature of the coasts, continental margins, and the deep-sea floor. Still another is the biologic ocean: the fishes, plants, and assorted other life forms that populate the deep blue sea. (There is also the nautical ocean, of course, the sailing routes used by pilots of boats and ships. See "Nautical Charts and Maps" for information on those maps.)

The earliest world maps were, in effect, maps of the oceans, which served as boundaries for earlier civilizations' perspectives of their planet. Indeed, some of the earliest surviving maps are European nautical charts that date back to the second half of the 13th century. Today's maps of the seas are somewhat more accurate, based on the digital recordings made by satellites instead of visual recordings made by seamen.

Scientists have been fascinated with the bottom of the ocean almost as long as they have studied the lands ashore. Geologists in the early 19th century speculated that the ocean floors were dull expanses of mud—featureless and flat. For centuries, naturalists thought that the oldest rocks on Earth were on the ocean floors. They believed that the present-day ocean basins formed at the very beginning of the Earth's history and that throughout time they had slowly been filled by a constant rain of sediment from the lands. Data gathered since the 1930s have enabled scientists to view the sea floor as relatively youthful and geologically dynamic, with mountains, canyons, and other topographic forms similar to those found on land. The sea floor, they found, is no more than 200 million years old—a "young" part of the globe's crust compared to the continents that contain rocks nearly twenty times that age.

Research conducted since World War II has produced an ocean of data on the sea floor, much of it based on the studies of the federal **National Ocean Service** conducted by the National Oceanic and Atmospheric Administration, an agency of the U.S. Department of Commerce. Actually, it was the war itself that produced the technology for this research. In testing different sound frequencies to help locate submarines, scientists found that certain frequencies were capable of sending sound waves through the seafloor and getting reflections from deeply buried layers of rock. This revolutionized the study of marine geology and the quality of maps of the ocean floor.

Among the most spectacular of cartographic products of this technology is the glow-in-the-dark "Map of the Ocean Floor" published by **Celestial Arts** (P.O. Box 7327, Berkeley, CA 94707; 415-845-8414; 800-841-2665). The 36-inch-square map provides accurate and highly illustrated details of the mysterious terrain that underlies our oceans—in effect, what the Earth would look like without water—showing all four hemispheres from several different viewpoints. With the lights out, areas of volcanic activity glow in the dark. The map ($9.95 plus $2 postage) comes with a helpful 28-page "guide and tourbook."

NOS produces a variety of ocean maps (see "Nautical Charts and Maps" for other NOS products). Among them are:

■ **Bathymetric maps**, which are topographic maps of the sea floor. Through the use of detailed depth contours and other data, the size, shape, and distribution of underwater features are vividly portrayed. These serve as the basic tool for performing the scientific, engineering, marine-geophysical, and marine-environmental studies that are often required for development of energy and marine resources.

■ **Topographic/bathymetric maps** are multipurpose maps showing the topography of the ocean floor and the land nearby. These are cooperatively produced by NOS and the U.S. Geological Survey to support the coastal-zone-management and environmental-impact programs. They may also be used by land-use planners, conservationists, oceanographers, marine geologists, and others interested in the coastal zone and the physical environment of the Continental Shelf.

■ **Bathymetric fishing maps** are topographic maps of the sea floor designed primarily for use by commercial and sport fishermen. This series of maps, produced at a 1:100,000 scale, includes information about the type and distribution of bottom sediment and known obstructions on the sea floor, in addition to the basic information found on a standard bathymetric map. It is intended to aid fisher-

Portion of the "World Ocean Map (Spilhaus Projection)," from GeoLearning Corp., showing "the whole world ocean and all the continents uncut by the edge of the map."

men in identifying where the "big ones" are biting.

■ **Geophysical maps** consist of a base bathymetric map, a magnetic map, a gravity map, and, where possible, a sediment overprint. The bathymetric map, when combined with the others, serves as a base for making geological-geophysical studies of the ocean bottom's crustal structure and composition. There are two series of geophysical maps. The 1:250,000-scale series contains the geophysical data for the Continental Shelf and Slope. The 1:1,000,000-scale series covers geophysical data gathered in the deep-sea areas, sometimes including the adjacent Continental Shelf.

GOVERNMENT SOURCES, UNITED STATES

A free NOAA publication, *Map Catalog 5* (Distribution Div., National Ocean Service, 6501 Lafayette Ave., Riverdale, MD 20737; 301-436-6990), provides a U.S. map overlaid with a grid showing the availability of each type of map for all of the U.S., including Alaska and Hawaii, along with ordering instructions. Map prices range from $2.50 to $4.30.

National Geophysical Data Center. The NGDC, part of NOAA, distributes a full-color "Gravity Field of the World's Oceans," a 35″ X 46″ map portraying the gravitational effect of sea floor topography such as sea mounts, ridges, and fracture zones ($10).

U.S. Geological Survey. The USGS is mapping the ocean floor in the Exclusive Economic Zone (EEZ), using imaging sonar along with seismic, magnetic and bathymetric data. (The EEZ, established by presidential proclamation in 1983, gives the U.S. jurisdiction over the living and natural resources in the ocean area extending 200 nautical miles seaward from the offshore three-mile limit off the coast of the U.S. and its territories.) To date, two atlases have been published that show 2-degree X 2-degree computer-enhanced sonar mosaics of the ocean floor, at a scale of 1:500,000. Facing each sonar mosaic is a matching image map containing preliminary geologic interpretation,

names of major features and bathymetric contours. *Atlas of the Exclusive Economic Zone Western Conterminous United States* (I-1792; 152 pages) and *Atlas of the Exclusive Economic Zone, Gulf of Mexico and Eastern Caribbean Areas* (I-1864 A,B; 162 pages) are both available for $45 each. Atlases of the East Coast, Alaska and Hawaii are in preparation.

GOVERNMENT SOURCES, CANADA

Canadian Hydrographic Service (Dept. of Fisheries and Oceans, Ottawa, Ontario K1A 0E6; 613-998-4931) produces a "Natural Resource Maps" series of the ocean floor adjacent to Canada. The maps are used to identify areas of geological and geophysical significance in the search for new resources. They are available in 1:250,000 scale; maps drawn in 1:1,000,000 scale are available for some areas. CHS also publishes 1:1,000,000-scale bathymetric charts of Canadian areas as part of a "General Bathymetric Chart of the Oceans," sponsored by the International Hydrographic Organization and the United Nations Educational, Scientific, and Cultural Organization (UNESCO). Special-purpose maps are also produced along with descriptive text in the "Marine Science Paper" series.

COMMERCIAL SOURCES

AAPG Bookstore (P.O. Box 979, Tulsa, OK 74101; 918-584-2555), affiliated with the American Association of Petroleum Geologists, distributes geologic maps of ocean floors such as: "Magnetic Lineations of the World's Ocean Basins" ($18), depicting sea-floor-spreading magnetic anomalies in the world's ocean basins; and "Sediment Thickness Map of the Indian Ocean" ($18), showing total sediment thickness between the sea floor and oceanic basement.

Circum-Pacific Map Project (345 Middlefield Rd., MS 952, Menlo Park, CA 94025; 415-329-4002) produces various maps of the Pacific Ocean area. See "Geologic Maps."

GeoLearning Corp. (555 Absaraka St., Sheridan, WY 82801; 307-674-6436) distributes the "World Ocean Map," drawn with the Spilhaus projection, showing continents and oceans uncut by map edges; text provides facts about the oceans of planet Earth. The map measures 24″ X 36″ and costs $8.95.

Marie Tharp, Oceanographic Cartographer (1 Washington Ave., South Nyack, NY 10960; 914-358-5132) produces beautiful color maps of the world ocean floor, including "Floor of the Oceans" ($20), a 27″ X 41″ map created in 1975; "World Ocean Floor Panorama" ($20), measuring 24″ X 38″, available laminated or in Kimdura plastic; "Seismicity of the Earth" ($25), showing areas that experience frequent seismic activity; and "Ocean Floor Sediment and Polymetallic Nodules Map" ($15), measuring 44″ X 76″. Also available are physiographic diagrams of the South Atlantic, the Indian Ocean, and the Western Pacific ($7.50 each).

National Geographic Society (17th & M Sts., NW, Washington, DC 20036; 202-921-1200)

sells four double-sided ocean-floor maps:

■ "Arctic Ocean/Ocean Floor" ($3; 02287; 24″ X 18 1/4″).
■ "Indian Ocean/Ocan Floor" ($4; 02025; 24″ X 18 1/4″).
■ "Pacific Ocean/Ocean Floor" ($4; 02024; 24″ X 18 1/4″).
■ "The Mediterranean Seafloor" ($4; 02006; 23″ X 37″), and includes descriptive notes.

Perigee Books (200 Madison Ave., New York, NY 10016; 201-933-9292; 800-631-8571) distributes Van Dam's "Environment Unfolds" map series, pocket-size maps that unfold to 12 times the original size, then refold automatically. "The Ocean Unfolds" ($9.95) features two four-color maps and a 16-page booklet exploring the historic and scientific aspects of the ocean.

■ **See also: "Antique Maps," "Nautical Charts and Maps," "Selected Atlases," "Space Imagery," "Tide and Current Maps," and "World Maps."**

River, Lake, and Waterway Maps

For mapping purposes, rivers, lakes, and waterways fall primarily into the jurisdictions of three federal agencies: the Army Corps of Engineers, the U.S. Geological Survey, and the Tennessee Valley Authority, all of which are prolific cartographers. Some of their products are covered in two other sections of this book, "Nautical Charts and Maps" and "Recreation Maps."

GOVERNMENT SOURCES, UNITED STATES

U.S. Geological Survey. General information about individual major U.S. rivers may be found in a series of free brochures, "River Basins of the United States" produced by USGS. The series, which covers the Colorado, Delaware, Hudson, Potomac, Suwannee, and Wabash Rivers, includes general rather than highly detailed maps of the river basins, each of which spans two or more states. Also included is information about each river's early exploration; its headwaters and mouth; its major tributaries; a description of its course, length, width, depth, and rate of flow; the river's dams, reservoirs, and canals; its geologic setting and drainage area; its water quality and use; and the major cities it passes along its route. These brochures are available from USGS publication-distribution centers (see Appendix B) and are listed in a free catalog, *Popular Publications of the U.S. Geological Survey.*

Another USGS series is its State Hydrologic Unit Maps, highly detailed maps showing the hydrographic boundaries of major U.S. river basins that have drainage areas greater than 700 square miles. (Hydrology is the study of the water cycle, from precipitation as rain and snow through evaporation back into the atmosphere.) The four-color maps provide information on drainage, culture, hydrography, and hydrologic boundaries for each of 21 regions and 222 subregions. The maps, which cost from $1.75 to $5 each, are used primarily by water-resource planners for managing water resources and flood potential, and by land-resource planners for managing natural resources and recreation areas. The maps are available from USGS Distribution Centers (see Appendix B). A free brochure, "State Hydrologic Unit Maps," also is available from USGS.

Army Corps of Engineers. The big picture of U.S. waterways is contained in a 15″ X 22″ black-and-white Corps of Engineers map, "Major Waterways and Ports of the United States." On the map, navigable waterways are shown as heavy black lines, with other rivers shown with less emphasis. For the shipping rivers, and waterways, there are symbols indicating locks, ports, and principal cities.

The Corps also produces maps of major rivers—considerably more colorful than the USGS versions—containing much the same types of information, with one major exception: a detailed chart of recreational facilities along each river, such as boat ramps, sanitary facilities, camping facilities, and picnic tables. The sites on the charts are keyed to the map.

The Corps of Engineers' series of maps of U.S. lakes provides considerably more detail, right down to the lakes' boat ramps, campgrounds, and concessions stands. Most helpful is a series of nine regional maps: "Lakeside Recreation in New England," "Lakeside Recreation in Mid-Atlantic States," "Lakeside Recreation in the Southeast," "Lakeside Recreation in the Great Lakes States," "Lakeside Recreation in the Upper Mississippi Basin," "Lakeside Recreation in the South Central States," "Lakeside Recreation in the Great Plains," "Lakeside Recreation in the Northwest," and

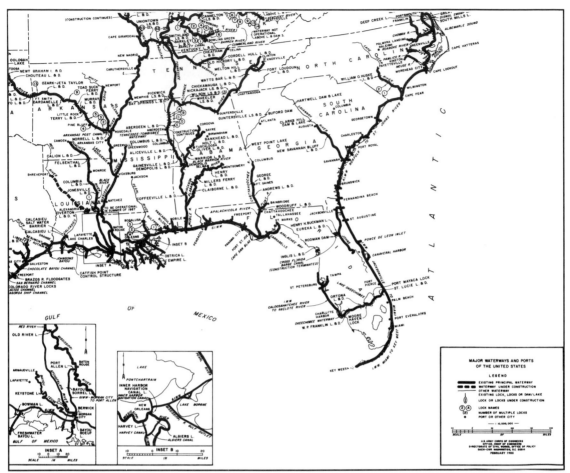

Section from the U.S. Army Corps of Engineers map "Major Waterways and Ports of the United States," showing existing and planned waterways and ports.

"Lakeside Recreation in the Southwest." Each contains a map of the states covered along with a detailed chart, showing the facilities available at each lake—everything from the type of campsites (developed or primitive) to boat-launching ramps, showers, hunting, ice fishing, and nearby hotels and restaurants. In addition to the regional maps, there are detailed maps covering dozens of individual lakes. These show recreation areas in considerable detail, including information about all facilities. Both river and lake maps are available at all Corps offices and from the U.S. Army Corps of Engineers' Washington, D.C. office.

Unfortunately, each of the Corps' divisions and districts publishes its own maps and

publications, and there is neither a central distribution center nor a single listing of all Corps of Engineers maps. The Missouri River Division, for example, distributes a "Descriptive List of Maps and Charts for Sale to the Public," available from that division's two offices, in Omaha and Kansas City, Missouri. Similarly, the Ohio River Division distributes maps and charts of that river, and the Chicago District distributes maps of the middle and upper Mississippi River and the Illinois Waterway to Lake Michigan; lower Mississippi maps come from the Corps' Vicksburg (Mississippi) District. And then there are the Black Warrior, Alabama, Tombigbee, Apalachicola, and Pearl Rivers, all of whose maps come from the Corps' Mobile

(Alabama) District. When seeking maps of a particular area, your best bet is to contact the Corps of Engineers office closest to that area (see Appendix B).

Tennessee Valley Authority. TVA's Mapping Services Branch produces a wide range of maps and publications, all listed in a free TVA *Maps Price Catalog*. In addition to its recreation maps of TVA lakes (see "Recreation Maps"), there is also a series of full-color Tributary Watershed Maps, showing drainage basins, topography, highways, railroads, and nearby cities; and a series of full-color Reservoir Area Maps, showing reservoirs and surrounding regions. TVA also publishes navigation maps of its many rivers and tributaries.

Government Sources, Canada

Canada Map Office (615 Booth St., Ottawa, Ontario K1A 0E9; 613-952-7000) distributes "Inland Water Directorate Maps" for Canadian areas. IWD-series maps available are Peyto Glacier Map (IWD 1010), and Columbia Icefield (IWD 1011). The maps are $3 Canadian each. A map index/price list will be sent upon request.

Saskatchewan Property Management Corp. (Central Survey and Mapping Agency, Distribution Ctr., 2045 Broad St., 1st Fl., Regina, Saskatchewan S4P 3V7; 306-787-6911) produces two special 1:50,000-scale waterway maps: "Churchill River" from Otter Lake to Nistowiak Lake, and "Churchill River—MacKay Lake" ($3 Canadian each). The agency also produces hydrographic charts for Lake Athabasca, Lac La Ronge, Poplar Point to Stony

Rapid, and Slave River to Mackenzie River ($10 Canadian each). Bathymetric maps are also available for more than 525 lakes in the province. Map sizes and prices vary; a list is available upon request.

Commercial Sources

Clarkson Map Co. (1225 Delanglade St., P.O. Box 218, Kaukauna, WI 54130; 414-766-3000) sells maps for lakes in Minnesota, upper Michigan, and Wisconsin (90 cents each). A free map index is available.

Map Works Inc. (2125 Buffalo Rd., Ste. 112, Rochester, NY 14624; 716-426-3880) publishes "Finger Lakes Map," measuring 38" X 45" and covering the 14-county Finger Lakes region of New York, with a special emphasis on fishing in these lakes ($3.25).

Northwest Map Service (W. 713 Spokane Falls Blvd., Spokane, WA 99201; 509-455-6981) publishes nautical and facilities maps of Roosevelt Lake in Washington ($5.50 each; 24" X 35"). Also available is "Banks Lake Bathymetric/Topographic and Facilities Map" ($5.50; 24" X 30").

Ozark Map Co. (RR2, Box 700-B, Gravois Mills, MO 65037; 314-374-6553) produces a map of the Truman Reservoir in South Central Missouri ($1.95).

■ **See also: "Nautical Charts and Maps," "Recreation Maps," and "Tide and Current Maps."**

Tide and Current Maps

Among the many remarkable qualities of Mother Earth is her ability to perform certain rituals with uncanny regularity. Tides, for example, can be predicted with astounding accuracy and plotted on tidal-current charts. These are available from the **National Ocean Service**, part of the Commerce Department's National Oceanic and Atmospheric Administration. The NOS tidal-current charts each consist of a set of 11 charts that depict, by means of arrows and figures, the direction and velocity of the tidal-current for each hour of the tidal cycle. The charts, which may be used for any year, present a comprehensive view of the tidal-current movement in the respective waterways as a whole and also supply a means for rapidly determining, at any given moment, the direction and velocity of the current at various points throughout the water areas covered.

There are twelve available charts ($4 each), covering Boston Harbor, Charleston Harbor, Narragansett Bay (two charts), Long Island and Block Island Sounds, New York Harbor, Delaware Bay and River, Upper Chesapeake Bay, Tampa Bay, San Francisco Bay, and Puget Sound (two charts). All but the Narragansett Bay chart require that you also purchase of one of the NOS Tidal Current Tables ($5.50). There are two: one covering the Atlantic Coast, the other covering the Pacific Coast and Asia. The Narragansett Bay chart requires purchase of the NOS table ($6.75) covering the East Coast. All NOS charts, tables, and publications are available directly from NOS (see Appendix B) or one of its authorized dealers.

Maps Alberta (Land Information Services Div., Main Fl., Brittania Bldg.,

703 6th Ave. SW, Calgary, Alberta T2P 0T9; 403-297-7389) distributes the publication, "Tides in Canadian Waters" ($2 Canadian).

■ **See also: "Nautical Charts and Maps," "Ocean Maps," and "River, Lake, and Waterway Maps."**

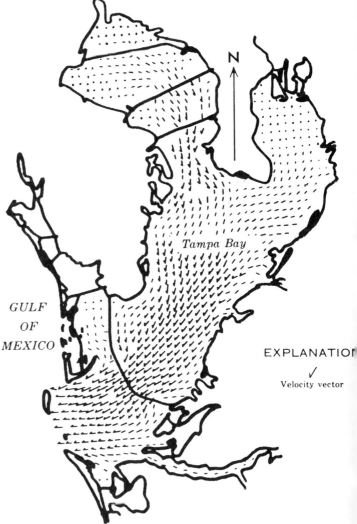

Map showing current flow and velocity in Tampa Bay, Florida, from information compiled from Landsat satellite imagery.

SKY MAPS

Aeronautical Charts

Aeronautical charts provide a wealth of fascinating information, regardless of whether you intend to take over a Beechcraft or a Boeing. Granted, these are not simple maps to read. At first glance, they seem filled with colorful circles, arrows, and other strange markings, and the little text that exists consists primarily of cryptic letters and numbers— "Picket 3 MOA 4000' to and incl. 10,000'" is one example.

But with a bit of patience, you can spot areas that are prohibited or restricted to fly over—usually military bases and other high-security locations. And you are likely to find dozens of heretofore unknown landing strips in your region—possibly belonging to a tycoon down the road who may have quietly constructed a runway capable of handling a jumbo jet. You'll find out exactly where low-flying aircraft are permitted and where they're banned, perhaps giving you the informational ammunition to inform the authorities about a commercial airliner that has been repeatedly flying over your home at an altitude a bit too close for comfort. It's all there, if you know how and where to look.

Aeronautical charts are published by both government and private publishers. Both create charts covering the skies over the United States as well as the rest of the world.

GOVERNMENT SOURCES, UNITED STATES

The **National Ocean Service** (NOS), part of the Commerce Department's National Oceanic and Atmospheric Administration, publishes and distributes U.S. aeronautical charts. Charts of foreign airways are published by the Defense Mapping Agency. In addition to the NOS's two distribution centers, domestic aeronautical charts and related publications are available through a network of several hundred sales agents, usually located at or near airports. A free publication, *Aeronautical Charts and Related Products*, includes a list of such dealers, as well as detailed descriptions of the various charts and publications distributed by the NOS.

According to the NOS, the date of an aeronautical chart is important if you are using it for aviation purposes. The agency notes that when charted information becomes obsolete, **using the chart or publication for navigation may be dangerous**. Critical changes occur constantly, and it is important for pilots to purchase up-to-date charts. To ensure that only the latest charts are used, NOS publishes *Dates of Latest Edition*, available free from NOS distribution centers.

Available NOS charts include:

Aeronautical planning charts, used for preflight planning of long flights. Portions of the flight route can then be transferred to more detailed charts for actual use in flight. NOS publishes two types of flight-planning charts: "VFR/IFR Wall Planning Charts," at a scale of 1:2,333,232, a large (82" X 56") chart in two pieces, which can be assembled to form a composite visual flight rules (VFR) planning chart on one side and an instrument-flight-rules (IFR) chart on the other. The chart is revised every 56 days. A one-year subscription is $45.50; single copies are $6.50. Another planning chart, the "Flight Case Planning Chart," is a somewhat smaller (30" X 50"), folded chart (scale 1:4,374,803) designed for pre- and in-flight use. It contains the same information as the VFR/IFR chart, with the addition of selected flight-services stations and National Weather Service offices at airports, parachute-jumping areas, a tabulation of special-use airspace areas, a mileage table listing distances between 174 major airports, and a city-airport location index. This chart is revised every 24 weeks. A two-year subscription is $21.60; single copies are $3.60.

Portion of a NOAA sectional chart for the New York City area, showing airports, controlled airspace, restricted flying areas, and other vital information for pilots.

Visual aeronautical charts, multicolor charts designed for visual navigation of slow- to medium-speed planes. The information featured includes selected visual checkpoints, including populated places, roads, railroads, and other distinctive landmarks. There are three types: Sectional Aeronautical Charts (1:500,000; 20″ X 60″) show the airspace for a large region of several hundred square miles; Terminal Area Charts (1:250,000; 20″ X 25″) show the airspace designated as Terminal Control Areas around airports; and World Aeronautical Charts (1:1,000,000; 20″ X 60″) cover much larger areas with much less detail. Two-year subscriptions are $5.50 to $21; single-copy prices are $5.25 and $2.75. One helpful publication is VFR *Chart User's Guide* ($4), a colorful 36-page booklet intended as a learning tool for all three types of charts. It includes detailed definitions of the dozens of symbols used on these charts and can help any aeronautical chart reader understand the symbols.

Instrument navigation charts, providing information for navigation under instrument flight rules. There are different series for low-altitude flights (below 18,000 feet), high-altitude flights, and instrument-approach-procedure (IAP) charts.

The **Defense Department** also publishes aeronautical charts, which primarily provide information about flying over airspace outside the U.S. These can be ordered through Defense Mapping Agency Distribution Services (Washington, DC 20315; 202-227-2495).

■ **FLIP (Flight Information Publications) charts** are available for most parts of the world, including Africa, Asia, and Antarctica. Each set of maps provides the information needed for flying in foreign airspace.
■ **Operational navigation charts** provide information on high-speed navigation requirements at medium altitudes.
■ **Tactical pilotage charts** provide information on high-speed, low-altitude, radar, and visual navigation of high-performance tactical and reconnaissance aircraft at very low through medium altitudes.
■ **Jet navigation charts** are used for long-range, high-altitude, high-speed navigation.
■ **Global navigation and planning charts** are suitable for flight planning, operations over long distances, and en route navigation in long-range, high-altitude, high-speed aircraft.

GOVERNMENT SOURCES, CANADA

Canada Map Office (615 Booth St., Ottawa, Ontario K1A 0E9; 613-952-7000) distributes aeronautical charts for Canada, including VFR Navigation Charts (VNC) ($5.50 Canadian each) drawn on a 1:500,000-scale; Canadian Pilotage Charts (CPC) ($4.50 Canadian each); World Aeronautical Charts (WAC) ($5.50 Canadian each); and miscellaneous charts. Also available are the booklets *Air Tourist Information* (free), providing information on customs, licensing requirements, Canada's airspace, and major differences between Canadian and U.S. flight

procedures; and *Flying the Alaska Highway in Canada*, giving useful information on preparing for an air trip along this scenic route.

Manitoba Natural Resources (Surveys and Mapping Branch, 1007 Century St., Winnipeg, Manitoba R3H 0W4; 204-945-6666) distributes World Aeronautical Charts, Canadian Pilotage Charts, and VFR Navigational Charts for Canada. A pamphlet is available.

Maps Alberta (Land Information Services Div., Main Fl., Brittania Bldg., 703 6th Ave. SW, Calgary, Alberta T2P 0T9; 403-297-7389) distributes World Aeronautical Charts, Canadian Pilotage Charts, Aeronautical Planning and Plotting Charts, and VFR Terminal Area Charts for Canada. A catalog is available.

Saskatchewan Property Management Corp. (Central Survey and Mapping Agency, Distribution Center, 2045 Broad St., 1st Fl., Regina, Saskatchewan S4P 3V7; 306-787-6911) distributes World Aeronautical Charts ($5.50 Canadian each) for all areas in Canada.

COMMERCIAL SOURCES

One of the oldest and largest producers of aeronautical charts is **Jeppesen Sanderson** (55 Inverness Dr. E., Englewood, CO 80112; 303-799-9090), which creates a wide range of charts and other publications for pilots. "Jeppesens," as they are known among pilots, are available only by subscription, which includes the initial charts for a particular area plus updates (biweekly for the U.S. and Canada, weekly for other areas). In addition to the navigation charts, Jeppesens include information pages covering topics such as chart terminology, a glossary, and standard abbreviations.

Jeppesen offers a variety of subscription services, which are described in a free catalog available by writing or calling the company. Also available is a set of sample charts—a U.S. Low Flight Planning Chart, an Enroute Chart, an Area Chart, and the Terminal Charts for an airport in your local area that has a published

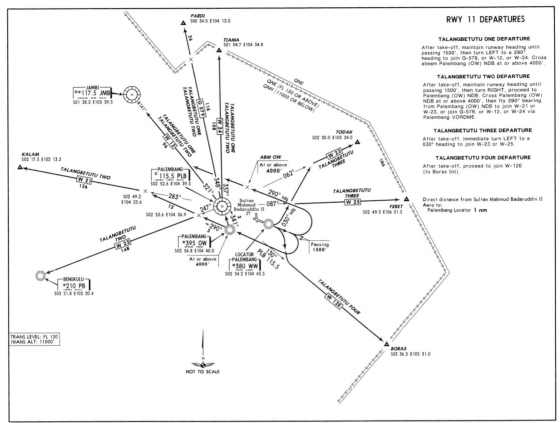

Jeppesen chart showing departure routes at Sultan Mahmud Badaruddin II Airport, Palembang, Indonesia. Copyright 1989 Jeppesen Sanderson, Inc.

approach. To obtain the samples, send $1 to cover postage and handling.

WHERE TO BUY CHARTS

The best place to purchase government and nongovernment aeronautical charts is at almost any general aviation airport—that is, an airport, or portion of an airport, that serves private, noncommercial flights. The "fixed-base operators" (or FBOs) that run these private airport facilities typically sell charts covering the immediate region surrounding the airport; some also sell other domestic or international charts. In addition, there are many other retailers of the charts, including many map stores; a list of select retailers is below.

Government-produced charts, besides being available at many airports, may also be purchased through three NOS offices:

■ 439 W. York St., Norfolk, VA 23510; 804-441-6616.
■ 1801 Fairview Ave. E., Seattle, WA 98102; 206-442-7657.
■ Federal Bldg. and Courthouse, 701 C St., Box 38, Anchorage, AK 99513; 907-271-5040.

These three NOS offices also can send you NOS's free catalog mentioned earlier.

Some of the many commercial aeronautical chart dealers are:

Aero Supply Inc.
1624 Aviation Center Pkwy.
Daytona Beach, FL 32014

Air Delphia
1834 E. High Ave.
New Philadelphia, OH 44663

Atlanta Pilot Center
1954 Airport Rd., Ste. 66
Atlanta, GA 30341

Aviation International Corp.
5555 NW 36th St.
Miami, FL 33166

Aviation Management
195th & Bernham
P.O. Box 553
Lansing, IL 60438

Bodas Aero Mart
3930 Campus Dr.
Newport Beach, CA 92660

Cableair Inc.
Cable Airport
1749 W. 13th St.
Upland, CA 91786

Centerline Flight
P.O. Box 328
Valley City, ND 58072

Chief Aircraft Parts
345 Whispering Pines
Grants Pass, OR 97527

D & M Pilot Supply
4N 116 Wyant Rd.
West Chicago, IL 60185

El Cajon Flying Service Inc.
1825 N. Marshall Ave.
El Cajon, CA 92020

International Aviation
4135 Donald Douglas Dr.
Long Beach, CA 90808

J-Mar Aero Service
R.R. 2, Box 6C
Litchfield, IL 62056

J. C. Air
400 Industrial Pkwy.
Industrial Airport, KS 66031

Jem Aero
2003 Quail St.
Newport Beach, CA 92660

Kimco Flight Crew
107 Millwood Dr.
Warner Robbins, GA 31088

Marv Golden Discount Store
8690 Aero Dr., Ste. 102
San Diego, CA 92123

The Pilot Shop
Norwood Municipal Airport
Norwood, MA 02062

Plane Things
1875 W. Commercial Blvd., Ste. 180
Ft. Lauderdale, FL 33304

Prescott Pilot Shoppe Inc.
2054 Old Kettle Dr.
Prescott, AZ 86301

San-Val Aircraft Parts
7456 Valjean Ave.
Van Nuys, CA 91406

Sportsman Market
Clermont County Airport
Batavia, OH 45103

Sporty's Pilot Shop
Clermont Airport
Batavia, OH 45103

Wings Inc.
501 N. Daleville Ave.
Daleville, AL 36322

Astronomical Charts and Maps

For thousands of years observers have tried to unlock the secrets of the heavens. And with that effort has come a never-ending stream of celestial charts and maps. Today there are maps showing the landscapes of the planets and satellites that our spacecraft have visited or flown by, and charts that show the bright stars of the Milky Way galaxy; some catalog the positions of other galaxies. The charts are redrawn every 50 years (a period known as an *epoch*), with additional updating done in the intervening years.

There are several kinds of star charts. Some are decorative, made to be put on a wall for classroom use or to pique the interest of curious passersby. The most stunning chart of this variety is the *Map of the Universe*, published by **Celestial Arts** (P.O. Box 7327, Berkeley, CA 94707; 415-845-8414; 800-841-2665). Printed on black paper, this brightly colored map glows in the dark. Major constellations and zodiacal signs are vividly illustrated, as are comet paths, meteor showers, and other celestial wonders. A 16-page booklet accompanies the map.

For beginners, a novel and effective way to get to know the sky is an audio star chart. Called *Tapes of the Night Sky* ($19.95 from **Astronomical Society of the Pacific** (390 Ashton Ave., San Francisco, CA 94112; 415-337-1100), the two cassette tapes feature four "guided tours" of the bright stars and constellations of each season and come with a 60-page book of instructions, transcripts, maps, and readings.

Young astronomers may find it easier to work with a planisphere, a wheel that is dialed to the time and date of the visible stars and constellations in the sky. Planispheres may be the most basic and helpful astronomical tool for the novice.

Many amateur astronomers interested in using star charts for outdoor gazing turn to star-atlas field guides. Often these are printed on vinyl or on a durable paper that can withstand inclement weather and other hazards that may be encountered outdoors. As you peruse what's available, think about how much information you need. A weekend observer may need charts that show only stars visible with the naked eye. Such charts are less cluttered and a lot easier to use. If you plan to take your chart outside to use while viewing, make sure that it is visible at night (stars printed in white on a black field) and that it is compact enough to accompany you, your telescope, and other necessities. If you plan to use the chart to make observations, make sure you bring along a flashlight with a red filter so you can use the atlas without being blinded by white light.

Sky Publishing Corp., **Kalmbach Publishing Co.**, and **Willmann-Bell Inc.** are the three major companies publishing reputable star charts used by amateur astronomers. Government agencies also publish star charts, as well as maps of planets and the solar system. The **National Geographic Society** offers individual maps and two beautiful star charts in its comprehensive *Atlas of the World*. Finally, a few additional companies sell planispheres and other unique types of star charts. Following are descriptions of what's out there to make your journey through the stars easier and more enjoyable.

GOVERNMENT SOURCES

The **Government Printing Office** has the following selection in astronomical charts and maps:

Night-sky planisphere from David Chandler Co. allows you to dial in on the constellation of your choice according to the current date and time.

■ Astronomical Almanac for the Year comes in two volumes, each more than 500 pages long, that cover 1985 through 1990. These rather technical almanacs include charts, maps, graphs, tables, and other statistics relating to astronomical events during the years covered (1985 and 1986: S/N 008-054-00103, $25; 1987: S/N 008-054-00124-6, $18; 1989: 008-054-00135; $24; 1990: 008-054-00129-7; $23).
■ Astronomical Phenomena for the Year comes in three volumes covering 1986 through 1988. Each volume contains information on the sun, the moon, planets, eclipses, Gregorian calendar and Julian day, time zones, and related phenomena (1986: S/N 008-054-00112-2; $2.75; 1987: S/N 008-054-00118-1; $3; 1988: S/N 008-054-00123-8; $3).

■ Atlas of Mercury contains photographs of the planet Mercury taken by the Mariner 10 spacecraft (S/N 033-000-00695-1; $21).
■ Stars in Your Eyes: A Guide to the Northern Skies explains how to locate several summer constellations of the northern sky and includes famous mythological lore (S/N 008-022-00155-7; $1.50).

U.S. Geological Survey. The USGS maintains an extensive stock of moon and planetary maps. Titles range from "Geologic map of the Theophilus quandrangle of the Moon" ($3; I-546, 1:1,000,000; 29″ X 48″), produced in 1968, to a "Reference Mosaic of Mercury" ($2.40; I-903, 1:10,000,000; 23″ X 26″), made in 1974, to a "Preliminary pictorial map of Io" ($2; I-1240, 1:25,000,000; 22″ X 25″).

COMMERCIAL SOURCES

Abbeville Press (488 Madison Ave., New York, NY 10022; 212-888-1969; 800-227-7210) publishes *Maps of the Heavens*, a 148-page book featuring full-color maps of the the zodiac and the constellations ($59.95).

American Map Corp. (46-35 54th Rd., Maspeth, NY 11378; 718-784-0055) offers Hallwag's highly detailed full-color astronomy maps with heavy-duty covers ($9.95 each). The following maps, each of which comes with an illustrated explanatory booklet, are available:
■ "The Moon," a two-sided photographic map showing both faces of the moon.
■ "The Stars," which depicts zodiac signs, constellations, and a variety of star types.
■ "The Solar System," depicts orbits as well as the planets' relative positions and sizes.

Celestial Arts (P.O. Box 7327, Berkeley, CA 94707; 800-841-2665; 415-845-8414) offers four 36″ X 36″ posters showing a map of the universe, a map of the solar system, the phases and faces of the moon, and the cosmos as perceived by the Aztecs. The two maps have the added benefit of glowing in the dark. The maps are $9.95 each.

Champion (200 Fentress Blvd., Daytona Beach, FL 32014; 904-258-1270; 800-874-7010; 800-342-1072 in Florida) distributes Replogle's celestial globes, including the 12″ "NASA Moon Globe" ($26.95) and "The Apollo," a 12″ model for $29.95.

David Chandler Co. (P.O. Box 309, La Verne, CA 91750; 714-946-4814) has created a night-sky planisphere that allows you to dial in on the constellations of your choice according to the current date and time. Chandler's 10″ chart is double-sided to accommodate both the northern and southern sky views, and four different versions are available for the following latitudes: northern United States/Canada/Europe; southern United States; Mexico/far southern United States; and South America/Australia/New Zealand. Chandler's book

Exploring the Night Sky with Binoculars provides additional information about stargazing. Price for the planisphere is $3.25; the book is $3.95. These are also available through **Sky Publishing Corp**. and the **Astronomical Society of the Pacific**.

Exploration USA (P.O. Box 456, College Park, MD 20740; 301-490-2236) produces a 35″ X 37″ "Mars Map," based on USGS and NASA information, which includes the latest official names approved by the International Astronomical Union (IAU). The 1:25,000,000-scale map comes with a glossary and is available directly from Exploration USA for $5 plus $2 shipping.

Hubbard (P.O. Box 104, Northbrook, IL 60062; 312-272-7810) produces several teaching charts and tools. Among the products are:

■ "Astrocharts" ($52.50 per set of six). These 29″ X 23″ charts are enlarged reproductions of astronomical-object photographs taken from the Mt. Wilson and Mt. Palomar observatories. The color, quality, and subjects—which include depictions of Orion's Great Nebula, Andromeda's Great Galaxy, and Sagittarius's Frifid Nebula—make them suitable for classroom reference or decorative use.
■ "Astronomy Study Prints" ($16.50) is a set of twelve 9″ X 15″ charts for teaching the basics of astronomy. The set includes prints of star charts of the Northern and Southern Hemisphere, as well as prints of the planets, the sun, the moon, and the galaxies.
■ "Star Charts" ($29.95) are reversible 44″ X 44″ sky maps of the Northern and Southern Hemispheres. The stars, nebulae, planets, constellations, and phenomena are depicted with light colors against a dark background for easy reference.
■ "Star Finder/Zodiac Dial" ($3.95) is a reversible wheel for finding stars and identifying zodiac constellations.
■ "Universal Celestial Globe" ($109) is a clear model of the celestial sphere that provides a physical demonstration of star and planetary

movement. It comes with a meridian ring and horizontal mount and is also available without these features ($65).

Kalmbach Publishing Co. (P.O. Box 1612, Waukesha, WI 53187; 414-796-8776) produces a wide range of astronomy materials, including magazines, books, and posters. Here are a few samples from Kalmbach's list:

■ *The Star Book*, by Robert Burnham ($8.95), is a good basic field guide to the constellations of the Northern Hemisphere. Half of the 18 pages illustrate the changes in the night sky for both the early and late halves of the four seasons. The other half describe the constellations and their history. Illustrations are white on blue for easier night-time reference. The accompanying text and illustrations explain clearly how to identify stars and constellations.

■ *Leslie Peltier's Guide to the Stars: Exploring the Sky With Binoculars* ($11.95) is a detailed guide on how to use binoculars to see constellations, stars, variable stars, the sun, the planets, the moon, and comets and meteors. Easy-to-read seasonal star charts and diagrams are included.

■ *Our Galaxy: The Milky Way Poster* ($16.95) shows the entire sky visible from Earth. What sets this star chart apart is that everything in the sky is shown in relation to the Milky Way. The poster shows more than 9,000 stars, all 88 constellations, and more than 250 of the brightest and most significant deep-sky objects.

Millennium Enterprises (P.O. Box 4730, Fresno, CA 93744; 209-266-2239) publishes a "World Star Atlas" ($12.50 postpaid), an attractive 23″ X 35″ two-color star chart. Although not drawn to scale, it provides a clear perspective of the key celestial objects.

National Geographic Society (17th and M Sts. NW, Washington, DC 20036; 202-921-1200) offers the *Atlas of the World*, which includes two magnificent star charts of both the northern sky and the southern sky. Magnitudes up to 6 are included, as well as quasars, pulsars, stars, nebulae, and even a possible black hole. Also

Using the NightStar star globe.

included is a map of the solar system and maps of both the near side and far side of the moon. Other astronomical maps sold separately are:

■ *The Earth's Moon* ($3 paper, $4 plastic; 42″ X 28″), with descriptive notes and diagrams.

■ *Mars* ($3; 38″ X 23″), revealing its cratered topography.

■ *The Solar System/Saturn* ($3; 22 1/2″ X 17″), printed on both sides and including illustrations and text.

■ *The Universe/Sky Survey* ($3; 22 3/4″ X 34″), with illustrations and text on both sides.

The NightStar Co. (1334 Brommer St., Santa Cruz, CA 95062; 408-462-1049; 800-782-7122) manufactures three models of star globes that are made of soft plastic, then deflated, and sealed into the shape of a bowl. The models have a movable surface that can be adjusted for any time and place on Earth. They include the whereabouts of all 88 constellations. There are several models available: *NightStar Traveler* ($26) is compact enough to fit in your pocket or backpack. The package includes a full-feature foldable two-color (white on dark blue) map with simplified instructions. *NightStar Classic* ($39) is the top-of-the-line model; it includes a comprehensive activities handbook, two planet-finder overlays showing one full year of planetary movements, and two snap-on dials that offer the option of presetting the sky for any time, date, or latitude. *NightStar Traveler Deluxe* ($39) combines features of both models. It comes with the same handbook as the *Classic* and is as compact as the *Traveler*.

Two additional NightStar products are *Learning Astronomy With NightStar* ($7), an easy-to-follow activities booklet for beginning students of the stars, and *Astronomy and the Imagination* ($15), which is packed with fascinating astronomical facts that can be appreciated by viewing the sky with the naked eye.

Perigee Books (200 Madison Ave., New York, NY 10016; 201-933-9292; 800-631-8571) distributes Van Dam's "Environment Unfolds" map series, pocket-size maps that unfold to 12 times the original size then refold automatically. Three maps featuring space environments are available: "The Universe Unfolds," "The Moon Unfolds," and "Mars Unfolds" ($9.95 each).

Replogle Globes Inc. (2801 S. 25th Ave., Broadview, IL 60153; 313-343-0900), the world's largest manufacturer of globes, offers a 12″ moon globe and a 12″ celestial globe. Developed under the auspices of NASA, the moon globe ($28.95; 382455) has a three-dimensional look that reveals craters, seas, and mountains. The celestial globe ($32.95; 38848) locates 1,200 stars from 1st to 4th magnitude and shows constellations, star clusters, new stars, and double stars.

Sky Publishing Corp. (49 Bay State Rd., Cambridge, MA 02238; 617-864-7360) offers a wide range of publications for both the ama-

Replogle Globes's moon globe.

teur and the more advanced astronomer. Its monthly magazine, *Sky & Telescope*, includes timely articles on space research, telescope-making, astrophotography, planetariums, and more. Each issue contains a map of the sky that is especially useful for locating the constellations. Sky Publishing has a Book Faire division that specializes in quality astronomical books and publications from numerous other publishers. It also has its own publishing arm, which distributes star charts and atlases, among other things. A catalog is available upon request. Sky Publishing's products include:

■ *Norton's Star Atlas and Reference Handbook*, edited by Gilbert Satterthwaite, Patrick Moore, and Robert Inglis ($29.95; 06780), is a favorite among serious stargazers. Now in its 17th edition, the book serves as a handy reference book for explanations of unfamiliar terms, as well as maps that show more than 8,400 stars, 600 deep-sky objects, and constellations from pole to pole. In addition to the numerous charts, the book contains information about astronomy concepts, observation techniques, mysteries of the solar system, the composition of stars, and telescopes and accessories.
■ *Sky Atlas 2000.0*, by Will Tirion (black-and-white version $15.95; 4631x; color version, $34.95; 46336) is considered a necessity for

the practicing astronomer. A 27-page collection of 13-by-18-inch charts, the book includes more than 43,000 stars and 2,500 deep-sky objects. The atlas is also available in a field version, which reverses the normal black-on-white format for easier night-time stargazing.

■ Atlas Borealis and Atlas Australis, by Antonin Becvar ($49.95 each), plot the location, spectral class, and brightness of more than 320,000 stars in the northern and southern skies and in the equatorial region. Comets, asteroids, and other deep-sky phenomena are also included. Each volume contains 24 13 1/2" X 19" charts.

■ Atlas Stellarum 1950.0, by Hans Vehrenberg, comprises two volumes: Stellarum North ($145; 49051) and Stellarum South ($80; 4906X). Both include a collection of photo-offset black-and-white images and illustrate star positions in the northern and southern skies, respectively. The charts are scaled to two arc-minutes per millimeter on a transparent grid and magnify the sky 14 times.

■ Atlas of Selected Areas ($33; 49000) is a compilation of 206 regions in the celestial sphere. Each region contains one star of 8th or 9th magnitude, and each page maps a general field as well as a magnified version of the same region. Selected areas are spaced at every half-hour of right ascension.

■ Photographic Star Atlas—The Falkau Atlas, by Hans Vehrenberg, is available in several versions: white-on-black volume (A version) of the northern and southern skies and black-on-white volumn (B version) of the northern skies. All volumes include both photographs and charts (northern sky A version, $48 each; southern sky B version, $40 each).

■ Handbook of Constellations, by Hans Vehrenberg and Dieter Blank ($39.45; 49086), presents 55 dark-gray maps of the entire sky. Additional information includes maximum and minimum brightnesses, spectral types, periods, and some epochs.

■ Sky Catalogue 2000.0, edited by Alan Hirschfeld and Roger Sinnott, is available in two volumes. Volume 1 lists stars to Magnitude 8, and Volume 2 presents data on double stars, variable stars, and nonstellar objects. Both volumes use data from the NASA-Goddard SKYMAP project, with Vol. 1 listing timely star positions, magnitudes, and spectral classes, and Vol. 2 offering a variety of data on more than 3,100 galaxies. Both volumes are available in paperback and cloth for $29.95 and $49.95, respectively.

■ Sky & Telescope Guide to the Heavens, published annually, weaves a chart of astronomical events into the astronomical year. Planetary movements, the waxing and waning of the moon, meteor-shower activity, and star magnitude and location are charted on an easy-to-read color fold-out. Also included is a day-by-day guide to celestial activity, which is $5.95.

Spherical Concepts (2 Davis Ave., Frazer, PA 19355; 215-296-4119) produces silk-screen clear-acrylic celestial globes, including:

■ "Star Ball" ($27.50), a five-inch celestial globe featuring all 88 constellations with names and patterns, and including a booklet of interesting "sky facts."

■ "The Stars Above" ($69 to $89), designed to show the actual sky at any time of any night of the year by matching the date and time on the rims of the two 14-inch hemispheres. Constellations and stars of first and second magnitude are shown, along with the Milky Way and the path of the planets. It comes on a clear base with an eight-page booklet and wax marking pencil, available in either blue or gold, illuminated or not.

■ "Starship Earth," a fascinating clear celestial globe with a small inner globe of the Earth and a "sun ball." With the "Starship," it is possible to see planetary movements and deep-sky objects, and to trace the movements of the sun. The globe comes with a "Certificate of Authenticity," a wax marking pencil, and a 36-page booklet written by astronomical columnist George Lovi. The globe is available in a 16-inch diameter ($249 to $335, depending on base) or 30-inch diameter ($1,250).

■ "S.C.I. Celestial Globe," a six-color globe showing most naked-eye stars in their true colors (temperatures) and magnitudes (bright-

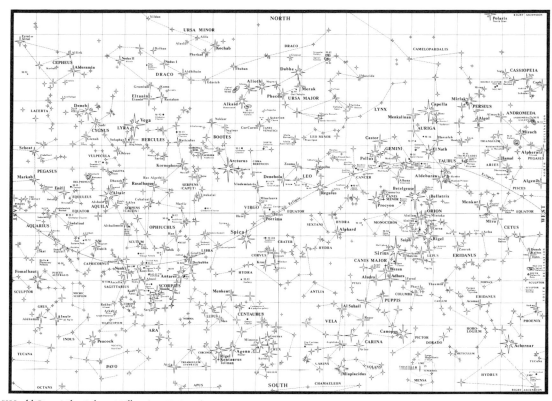

"*World Star Atlas*," *from Millennium Enterprises.*

ness), and constellations, star cluster, nebulae, and the like. The globe comes with an instruction booklet written by George Lovi ($289 to $325, depending on base).

Star Finders Inc. (2406 Lawrence St., Eugene, OR 97405; 503-686-6754) produces a glow-in-the-dark star chart ($9; 22" X 28") of the northern sky for learning the stars indoors. It shows all the constellations of the Northern Hemisphere. Star Finders also produces a brass-and-vinyl planisphere ($12) for learning the same stars and constellations outdoors. The planisphere comes with a red plastic filter for use with a flashlight and a primer to stargazing.

Sunstone Publications (R.D. 3, Box 100A, Cooperstown, NY 13326; 607-547-8207) offers ASTRO-DOME R ($9.95), a three-dimensional map of the night sky developed by schoolteacher Klaus Hunig. With easy-to-follow directions, the precut pages transform into an

attractive miniature planetarium depicting the stars and constellations of the Northern Hemisphere. ASTRO-DOME measures 20 inches in diameter when assembled and includes a 24-page *Constellation Handbook* explaining the history and details of the major constellations. The map glows in the dark and is intended for children ages 12 and up.

Willmann-Bell Inc. (P.O. Box 35025, Richmond, VA 23235; 804-320-7016) is one of the largest retail book dealers in astronomy-related materials. It carries 1,500 titles, many of which are also carried by Sky Publishing and Kalmbach Publishing. The catalog costs $1. The company's own star chart, *Uranometria* 2000.0, by Will Tirion, Barry Rappaport, and George Lovi ($34.95), charts the heavens in two volumes. Volume 1 shows the Northern Hemisphere to –6 degrees. Volume 2 shows the Southern Hemisphere to +6 degrees. Each volume contains 259 charts.

Weather Maps

Weather has always been a provocative and mystical element of the natural world. From the songs and poems it has inspired to the conversations it has started (or saved), our romance with blue skies and blustery winds is as old as weather itself. Because weather is often unpredictable and erratic, the maps, charts, and tables that trace its patterns and behavior can be lifesaving tools for pilots, sailors, and others.

Weather maps, by their very nature, are usually out of date within a few days or even hours of their creation. There are exceptions—including historical climatology maps, general tracking guides, do-it-yourself prediction kits, and National Weather Service charts—but most specific weather maps are useful only for an extremely limited period of time.

The federal government produces a number of weather maps, charts, atlases, and guides aimed at educating and informing the public. The **National Weather Service** (part of the Commerce Department's National Oceanic and Atmospheric Administration), which provides the information for the telephone recordings and radio and TV station forecasts upon which many people rely, also issues daily weather charts and maps.

The Weather Service's daily maps (available by subscription from NOAA and the **Government Printing Office**) show the surface weather of the United States as it is observed at 7 a.m. Eastern time each morning. A weekly series, also available by subscription, includes each day's maps for the seven-day period, plus smaller maps showing high and low temperatures, precipitation, and wind patterns. These maps are widely used by airplane pilots and boat captains.

Another set of useful weather maps produced by the Weather Service are the 15 Marine Weather Services Charts covering the waters of the United States and Puerto Rico, which illustrate the locations of visual storm-warning display sites, and of weather radio stations broadcasting Weather Service information.

Other weather maps and charts produced by the federal government include the historical climatology series of maps and atlases published by the **National Climatic Data Center**, the sunshine temperature maps that are included in the **U.S. Geological Survey's** National Atlas Program, and a variety of climatic, storm, and weather-pattern data and maps available from the Government Printing Office.

In addition to these printed materials, there are several new technologically advanced methods of getting up-to-date weather maps. For example, through on-line weather services, you may tap into data bases via a modem.

GOVERNMENT SOURCES

Government Printing Office. The GPO sells a variety of weather-related publications. A sampling is:

■ "Cloud Code Chart" ($2.25; S/N 003-018-00050-4), created in 1972 and reprinted in 1980, includes a picture of each cloud type, the name, a brief description, and the cloud code figure according to the International System of Cloud Classification.
■ "Daily Weather Maps Weekly Series" (subscription prices: $60 per year in the U.S., $75 per year elsewhere; single-copy prices: $1.50 in the U.S., $1.88 elsewhere; S/N 703-021-00000-0) offers daily weather maps created by the National Weather Service. Explanations of the maps and symbols are included.
■ "Explanation of the Daily Weather Map" ($2; S/N 003-017-00505-4), created in 1982, explains symbols and other details of daily weather maps.

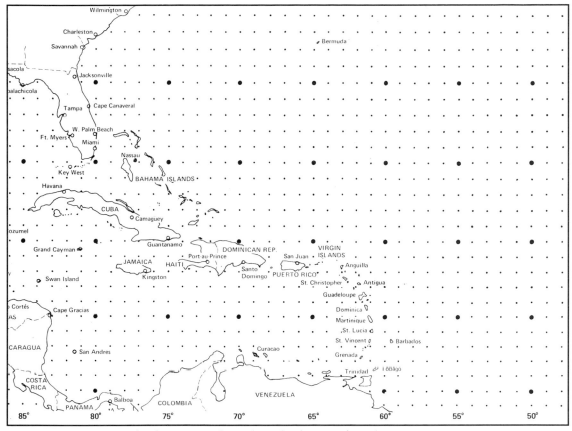

From NOAA's *"Storm Search and Hurricane Safety with North Atlantic Tracking Chart."*

■ "World Weather Extremes" ($4.50; S/N 008-022-00230-8), created in 1985, provides maps of weather extremes on a worldwide basis and for the U.S. and Canada.

■ *Climatological Atlas of the World Ocean* ($11; S/N 003-017-00509-7). This 189-page atlas has maps, charts, graphs, and tables that synthesize National Ocean Survey data about the temperature and other climatological information.

■ *United States Navy Hindcast Pectral Ocean Wave Model Climatic Atlas: North Atlantic Ocean* ($39; S/N 008-042-00074-8). This 393-page atlas, created in 1983, is filled with maps, charts, and other historical data about the wind and wave climatology of the North Atlantic Ocean.

National Climatic Data Center. The National Climatic Data Center (Federal Bldg., Asheville, NC 28801; 704-259-0682) part of the **National**

Oceanic and Atmospheric Administration, has a Historical Climatology series that includes 11 weather-related atlases. The maps in the atlases depict local, regional, or continental climatological parameters, showing climate changes over a relatively long time. A service charge of $5 per order must be included when purchasing the atlases by mail (a $7 minimum order, in addition to the $5 service charge, is required). Available atlases include:

■ *Atlas of Mean Winter Temperature Departures From the Long-Term Mean over the Contiguous U.S., 1895-1983* ($3; Series Title 3-1), with seasonal maps of departures from mean winter temperatures in the U.S. from the late 19th to late 20th centuries.

(continued on page 237)

How to Read a Weather Map

All weather maps use common symbols. Here is a description of the symbols used to represent different weather phenomena:

■ **Sky conditions** are indicated by small circles (representing key cities) that are completely filled in (overcast skies), partially filled in (partly cloudy conditions), or empty (clear conditions). If it is raining or snowing, the circle may have an R or an S inside. On professional maps, symbols are used—jagged lines for lightning and snowflakes for snow, for example.

■ **Wind** is represented by a line drawn from a sky-condition circle. To determine wind direction, assume that the top of the circle is north, the bottom south, the right east, and the left west. The line is drawn in the direction from which the wind originates. Barbs representing 10 miles per hour each are drawn horizontally from the wind direction line to indicate wind speed. So a 30-mph wind would have three barbs. Some maps simply use arrows to indicate wind direction. The arrows point in the direction the wind is blowing.

■ **Temperature** is usually written directly on the weather map in degrees Fahrenheit, although Celsius may also be used.

■ **Barometric pressure** is represented by isobars — solid, curved black lines. High-pressure weather systems (clear weather) are drawn as ascending isobars. Low-pressure weather systems (cloudy weather) are drawn as descending isobars.

■ **Fronts** can be shown in three ways. A *cold front*, a leading edge of a moving mass of cold air, is drawn as a solid black line with small triangles. The points of the triangles face the direction in which the front is moving. A *warm front*, the leading edge of an advancing mass of warm air, is drawn as a solid black line with solid black semicircles. A *stationary front*, the separation of two air masses, neither of which is moving, is drawn by a solid black line that has alternate triangles and semicircles.

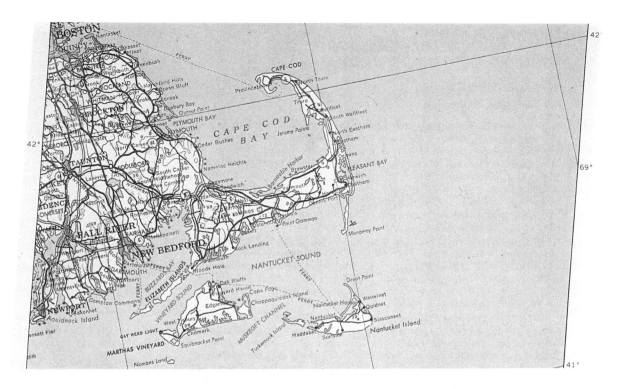

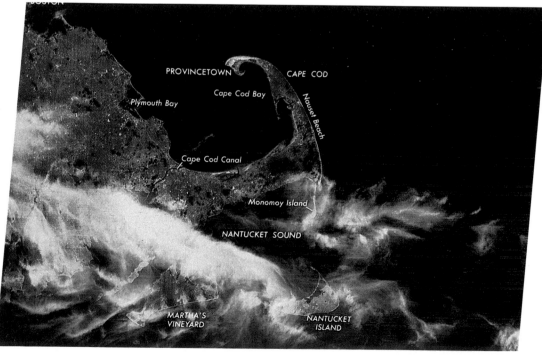

Two views of Cape Cod, Massachusetts. Top: Part of the Boston sheet of the Defense Mapping Agency's "International Map of the World." Bottom: A Landsat image covering the same region. Courtesy U.S. Geological Survey.

A polar projection of the North Pole from the Nimbus-5 weather satellite. Colors indicate variations in microwave radiometric temperature.

"Map of the Universe," compiled and designed by Tomas J. Filsinger, showing astronomical and astrological phenomena. *The Milky Way and other highlights glow in the dark on this poster-sized map. ©1981. Used with permission of Celestial Arts, P.O. Box 7327, Berkeley, CA 94707.*

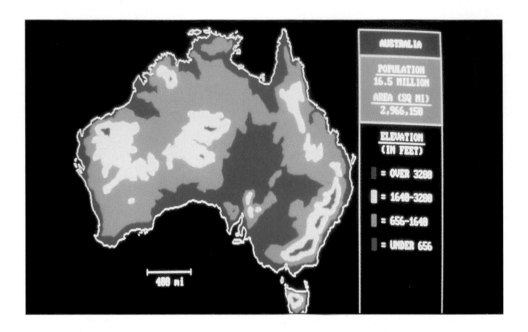

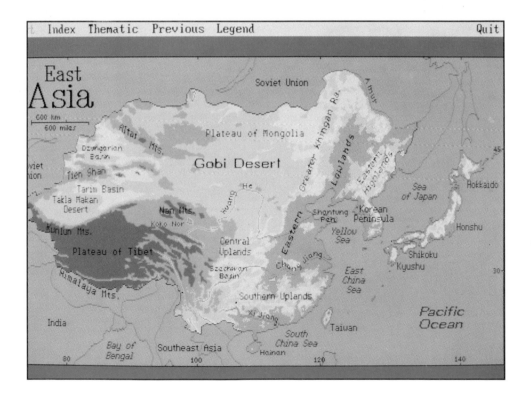

Two samples of geography education software. Top: *Screen from PC Globe,* courtesy Comwell Systems Inc. Bottom: Electromap World Atlas, *courtesy Electromap, Inc.*

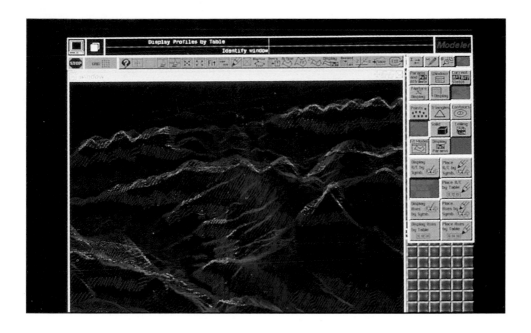

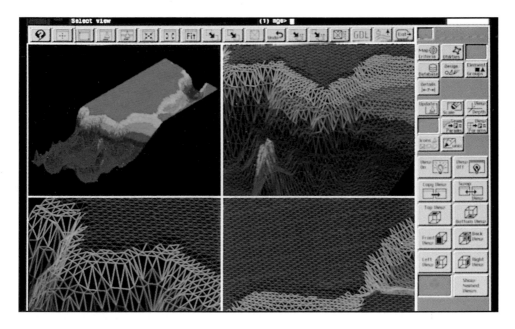

Two maps created with Intergraph Corporation's MicroStation computerized map-making system. Top: Color-coded surface mesh showing the Blue Ridge Mountains, created with TIGRIS Modeler, using a dataset from the U.S. Geological Survey. Bottom: A bathymetric model of the Straits of Gibraltar.

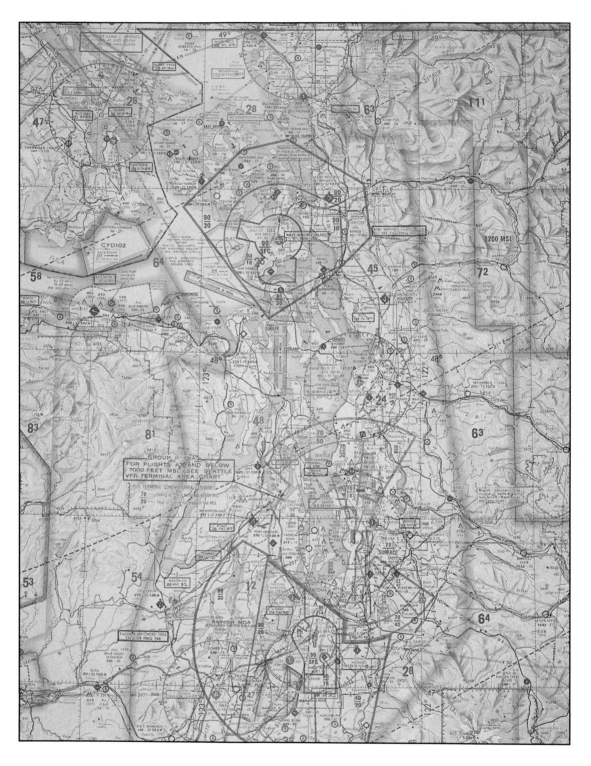

Aeronautical chart of western Washington, including Seattle, the San Juan Islands, and part of Vancouver, B.C.

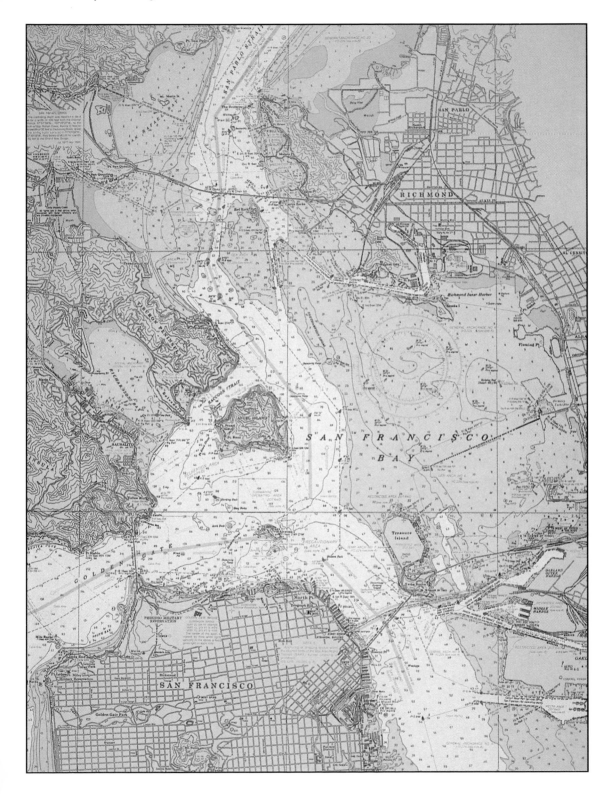

Nautical chart of San Francisco Bay, showing depth markings and other navigation information for sailors.

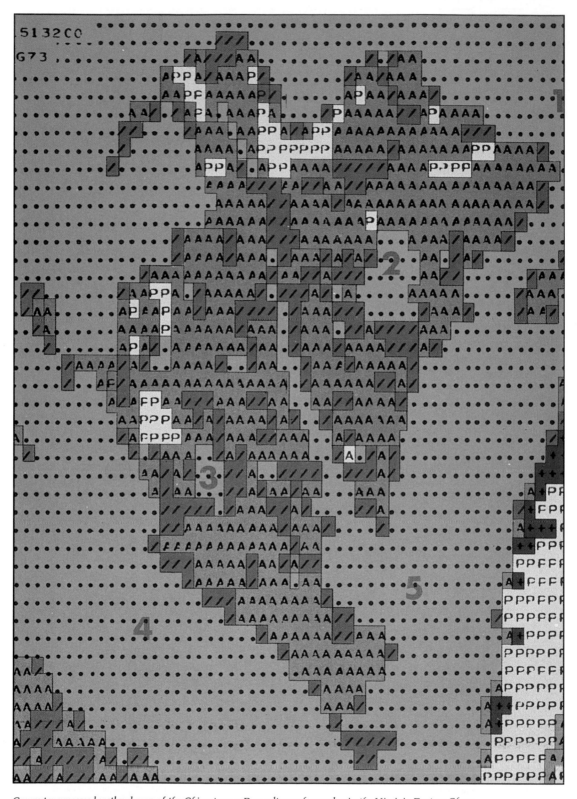

Computer-processed wetland map of the Chincoteague Bay salt-marsh complex in the Virginia Eastern Shore.

(continued from page 227)

■ *Atlas of Monthly and Seasonal Temperature Departures*, 1895-1983 (four volumes, $10 each or $36 for all four; Series Titles: 3-2 [Winter]; 3-3 [Spring]; 3-4 [Summer]; and 3-5 [Fall]). This series includes maps of departures from long-term, statewide, average monthly and seasonal temperatures.

■ *Atlas of Monthly Palmer Hydrological Drought Indices* (1895-1930) *for the Contiguous United States* ($12; Series Title 3-6), with maps showing drought conditions in the U.S. according to the Palmer Hydrological Drought Index, an objective measure of moisture conditions. A companion volume, for years 1931-1983 ($17; Series title 3-7), is also available.

For a complete list of atlases and other products in the Historical Climatology Series, write to the NCDC and ask for the *Environmental Information Summary* (C-24).

The NCDC also produces "Annual Average Climatic Maps of the United States" series, including "Temperature, 1941-70," "Total Precipitation, 1941-70," "Relative Humidity, generally 1948-77," "Dew Point Temperature, generally 1946-65," "Annual Average Daily Global Solar Radiation on a South Facing Surface, generally 1953-75," "Mean Annual Number of Thunderstorms, generally 1948-77," and "Normal Annual Heating Degree Day (contiguous U.S. and Alaska), generally 1951-80." The climatic maps are free from NCDC.

National Oceanic and Atmospheric Administration. NOAA publishes "Storm Search and Hurricane Safety with North Atlantic Tracking Chart" (NOAA/PA 78019), a brochure with chart included. It is free from the NOAA Logistics Supply Center (1500 E. Bannister Rd., Bldg. 1, Kansas City, MO 64131; 816-926-7993; 800-669-1099).

U.S. Geological Survey. USGS's National Atlas Program includes several weather-related maps. The maps are colorful and suitable for framing, albeit a bit out of date. Prices shown include surface mailing within the U.S. The following maps are available:

■ "Monthly Sunshine" ($3.10; 00478, Ref. Code 38077-AI-NA-17M-00), created in 1965.
■ "Annual Sunshine, Evaporation, and Solar Radiation" ($3.10; 00565, Ref. Code 38077-AJ-NA-17M-00), created in 1969.
■ "Monthly Average Temperature" ($3.10; 00661, Ref. Code AK-NA-17M-00), created in 1965.
■ "Monthly Minimum Temperatures" ($3.10; 00662-38077-AL-NA-17M-00), created in 1965.

GOVERNMENT SOURCES, CANADA

Canada Map Office (615 Booth St., Ottawa, Ontario K1A 0E9; 613-952-7000) distributes 11 weather-related maps as part of the *National Atlas of Canada*, 5th Edition series: "Annual Solar Radiation" (MCR 4076), "Length of Day" (MCR 4068), "Last Frost in Spring" (MCR 4035), "Growing-Degree-Days" (MCR 4034), and "Temperature—January and July" (MCR 4058). The maps are $5.50 Canadian each.

Maps from the *National Atlas of Canada*, 4th Edition (1973) are available for $2 Canadian each. Titles include: "Soil Climate" (MCR 1132), "Weather Stations and Average Annual Precipitation" (MCR 1134), "Frost" (MCR 1137), "Precipitation-Regions" (MCR 1139), "Precipitation-Monthly Average, April-Sept." (MCR 1140), "Precipitation-Seasonal Average" (MCR 1141), "Temperature-Winter" (MCR 1142), "Temperature-Spring" (MCR 1143), "Temperature-Summer" (MCR 1144), "Temperature-Autumn" (MCR 1145).

Maps Alberta (Land Information Services Div., Main Fl., Brittania Bldg., 703 6th Ave. SW, Calgary, Alberta T2P 0T9; 403-297-7389) distributes various weather maps of the province, which are part of the *Alberta Resource and Economic Atlas*. Part of the "Physical Features" series of the atlas, the weather maps (75 cents Canadian each) are drawn at a scale of 1:5,000,000 and include: "Sunshine" (D10),

"Precipitation" (D11), "Snowfall" (D12), "Mean Temperature" (D13), and "Average Frost Free Days" (D14).

COMMERCIAL SOURCES

American Association of Weather Observers (401 Whitney Blvd., Belvidere, IL 61008; 815-544-9811) founded in 1984, acts as a clearinghouse for people interested in pooling weather data. Information submitted by members is published monthly in the organization's newsletter, *American Weather Observer*. It also becomes part of the organization's permanent data base, available to members. The newsletter includes other information close to weather watchers' hearts: new tools of the trade, reports from other organizations, and a weather map based on members' data, among other things. The group has about 1,500 members, who pay $12 a year ($18 for those not members of AAWO-affiliated groups), which includes membership and a dozen issues of the *Observer*.

American Weather Enterprises (P.O. Box 1383, Media, PA 19063; 215-565-1232) sells books, software, and educational materials, in addition to weather instruments. Some choice items from the catalog include a sturdy wall chart (55" X 33") that details global atmospheric influences, meteorological motion, severe weather, clouds and precipitation, examples of pollution, aviation meteorology, and more in an easy-to-understand flow-chart design ($16.50) and Geochron, an attractive chalkboard-sized electronic map that continually shows the exact time of day and amount of sunlight anywhere in the world (basic unit is $1,200; deluxe models up to $2,000).

Blue Hill Observatory (P.O. Box 101, East Milton, MA 02186; 617-698-5397) sells a hurricane tracking chart that includes two 11" X 24" maps and data on hurricanes covering the Atlantic, Gulf, and Pacific coasts. Included are instructions on how to plot a hurricane's course using longitude and latitude, a hurricane-

survival checklist and a list of current hurricane names. The observatory offers weather-related publications, including a quarterly bulletin, available to all members; membership is $10 a year. Write for a free brochure.

Geoscience Resources (2990 Anthony Rd., P.O. Box 2096, Burlington, NC 27216; 919-227-8300; 800-742-2677) distributes various weather-related maps of foreign countries, such as the "China—Annual Rainfall Map" ($10.95).

Hammond Inc. (515 Valley St., Maplewood, NJ 07040; 201-763-6000; 800-526-4953). Of special interest to weather watchers is the Hammond Weather Kit, complete with chart and wheel. The 25" X 28" weather chart offers graphic explanations of popular weather terms, including cloud types, weather fronts, tornado formation, and more. The weather wheel helps amateurs create their own forecasts.

Hubbard Scientific (P.O. Box 104, Northbrook, IL 60065; 708-272-7810; 800-323-8368) distributes a 44" X 50" fluorescent-green weather map and plotting chart of the U.S. The map and chart are laminated, so weather conditions can be written on and wiped off later ($30).

New Orleans Map Co. (3130 Paris Ave., New Orleans, LA 70119; 504-943-0878) distributes a 29" X 35" "Hurricane Tracking Chart" that covers the Gulf of Mexico, the Yucatan Peninsula, the northern coast of South America, the West Indies, and the Atlantic coast of the U.S. north to Baltimore ($10 plus tax and shipping). The map is laminated in flexible plastic, so hurricanes can be tracked with a grease pen; when the hurricane is over, the markings can be wiped off and the map rolled up and stored away until the next one.

COMPUTER SOFTWARE AND ON-LINE SERVICES

The **National Weather Service** offers various data lines that can be subscribed to for up-to-

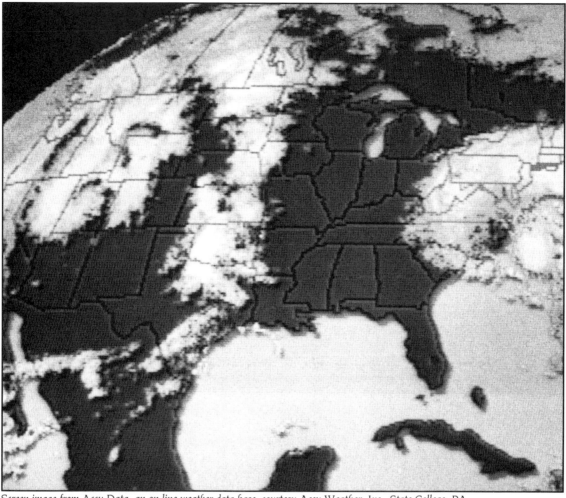

Screen image from Accu-Data, an on-line weather data base, courtesy Accu-Weather, Inc., State College, PA.

the-minute weather information. The problem is that most of this data is in coded form. You can't really read it without computer software and hardware to help you decipher what is being sent. A number of companies offer software packages to help translate the data. Other companies offer menu-driven systems that are not dependent on NWS data. Each of these systems has its own set of strengths and limitations.

Accu-Data (Accu-Weather Inc., 619 W. College Ave., State College, PA 16801; 814-237-0309) is a complete, real-time weather data base that interfaces with most terminals, personal computers, and business computers. Available data from Accu-Data include hourly surface observations; digitized radar signals; upper-air data; worldwide observations; DIFAX, LFM/NGM, MOS, and other modal output; plain-language and coded forecasts; severe-weather bulletins; watches and warnings; climatological summaries; and marine data. Data are transmitted as both charts and maps. The data base is compatible with both IBM and Apple computers. There are three different prices for Accu-Data. Commercial users pay $69 an hour and secondary schools $42 an hour, both with access to a toll-free "800" number. Hobbyists pay $16.95 an hour, but must also pay long-distance charges. Prices are for 1200 baud.)

National Geographic Society (17th & M Sts. NW, Washington, DC 20036; 202-857-7000, 800-368-2728) distributes the "Weather Machine." The Weather Machine allows students to select a packet of weather data culled from the National Weather Service's Domestic Data Service (held in a National Geographic computer) which are translated into color computer maps superimposed over a map of North America. Readings include temperature, air pressure, wind speed and direction, dew point, cloud cover, and more. The Weather Machine can be used on the Apple IIc, IIe, or IIgs; a modem is needed if users plan to tap into the daily updates offered. A detailed curriculum is part of the package.

Weathervane Software (P.O. Box 277, Grafton, MA 01519; 508-839-6777) is the publisher of *Atlantic Hurricane Watcher* ($30), a menu-driven program that allows you to plot and compare the tracks of hurricanes on a color or black-and-white map of the U.S., western Atlantic Ocean, Caribbean, and Gulf of Mexico. New storms can be included easily in the 1985 data base. IBM-PC compatible.

WEATHER BY FAX

It's not generally known, but you can use a short-wave radio to pick up much more than just voice transmissions or Morse code. For example, some weather stations transmit in a facsimile, or fax, mode, which means that they send photos and maps over short-wave radio bands. These are picked up by news and weather services around the country, but anyone with a little know-how can tap into the system. For some short-wave radios, you won't need a special antenna. Two sources are:

Stephens Engineering Associates Inc. (7030 220th St. SW, Mountlake Terrace, WA 98043; 206-771-2182) offers the SEAFAX weather facsimile digital signal processor, with high-resolution printouts. The system must be attached to a single-side band radio receiver and a printer. Price is $995, or $1,145 with IBM-compatible serial output.

Universal Amateur Radio Inc. (1280 Aida Dr., Reynoldsburg, OH 43068; 614-866-4267). If you'd like to tap into the weather map scene, take a look at the Information-Tech M-800 ($349), which can be hooked up to a short-wave radio and an Epson printer for a crisp, clear map. A variety of other electronic gear is also offered in the Universal catalog.

■ **See also: "Aeronautical Charts," "Emergency Information Maps," "Energy Maps," "Nautical Charts and Maps," and "Tide and Current Maps."**

IMAGES AS MAPS

Aerial Photographs

Ever since humans have been able to fly and hold a camera at the same time, aerial photographs have been an important part of the mapping process, as well as a means of preserving images of the Earth and its development. Images of the land taken from above are valuable tools, suitable for use as finished products, but they are also the starting points for creating most types of land maps. Cartographers often use stereoscopic aerial photos—two aerial photos of the same site taken from two different aerial camera positions—to create the three-dimensional perspective needed for some map-making. Aerial photos are also used in such fields as aeronautics, agriculture, engineering, and land planning.

Cameras in airplanes photograph land section by section. The resulting photographs must first be corrected to eliminate distortion, camera-tilt, and optical effects created by the land itself. Through a series of enhancements or overlays, the photos are then transformed into a variety of map types:

■ Cartographers often use aerial photographs to create or update **planimetric** or **topographic** maps, for example.
■ The addition of mapping symbols (such as major roads, borders, and selected place names) to photographs results in **photoimage** maps, also known as **pictomaps**.
■ **Orthophotoquads** and **orthophotomaps** result from further enhancements, such as eliminating most distortion, adding color and enhanced definition of coastlines, rivers, mountains, and other land forms.

GOVERNMENT SOURCES, UNITED STATES

Nearly every federal agency involved with land preservation, planning, or management maintains collections of aerial photographs that span the nation from picture-perfect coast to picture-perfect coast (see Appendix B for addresses).

The **U.S. Geological Survey's** Aerial Photography Summary Record System, established in 1976 as the first Earth Science Information Center data base, catalogs the in-progress and completed aerial photographs for the U.S. to assist in locating a desired photograph. The USGS, through its ESIC offices, can sell prints from photography from a number of federal agencies, such as the Bureau of Reclamation, the U.S. Navy, and the Environmental Protection Agency. A pamphlet, *How to Obtain Aerial Photographs*, available from USGS, may be helpful in obtaining aerial photos.

The **National Archives** has a large stockpile of older aerial photographs of the U.S. The Archives also maintains a large collection of World War II-vintage allied-flown aerial photography of foreign areas, as well as World War II-era Luftwaffe-flown aerial prints covering much of Europe, the United Kingdom, and North Africa. These may be located through queries to Archives staff (see "Uncle Sam's Treasure Chests").

The **Earth Resources Observation Systems (EROS) Data Center**, near Sioux Falls, S.D., is the depository of the Department of Interior's aerial photography, radar imagery, and other federal-government-produced aerial photographs, all of which are available for sale (see "Space Imagery").

The **U.S. Department of Agriculture** Aerial Photography Field Office in Salt Lake City, part of the Agricultural Stabilization and Conservation Service, has thousands of aerial photographs covering most of the nation's major

Aerial photograph of the rapidly growing Tysons Corner, Virginia area. Courtesy Air Photographics Inc., Wheaton, Maryland.

cropland. USDA's National Forest Service has aerial photographs of the nation's forests.

The **Department of the Interior's** Bureau of Land Management has aerial photographs of most federally maintained land, through its offices or through the EROS Data Center.

The **Defense Mapping Agency** has aerial photographs of military installations and other defense-related areas around the world.

The **Department of Commerce's** National Ocean Service has aerial photographs of the nation's coastline.

The **Tennessee Valley Authority** has aerial photography covering the entire Tennessee River Drainage Basin with varying dates and scales.

GOVERNMENT SOURCES, CANADA

Manitoba Natural Resources (Surveys and Mapping Branch, 1007 Century St., Winnipeg, Manitoba R3H 0W4; 204-945-6666) maintains the Manitoba Air Photo Library, a collection of more than 1 million aerial photographs, some dating to 1923. Copies and enlargements of any photo in the collection can be ordered. There are photos for every area of Manitoba, although coverage is more extensive for some areas; it's best to visit, phone, or write the Air Photo Library before placing an order.

Maps Alberta (Land Information Services Div., Main Fl., Brittania Bldg., 703 6th Ave. SW, Calgary, Alberta T2P 0T9; 403-297-7389) maintains all the aerial photographs taken for the provincial government dating to 1949. Most of the photos are black and white, although some are in color. Contact prints, laser prints, color-contact prints, and enlargements of the aerial photos can be requested. "Flight Index Maps" and an "Aerial Photography Project Index," available from Maps Alberta, are handy resources when searching for aerial photographs of a specific area. A catalog explaining Maps Alberta's aerial photograph distribution is available upon request.

Quebec Ministère de l'Énergie et des Ressources (Photocartothèque Québecoise, 1995 Blvd. Charest Ouest, Sainte-Foy, Quebec G1N 4H9; 418-643-7704) produces aerial photographs for areas in Quebec in scales of 1:40,000, 1:36,000, 1:31,680, 1:30,000, 1:25,000, 1:15,840, 1:15,000, and 1:10,000. An index of available photographs, "Répértoire des Cartes, Plans et Photographies Aériennes," is free.

Saskatchewan Property Management Corp. (Central Survey and Mapping Agency, Distribution Center, 2045 Broad St., 1st Fl., Regina, Saskatchewan S4P 3V7; 306-787-6911) produces "Township Photomaps " at a scale of about 1:20,000, from aerial photography rectified to theoretical township corners.

Photomaps ($6 Canadian per township) are available for all areas of Saskatchewan.

COMMERCIAL SOURCES

Aerial maps are also available from the following sources and reference libraries:

Alabama
Atlantic Aerial Surveys Inc.
803 Franklin St.
Huntsville, AL 35804
205-722-0555

Alaska
Aero Map U.S.
2014 Merrill Field Dr.
Anchorage, AK 99501
907-563-3038

Arizona
Color Aerial Photography
1135 W. Las Palmas Dr.
Tucson, AZ 85704
602-887-0444

Kenny Aerial Mapping Inc.
1130 W. Fillmore St.
Phoenix, AZ 85007
602-258-6471

Arkansas
AMI Engineers
1615 Louisiana St.
Little Rock, AR 72206
501-376-6838

California
Ace Aerial Photo
P.O. Box 2040
Laguna Hills, CA 92654
714-830-2960

Aerial Map Industries
2012 S. Main St.
Santa Ana, CA 92702
714-546-7201

Foster Air Photo
9312 Florence Ln.
Garden Grove, CA
714-539-3890

Aerial photograph of Boston, courtesy USGS/EROS Data Center.

Western Economic Research Co. Inc.
8155 Van Nuys Blvd., Ste. 100
Panorama City, CA 91402
818-787-6277

Colorado
Intra Search Inc.
5351 S. Roslyn St.
Englewood, CO 80111
303-741-2020

Florida
Aerial Cartographics of America
7000 Lake Eleanor Dr.
Orlando, FL 32809
407-851-7880; 800-426-0407

Georgia
Georgia Aerial Surveys
4451 N. Log Cabin Dr., Ste. 124
Smyrna, GA 30080
404-434-2516

Hawaii
Air Survey Hawaii
677 Ala Moana St., Ste. 401
Honolulu, HI 96813
808-833-4881

Idaho
Moore's Photography
7128 Snohomish
Boise, ID 83709
208-362-3820

Valley Air Photos
4601 Aviation Way
Caldwell, ID 83606
208-454-1344

Illinois
Chicago Aerial Survey
2140 S. Wolf Rd.
Des Plaines, IL 60018
312-298-1480

Iowa
Aerial Services Inc.
2120 Center St.
Cedar Falls, IA 50615
319-266-6181

Kentucky
GRW Aerial Survey Inc.
801 Corporate Dr.
Lexington, KY 40503
606-223-3999

Park Aerial Surveys
606 Harding Ave.
Louisville, KY 40217
502-366-4571

Maryland
Aeroeco Inc.
232 Cardamon Dr.
Edgewater, MD 21037
301-261-8448

Air Photographics Inc.
11510 Georgia Ave., Ste. 130
Wheaton, MD 20902
301-933-5282

Massachusetts
Avis Airmap Inc.
454 Washington St.
Braintree, MA 02184
617-848-8000

Michigan
Abrams Aerial Survey Corp.
P.O. Box 15008
Lansing, MI 48901
517-372-8100

Nevada
Cooper Aerial of Nevada Inc.
3750 S. Valley View Dr.
Las Vegas, NV 89103
702-362-4776

New Jersey
Aerial Data Reduction Associates Inc.
P.O. Box 557
Pennsauken, NJ 08110
609-663-7200

New York
Aerial Cartographics of America
100 W. Main St.
Babylon, NY 11702
800-426-0407

Aerographics Inc.
P.O. Box 248
Bohemia, NY 11716
516-589-6045

Ohio
Cleveland Blueprint
1815 St. Clair Ave.
Cleveland, OH 44114
216-241-1815

Oregon
Insight Reconnaisance
525 SE Goodnight Ln.
Corvallis, OR 97333
503-754-6488

Pennsylvania
Eastern Mapping Co.
250 Freeport Rd.
Pittsburgh, PA 15238
412-828-3410

Texas
International Aerial Mapping Co.
8927 International Dr.
San Antonio, TX 78205
512-826-8681

Adams Aerial Surveys
P.O. Box 476
South Houston, TX 77587
713-946-0830

Virginia
Air Survey Corp.
460 Spring Park Pl., Ste. 1500
Herndon, VA 22070
703-437-6060

Space Imagery

Modern science has created new tools that have not only revolutionized techniques for making maps, but have also given rise to new types of map products. Among the most important of these tools is "remote-sensing," the process of detecting and monitoring chemical or physical properties of an area by measuring its reflected and emitted radiation.

The earliest remote-sensing experiments took place using photographs aboard the first manned Earth-orbiting missions: Mercury, Gemini, and Apollo. Astronauts used hand-held cameras to produce historic pictures that were examined by scientists worldwide, although no actual maps were made from the pictures.

In 1969, the first scientific space photographic experiment was performed on Apollo 9. Four 70-mm cameras were mounted on a metal frame that fit the spacecraft's command module hatch window. Taking a variety of pictures of the Phoenix, Arizona region, scientists later interconnected them—a process called "mosaicking"—and printed them as a single image to produce a standard line map. Although crude by today's standards, the experiment led the way for what would become highly sophisticated photo-mapping techniques using manned and unmanned spacecraft.

Landsat I (originally called the Earth Resources Technology Satellite, or ERTS), launched in 1972, was the first American spacecraft designed specifically to record images of the Earth. Since then, subsequent Landsat satellites have recorded thousands of images of the planet. At an altitude of 567 miles, each satellite circles the globe 14 times daily, scanning a particular scene every 18 days, or more than 40 times per year. Each image covers an area approximately 115 miles square, and Landsat can detect an image as small as 100 feet square. Since 1972, images from the five Landsat satellites have become valuable tools for farmers, oil companies, geologists, foresters, foreign governments, and others interested in land resource management. In 1985, ownership of Landsat was transferred by the federal government to the Earth Observation Satellite Co. (EOSAT), a partnership of Hughes Aircraft Co. and RCA Corp.

Although Landsat produces pictures of Earth, its principal viewing instruments are not cameras but two digital-sensor systems known as multispectral scanner systems. Digital cameras have several advantages over other cameras in terms of signal-to-noise ratio, weight, reliability, and simplicity of operation, and so have been flown on almost all planetary probes. Another advantage is that their digital information can be manipulated in a computer, allowing enhancement or suppression of images.

In multispectral scanner systems, satellites record information in two visible (red and green) wavelengths and two infrared wavelengths not visible to the human eye. The procedure results in four separate black-and-white digital images, combined by computer into a "false-color" portrait. In thematic mapper systems, satellites record information in three visible (blue, green, red) wavelengths, three infrared wavelengths, and one thermal (heat measuring) wavelength. Two of the visible and two of the infrared wavelengths are combined to make standard "false-color" images. Different colors describe the land below:

■ Healthy vegetation appears in shades of red and contrasts with unhealthy vegetation, which appears as blue-green.
■ Water appears as dark blue or black unless it is sediment laden, in which case it takes on a light-blue tone.
■ Most buildings, streets, and other "cultural features" appear as a steely blue-gray.

Another, newer space-imagery vehicle is

"Spot," a French-owned satellite launched in 1986, the first commercially owned satellite-sensing system. Spot's sensing abilities enable it to record images as small as 10 meters (about 33 feet) square—about half the size of a tennis court—and at a higher resolution than Landsat. Airplanes, bridges, ships, roads, even some houses can be detected by Spot with impressive clarity. One reason is that Spot uses a system of mirrors that can "look" to the side as well as straight down, enabling the satellite to "view" an object from two or more directions. Among other things, this allows for production of stereoscopic images that create a three-dimensional perspective of land images. Still, Landsat can produce bigger pictures than Spot, and in more wavelengths of light.

The uses of space imagery are widespread and growing. There is the sheer beauty of space maps, of course—some have been likened to French impressionist paintings—but there are myriad practical uses. Space imagery, for example, has been used to:

■ Analyze geologic structures, from earth-quake faults to mountain ranges, and to detect oil and mineral deposits.
■ Manage water resources and improve stream-flow characteristics and water quality in lakes, rivers, and bays.
■ Map, measure, and analyze glaciers and ice caps to predict reservoir inflow and potential flooding.
■ Create thematic maps of forested and cultivated land and coastal waterways.
■ Produce computer-aided mapping and land-use analysis.
■ Monitor forest fires, air pollution, oil-well fires, and changes in vegetation.
■ Analyze archaeological sites, detect road alignment, and monitor wildlife migration.

Ordering Space Imagery. The EROS Data Center (EDC), near Sioux Falls, South Dakota, operated by the U.S. Geological Survey, reproduces and sells copies of manned spacecraft photographs as well as aerial photographs (see

"Aerial Photographs"), NASA aircraft data, and other remote-sensing products. EDC reproduces and sells copies of imagery products of the USGS as well as photographs, geophysical data, and computer products collected by 16 different federal agencies.

Landsat Images. Landsat images are available for the 50 states and for most of the Earth's land surface outside the U.S. There are several products available for any given location, including:

■ Single black-and-white images, available as film negatives, film positives, or paper prints.
■ Complete sets of four black-and-white images and a false-color composite. All false-color composites are available as film positives or paper prints.
■ Computer tapes containing digital data.

Each Landsat image covers about 8 million acres. Images do not reveal outlines of small areas, like houses or small towns and villages, but provide views of broad areas and large features, such as mountain ranges and the outlines of major cities.

Landsat images are available through EOSAT offices located at the EROS Data Center, where orders are processed. When ordering, it is important to describe the exact area in which you are interested, including, if possible, the geographic coordinates or a map marked with the specific area. You should also indicate:

■ the type of product (black-and-white, false-color, or digital tapes);
■ the minimum image quality acceptable;
■ the maximum percent of acceptable cloud cover (10 percent to 90 percent); and
■ the preferred time of year.

A useful brochure, "Landsat Products and Services" is available from the EROS Data Center. This brochure includes a price list, order form, inquiry form, reference aids, and other information. To order specific images of

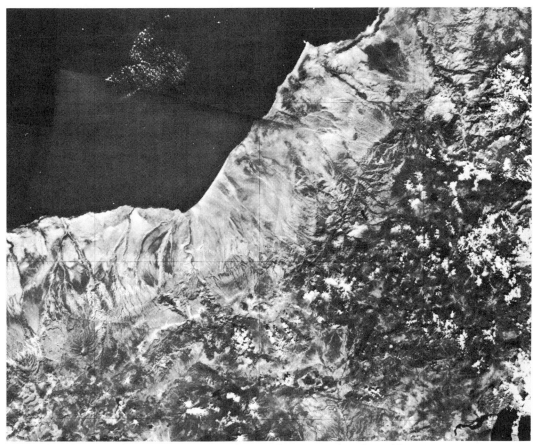

Portion of Landsat image showing parts of western Bolivia, southern Peru, and northern Chile.

the 48 contiguous United States, you can use the form *Selected Landsat Coverage* (NOAA Form 34-1205, available free from EROS and USGS Earth Science Information Centers), which includes a map of the U.S. showing the locations of individual Landsat images selected for their clarity and lack of cloud cover. USGS has also prepared a number of photomaps, mosaics, and other images produced from Landsat, SPOT, and other imagery, including an impressive view of the entire U.S. Prices for most of these range from $2.50 to $6, not including postage and handling charges for mail orders.

Ordering Spot Images. Spot's images may be obtained directly from Spot Image Corp. (1897 Preston White Dr., Reston, VA 22091; 703-620-2200), the wholly-owned subsidiary of the

French company created to market Spot's services. Spot data are available as digital information in computer-compatible tapes and as black-and-white or color prints and transparencies. Prices depend on several factors, including which of the three types of radiation that Spot records (two bands of visible light and one of near-infrared radiation) is desired, resolution quality (either 20-meter resolution or the sharper 10-meter resolution), and the size of the print or transparency. For example, a 9 1/2"-square black-and-white transparency showing all three bands at a scale of 1:400,000 and 10-meter resolution is $1,500—a color transparency at a scale of 1:400,000 (corrected for distortion to the point that it can be used to overlay on a printed map with high accuracy) and 20-meter resolution is $1,300. Prints are a bit less pricey ($250 each, black-and-white or color), but must accompany a tape or transparency order. At such prices, Spot images, however spectacular, are not intended for mere decoration.

Satellite-Image Maps. USGS has published satellite-image maps for selected areas in the U.S. and such areas as Antarctica, the Bahamas, and Iceland from multispectral scanner, thematic mapper, and Spot imagery. Most of these image maps are printed in false-color infrared. Notable examples are "Denali National Park, Alaska," at a scale of 1:250,000 with a standard topographic map on the reverse side ($7); "Washington, DC" at a 1:50,000-scale ($5.50); and "Point Loma, California," at a scale of 1:24,000 with a standard quadrangle map on the reverse side ($4). An order form listing these maps is available from USGS ESIC offices or from the Distribution Center in Denver, Colorado.

Other Space Imagery. Also available are older space-based images that predate Landsat and Spot, including those made from traditional photographic processes using both fixed on-board and hand-held cameras. A limited number of photographs from space are available from the Gemini and Apollo programs, for example, made between 1965 and 1970, and from the Skylab program in 1973 and 1974. The Gemini and Apollo images are limited by those spacecrafts' flight paths and cover primarily the Southwest, the Gulf Coast, and Florida. Skylab, however, includes extensive photographic coverage of most of the U.S., much of South America, and parts of Africa, Europe, and the Middle East. These photos are available from ESIC and EROS. A brochure may be helpful in ordering: "Manned Spacecraft Photographs and Major Metropolitan Area Photographs and Images."

COMMERCIAL SOURCES
Geoscience Resources (2990 Anthony Rd.,

P.O. Box 2096, Burlington, NC 27216; 919-227-8300; 800-742-2677) distributes many USGS satellite-image maps, many with topographic maps on the reverse side. Most titles it distributes portray areas of geologic interest. Titles available include "Anchorage, Alaska: Glacial Terrain" ($4.95); "Kansas City, Missouri: Meandering Rivers" ($4.95); "Pahute Mesa, Nevada: Eroded Mesas" ($4.95); and "Downtown Washington, DC" ($5.50).

National Air Survey Center Corp. (4321 Baltimore Ave., Bladensburg, MD 20710; 301-927-7177) sells a variety of satellite photography: "State Mosaics," false-color prints for all continental states and Hawaii; "River/Park/Lake Mosaics" ($25 to $150), available for "New Trinity River/Ft. Worth to Houston," "Lake Michigan," "Great Salt Lake," "Grand Canyon National Park," and "Great Smokey Mountains"; "Individual Scenes," made from Landsat imagery, available for Annapolis, Baltimore, Chesapeake Bay, Cincinnati, Des Moines, Detroit, Dulles Airport, Kansas City, New York City, Seattle, and Washington, D.C.; and "Mosaics of Countries," available for more than 30 areas, including "North Africa," "Greece," "Jordan," "Mexico," "Nile Delta," and "Venezuela." Prices range from $25 to $250 for color prints, and $60 to $350 for color transparencies. A brochure, "Satellite Photography—Space Portrait USA," is available.

Spaceshots Inc. (11111 Santa Monica Blvd., Ste. 210, Los Angeles, CA 90025; 213-478-8802; 800-272-2779) produces prints of satellite images of the Earth. More than 20 images are available for areas in North America, including "Cape Cod," "Michigan State," "Calgary/Banff," "Montreal," and "Finger Lakes, New York." Print size and area covered varies; prints are available on paper or laminated; prices range from $12.95 to $21.95.

■ **See also: "Aerial Photographs" and "Weather Maps."**

ATLASES AND GLOBES

How to Choose an Atlas
by June Crowe and James Minton

The atlas is a unique cartographic format valuable to students, teachers, travelers, and families who wish to have the world at their fingertips. The term *atlas* was first applied to a cartographic work by Gerhard Mercator, a Flemish geographer and map-maker, who in 1595 placed the Greek mythological figure of Atlas on the title page of a collection of maps in a book format.

Technically speaking, an atlas is a collection of maps that treat a specific subject in a systematic manner. Some atlases are broad in scope, such as the *Times Atlas of the World*, a general world atlas, while others are more specific, such as the *Atlas of Georgia*, which is a thematic atlas of that state. As with geographic area, subject matter also can vary, covering political, physical, economic, or social themes. Some examples of thematic atlases are *We The People: An Atlas of America's Ethnic Diversity*, *Cultural Atlas of Japan*, *Historical Atlas of Washington*, *Mariner's Atlas, Southwest Florida and the Florida Keys*, *Atlas of Oregon Lakes*, and *USSR Energy Atlas*. Most comprehensive world atlases focus on political and physical features with a limited selection of thematic maps.

Because there is a variety of possibilities to choose from, you must first decide which type of atlas will best satisfy your needs. Since many world atlases today cost more than $100, it is important that your selection process be deliberate and address the essential elements of quality atlas production. These criteria should cover the basic technical elements of maps (projection, scale, grids, etc.) and the physical properties of book design, layout, and printing and take into consideration your own individual preferences.

Here are some basic guidelines for selecting a world atlas for purchase:

Reputation and background of cartographer and publisher. The American Library Association's Reference and Subscription Books Reviews Committee (50 E. Huron St., Chicago, IL 60611; 312-944-6780) maintains information on a variety of atlas publishers and may answer your questions about a publisher's experience and reputation. A variety of library journals also contain reviews of atlases. Unfortunately, most atlas reviews appear many months after publication and are often difficult to locate. A selected list of journals that publish reviews on a regular basis is included at the end of this section. Ask the reference librarian at a local library for assistance in locating these reviews.

The cartographer is the person responsible for drafting the maps contained in an atlas. The cartographer (or cartographers) of individual atlas sheets is usually difficult to determine. The book's introduction may provide this information; if not, you may want to contact the publisher directly. General world atlases usually are not produced by a single cartographer, although thematic atlases often are. The cartographer's experience may be reflected in the accuracy, legibility, and clarity of the maps included in the atlas. Whether the cartographer or publisher is well known or relatively obscure, check the atlas for information on areas with which you are most familiar. Though omissions and errors are rare, they can be overlooked by even the best map-makers.

Your individual needs. Why are you buying an atlas? To plan your travels? To keep on your coffee table? Do you want the atlas to provide demographic and related information? Are you interested in the entire world or only a specific area? All these questions should be answered before selecting an atlas. In some situations, a geographical dictionary, almanac, gazateer, or encyclopedia may be better suited to your needs. Personal preference for size, binding, and other considerations should also be taken into account.

Technical aspects of cartographic materials. The type of *map projection* (the method of getting the spherical Earth on flat paper) used in map-making will affect the size, shape, distance, and directional characteristics of the world (see "Map Projections" for more on this). The intended use of a map determines the preferred projection. A clear explanation about the projections used in the atlas should be provided in the introduction. Moreover, the projection used should appear on each map. A variety of projections are normally used in a world atlas, including polar projections, equal-area azimuthal projections, Bonne projections, Albers conical projections, Winkel triple projections, and the Robinson projection.

The single most important technical element on a map is its *scale*. Scale determines the size of paper necessary to cover a particular geographical area and the amount of detail you can expect from a map. There are three widely used methods of indicating scale on a map: geographic or bar scale, a verbal scale, and the R.F. (representative fraction) or natural scale.

Maps contained in atlases may indicate scale with any one, or all, of these methods. (Better-quality atlases usually indicate scale in all three manners.) It is recommended that serious map users learn the R.F. method, because it allows you to work with any system of measurement—centimeters, inches, meters, feet, kilometers, or miles. Maps are referred to as either large-scale or small-scale. Small-scale (such as 1:35,000,000) maps show little detail and emphasize only major elements of world geography. Large-scale (such as 1:15,000) maps show greater detail and may even depict individual streets and buildings. If you are interested in such "big picture" things as global-transportation patterns or bird-migration routes, a small-scale map showing the entire world on one page may satisfy your needs. However, if you are interested in finding the name of a specific street, a much larger-scale map will be necessary. World atlases generally contain small-scale maps of continents and nations with a few larger-scale maps of selected cities. Don't expect a world atlas to contain a street index to your home town, for example.

Legends and symbols are also important. Maps use lines, points, and polygons (enclosed areas) to depict various physical, economic, social, and worldwide cultural features. These elements form the symbols used on maps that allow you to visually interpret information at a glance. Well-produced maps contain a legend or list of symbols. Most atlases contain a legend or symbol sheet in the front of the publication. Some publishers print a separate legend sheet that can be moved easily from page to page while using the maps without having to leaf to the specific legend page every time you need to consult it. All symbols used on the maps should be included and be clearly distinct from one another, especially if a range of colors is used to depict graduated changes.

Atlases are used most often in locating cultural and physical features—cities, lakes, mountains, and so forth. A variety of methods of *grids* is used to index these features. Some atlases contain lists of features (generally at the end of the publication) with a grid indicating its location on a page and specific spot within the map—"B-21," for example. Most good atlases, however, also list the latitude and longitude of the place along with the plate (or page) number. It is recommended that you become familiar with the systems of degrees, minutes, and seconds used to determine latitude and longitude. (Each degree of latitude or longitude is made up of 60 minutes; each minute consists of 60 seconds. Thus, 28° 40' 35"N refers to a latitude that is 28 degrees, 40 minutes, and 35 seconds north of the equator.) You shouldn't even consider an atlas whose index does not indicate latitude and longitude.

Publishing information. There are a number of considerations:

■ *Date of publication.* You should scan the title page and other introductory pages for the date of publication, place of publication, copyright, and revision information. Be aware that most atlases are revised or reprinted. However, some less-than-reputable publishers issue a

"new" atlas containing older maps without updating them. Select a particular section of the world that has undergone change and review the atlas to see if the changes have been incorporated. (For example: Do maps of China show the capital as "Beijing" or "Peking"?) Check for changes in names, roads, boundaries, and other pertinent information.

■ *Adequacy of coverage.* An important aspect in selecting an atlas is whether it contains detailed maps of an area important to you. If you are interested in detailed maps of Papua New Guinea, for example, it is unlikely that a world atlas will suffice. The place of publication is an important hint as to the depth of coverage you can expect. If the atlas were printed in Papua New Guinea or Australia, there would be a greater likelihood that it would contain more maps at larger scales of that region than those published elsewhere.

■ *Arrangement and organization of material.* Atlases have evolved into a somewhat standard format: title page, table of contents, general thematic maps, larger-scale maps, supplementary materials, and indexes. Atlas organization determines how efficiently it can be used. For example, you may want to quickly locate a place whose name you heard on the evening news. Is your atlas arranged to find it easily? Is there a recognizable geographic arrangement to the maps within the atlas?

■ *Indexes.* These are one of the most important parts of an atlas. Two types of indexes are common: a textual index to place names and a graphic index indicating map coverage and plat numbers. Textual indexes generally are included in an alphabetical arrangement at the end of the atlas along with the plate number (or page) and grid location (preferably latitude and longitude). Pay particular attention to the textual index. Do all names listed in the index appear on the maps? Are all names on the maps listed in the index? Are the spellings of names in both English and the native language of the country on the map? If not, you may have difficulty locating specific places. (Is it listed as Koln or Cologne?) Are physical features included in the index? Better-quality atlases have large indexes that include the map and the geographic location of each entry. The *Times Atlas of the World* and Rand McNally's *New International Atlas* each contain more than 200,000 entries. Select several place names from the index to determine whether these names appear on the map plates. According to *Kister's Atlas Buying Guide* (Oryx Press, 2214 N. Central Ave., Phoenix, AZ 85004; 602-254-1483; 1984), a general world atlas should contain between 30,000 and 50,000 entries to adequately cover the world's places and features.

Graphic indexes consist of outline maps indicating map coverage with plate or page number. These indexes provide quick access to most of the maps in an atlas. If you wish to locate a map of the island of New Guinea, for example, a quick glance at the graphic index should yield the plate number immediately, whereas the textual index will require locating the name in alphabetical order, noting the plate number, then locating the map. However, if you wanted to locate the city of Lae in Papua New Guinea, the textual index would provide a more efficient way to do so.

■ *Supplementary materials.* Most atlases contain material that is non-cartographic in nature. This may include bibliographies, charts, text, tables, population figures, and photographs. Many atlas publishers include such extraneous material to enhance marketing. But much of this type of information can be found in other publications providing better coverage and reliability than an atlas. Atlases should contain a limited number of pages devoted to supplementary material.

SELECTED JOURNALS THAT REVIEW ATLASES
■ *American Cartographer* (American Congress on Surveying and Mapping, 210 Little Falls St.,

Falls Church, VA 22046; 703-241-2446).
■ *Booklist: Including Reference Books Bulletin* (American Library Association, 50 E. Huron St., Chicago, IL 60611; 312-944-6780).
■ *Bulletin* (Special Libraries Association, Geography and Map Division, 1700 18th St. NW, Washington, DC 20009; 202-234-4700).
■ *Cartographic Journal* (British Cartographic Society, c/o Dept. of Geography, King's College, Strand, London, England W2CR 2LS; 01-836-5454).
■ *Choice* (Association of College and Research Libraries, 50 E. Huron St., Chicago, IL 60611; 312-944-6780).
■ *Geographical Journal* (Royal Geographical Society, 1 Kensington Gore, London, England SW7 2AR; 01-589-2173).
■ *Geographical Review* (American Geographical Society, 156 Fifth Ave., Ste. 600, New York, NY 10010; 212-242-0214).
■ *GeoTimes* (American Geological Institute, 4220 King St., Alexandria, VA 22302; 703-379-2480).
■ *Information Bulletin* (Western Association of Map Libraries, c/o Stanley D. Stevens, University Library-University of California, Santa Cruz, CA 95064; 408-429-2364).
■ *Library Journal* (R.R. Bowker, Bowker Magazine Group, Cahners Magazine Div., 249 W. 17th St., New York, NY 10011; 800-431-1713).
■ *Professional Geographer* (Association of American Geographers, 1710 16th St. NW, Washington, DC 20009; 202-234-1450).
■ *Surveying and Mapping* (American Congress on Surveying and Mapping, 210 Little Falls St., Falls Church, VA 22046; 703-241-2446).

June Crowe is Research Librarian with the Urban Land Institute in Washington, D.C., and former chair of the Geography and Map Division at the Special Libraries Association. James Minton is Map Librarian at the U.S. Geological Survey's National Center in Reston, Virginia.

Selected Atlases

Mention the word "atlas" and it usually brings to mind the word "world," although "world atlases" represent a mere fraction of atlases published. There are atlases of nearly every place and thing: agriculture atlases, Bible atlases, state atlases, country atlases, history atlases, plant and animal atlases, road atlases, celestial atlases, war atlases, and on and on. The Library of Congress, in Washington, D.C., has more than 11,000 books containing the world "atlas" in their titles—representing only books the library has cataloged since 1968!

WORLD ATLASES

Here are some of the world atlases currently available:

■ *Earthbook* (Boulder, CO: Graphic Learning International, 1987; $12.95) is a new world atlas, combining space-age technology and satellite photography with hundreds of full-color maps, and an encyclopedia section that chronicles Earth science, oceanography, and meteorology. It is available in a deluxe 327-page edition ($65) or a concise 215-page edition ($12.95).

■ *The Economist World Atlas and Almanac* (New York: Prentice Hall Press, 1989; $39.95) combines geographic, demographic, and political maps with lovely coverage of economics, politics, and current affairs, including regional and country profiles on each nation's geography, history, economy, people, politics, resources, and industry.

■ *Goode's World Atlas* (Chicago: Rand McNally, 1990; 384 pages; $22.95). This is one of the most popular world atlases used in schools (its 1922 through 1949 editions were titled *Goode's School Atlas*) because it provides more than 100 special-purpose maps typically used by students on subjects such as population,

economics, and agriculture. Goode's maps are particularly clear and readable, and the book's bindings were designed for rugged school-library use.

■ *Great World Atlas* (Maspeth, N.Y.: American Map Corp., 1989; 384 pages; $49.95). A revised and expanded edition, it includes chapters covering astronomy and Earth science, as well as a large section of satellite photos along with well-executed maps, several thematic and statistical maps, and an index.

■ *Hammond Gold Medallion World Atlas* (Maplewood, N.J.: Hammond Inc., 1988, 668 pages; $85). Hammond publishes several world atlases, but this is the most comprehensive, with more than 600 maps providing a wide range of data.

Hammond also publishes an *Ambassador World Atlas* (1988; 524 pages; $49.95); *Citation World Atlas* (1988; 388 pages; $19.95 soft, $29.95 hard); *Discovery World Atlas* (1988; 224 pages; $17.95 hard, $13.95 soft); *Nova World Atlas* (1988; 184 pages; $9.95); *International World Atlas* (1984; 200 pages; $18.95); *Headline World Atlas* (1986; 48 pages; $3.50); *Hammond Physical World Atlas* (1988; 40 pages; $5.95); and *Passport Travelmate and World Atlas* (1988; 128 pages; $4.95).

■ *National Geographic World Atlas* (Washington, D.C.: National Geographic Society, 1981; 383 pages; $44.95). Published by this venerable institution, the atlas is a comprehensive and attractive work, featuring clear, easily readable maps. Of particular note is the book's coverage of the oceans of Africa, both presented in detail not generally found in other world atlases. National Geographic also publishes *Atlas of North America: Space Age Portrait of a Continent* (1985; 264 pages; $39.95, or $49.95 for the "deluxe" hardcover edition with magnifier/scale).

■ *The Rand McNally New International Atlas* (Chicago: Rand McNally, 1989; 568 pages; $150). Compiled by more than 100 cartographers worldwide, this is known for its generally balanced coverage and attractive maps. It includes detailed maps of major world metropolitan areas at 1:1,000,000 and 1:3,000,000 scales, a helpful glossary giving translations of geographic terms in 52 languages, and a thorough 160,000-name index. It does not feature thematic or special-purpose maps.

Rand McNally also publishes the *Rand McNally Cosmopolitan World Atlas* (1987; 392 pages; $55); *Rand McNally Images of the World* (1984; 160 pages; $24.95); *Rand McNally Desk Reference World Atlas* (1987; 528 pages; $17.95); *Quick Reference World Atlas* (1989; 64 pages; $3.95); *Traveler's World Atlas and Guide* (1989; 224 pages; $6.95); *Family World Atlas* (1988; 256 pages; $13.95); and *Contemporary World Atlas* (1988; 256 pages; $9.95).

■ *Times Atlas of the World* (New York: Times Books, 1985; 520 pages; $149.95). The *Times of London* has produced world atlases since 1895, and they are consistently lauded for their comprehensive, balanced coverage and top-quality indexes. In addition to this "comprehensive" edition, there is also a "concise edition" of this fine book: *The New York Times Atlas of the World* (New York: Random House; 1987; 288 pages; $49.95).

Other world atlases include:

■ *Atlas of the Land* (New York: Ballantine Books, 1985; $9.95).
■ *The Bantam Illustrated World Atlas* (New York: Bantam, 1989; $39.95).
■ *Britannica Atlas* (Chicago: Encyclopedia Britannica, 1989; $99.50).
■ *Illustrated World Atlas* (New York: Franklin Watts Inc., 1988; $14.90).
■ *NBC News Rand McNally World News Atlas* (Chicago: Rand McNally, 1989; $7.95).
■ *Rand McNally World Atlas of Nations* (Chicago: Rand McNally, 1989; $34.95), lists countries alphabetically from A to Z.

■ *Reader's Digest Atlas of the World* (Chicago: Rand McNally, 1989; $39.95).
■ *Rand McNally Traveler's World Atlas & Guide* (Chicago: Rand McNally, 1989; $6.95).
■ *Signet-Hammond World Atlas* (New York: New American Library, 1982; $4.50).
■ *VNR Pocket Atlas* (New York: Van Nostrand Reinhold Inc., 1983; $15.95).
■ *World Atlas* (New York: Penguin Books Inc., 1979; $9.95).
■ *World Atlas* (New York: Random House, 1982; $10.95).

UNITED STATES ATLASES

Other atlases primarily covering the U.S. include:

■ *American Map Premier Resorts Edition, U.S. Road Atlas* (Maspeth, N.Y.: American Map Corp., 1990; $5.95), an annual.
■ *American Map U.S. Highways Atlas* (Maspeth, N.Y.: American Map Corp., 1990; $5.95), an annual.
■ *Atlas of the United States* (New York: Macmillan, 1986; $50).

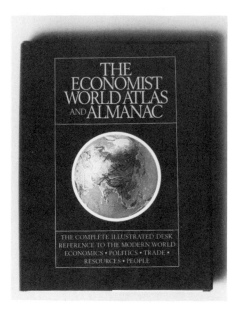

■ *Rand McNally Road Atlas & Vacation Guide* (Chicago: Rand McNally, 1989; $14.95), an annual.
■ *Rand McNally Road Atlas: United States, Canada, Mexico* (Chicago: Rand McNally, 1989; $6.95), an annual.
■ *Rand McNally Deluxe Road Atlas and Travel Guide* (Chicago: Rand McNally, 1989; $5.95), an annual.
■ *The United States Energy Atlas* (New York: Macmillan, 1985; $90).
■ *U.S. Atlas/New Road Atlas* (New York: Prentice Hall/Access Press, 1990; $29.95 hard, $12.95 soft).

SPECIALIZED ATLASES

These vary widely—in form, content, and availability. While there are thousands of specialized atlases, we've listed below some of the more interesting, comprehensive, or useful ones. Many, but not all, are still in print; a few are difficult to find outside of major metropolitan and university libraries.

STATE ATLASES

■ *Atlas of California* (Portland, Ore.: Professional Book Center Inc., 1979; $29.95).

■ *Atlas of Colorado* (Boulder, Colo.: Colorado Associated University Press, 1985; $17.50).
■ *Atlas of Georgia* (Athens, Ga.: Institute of Community & Area, University of Georgia, 1985; $47.50).
■ *Atlas of Illinois* (Madison, Wisc.: Madison Opportunity Center, 1976; $26.95).
■ *Atlas of Michigan* (Grand Rapids, Mich.: William B. Ferdmans Publishing Co., 1977; $5.95).
■ *Atlas of New York* (Madison, Wisc.: Madison Opportunity Center, 1978; $29.95).
■ *Atlas of the State of South Carolina: Prefaced with a Geographical, Statistical & Historical Map* (Easley, S.C.: Southern Historical Press, 1980 reproduction of 1825 edition; $50).
■ *Atlas of Wisconsin* (Madison, Wisc.: Madison Opportunity Center, 1976; $29.95).
■ *Historical Atlas of Kansas* (Norman, Okla.: University of Oklahoma Press, 1988; $24.95).
■ *Maine Atlas & Gazeteer* (Freeport, Me.: DeLorme Mapping Co., 1984; $9.95).
■ *New Jersey Road Maps of the Eighteenth Century* (Princeton, N.J.: Princeton University Library, 1981; $5).
■ *New Mexico in Maps* (Albuquerque, N.M.: University of New Mexico Press, 1986; $24.95).
■ *Trakker Maps Florida State Road Atlas* (Miami, Fla.: Trakker Maps Inc., 1989; $7.95), an annual.
■ *Vermont Atlas & Gazeteer* (Freeport, Me.: DeLorme Mapping Co., 1983; $9.95).
■ *Vermont Road Atlas and Guide* (Burlington, Vt.: Northern Cartographics, 1989; $12.95).
■ *Yankee Magazine's Travel Maps of New England* (Dublin, N.H.: Yankee Books, 1984; $4.95).

FOREIGN COUNTRY & REGIONAL ATLASES

■ *Atlas of Central America and the Caribbean* (New York: Macmillan, 1985; $50).
■ *Atlas of Ireland* (New York: St. Martin's Press Inc., 1980; $99.50).
■ *The Atlas of Israel* (New York: Macmillan, 1985; $185).
■ *Atlas of the Middle East* (New York: Macmillan, 1988; $60).
■ *Atlas of Peru* (New York: Ballantine Books, 1984; $9.95).

■ *Atlas of South Asia, Annotated* (Boulder, Colo.: Westview Press, 1987; $19.95).

■ *Atlas of Southeast Asia* (New York: Macmillan, 1988; $95).

■ *The Atlas of the Third World* (New York: Facts On File, 1989; $95).

■ *The Contemporary Atlas of China* (Boston: Houghton Mifflin Co., 1988; $39.95).

■ *Cultural Atlas of Africa* (New York: Facts On File, 1981; $40).

■ *Cultural Atlas of China* (New York: Facts On File, 1983; $40).

■ *The Cultural Atlas of Islam* (New York: Macmillan, 1986; $115).

■ *Cultural Atlas of Japan* (New York: Facts On File, 1988; $40).

■ *Cultural Atlas of Russia and the Soviet Union* (New York: Facts On File, 1989; $40).

■ *Hallwag Euro Guide* (Berne, Switzerland: Hallwag, 1989; $29.95), an annual.

■ *Hallwag Europa Road Atlas* (Berne, Switzerland: Hallwag, 1989; $19.95), an annual.

■ *Rand McNally Road Atlas & City Guide of Europe* (Chicago: Rand McNally, 1989; $14.95), an annual.

■ *Rand McNally Road Atlas of Britain* (Chicago: Rand McNally, 1989, $14.95), an annual.

■ *Rand McNally Road Atlas of Europe* (Chicago: Rand McNally, 1989; $7.95), an annual.

■ *Rand McNally Road Atlas of France* (Chicago: Rand McNally, 1989; $14.95), an annual.

■ *Third World Atlas* (Philadelphia: Taylor & Francis Inc., 1986; $65 hard, $26 soft).

HISTORICAL ATLASES

■ *Atlas of African History* (New York: Holmes & Meier Publishers Inc., 1978; $45 hard, $29.50 soft).

■ *Atlas of African History* (New York: Penguin Books Inc., 1980; $8.95).

■ *Atlas of American History* (New York: Charles Scribner's Sons, 1984; $55).

■ *Atlas of American History* (New York: Facts On File, 1987; $24.95).

■ *Atlas of Ancient America* (New York: Facts On File, 1986; $40).

■ *Atlas of Ancient Egypt* (New York: Facts On File, 1980; $40).

■ *Atlas of Ancient History* (New York: Penguin Books Inc., 1967; $6.95).

■ *Atlas of the British Empire* (New York: Facts On File, 1989; $40).

■ *Atlas of British Social and Economic History Since 1700* (New York: Macmillan, 1989; $85).

■ *Atlas of Classical History* (New York: Macmillan, 1985; $50).

■ *Atlas of Early American History: The Revolutionary Era* (Princeton, N.J.: Princeton University Press, 1976; $235).

■ *Atlas of the Greek World* (New York: Facts On File, 1981; $40).

■ *Atlas of Medieval Europe* (New York: Facts On File, 1983; $40).

■ *Atlas of Medieval History* (New York: Penguin Books Inc., 1968; $6.95).

■ *Atlas of Modern History* (New York: Penguin Books Inc., 1973; $6.95).

■ *Atlas of Nazi Germany* (New York: Macmillan, 1987; $55).

■ *Atlas of North American History* (New York: Penguin Books Inc., 1988; $6.95 soft, $19.95 hard).

■ *Atlas of Recent History* (New York: Penguin Books Inc., 1986; $6.95).

■ *Atlas of the Roman World* (New York: Facts On File, 1982; $40).

■ *Atlas of United States History* (Maplewood, N.J.: Hammond Inc., 1989; $10.95).

■ *Atlas of World History* (Chicago: Rand McNally, 1987; $17.95).

■ *Hammond Atlas of World History* (Maplewood, N.J.: Hammond Inc., 1987; $10.95).

■ *Hammond-Harwood Historical Atlas* (Baltimore, Md.: Johns Hopkins University Press, 1982; $47.50).

■ *Historical Atlas of Canada, Volume I* (Toronto: University of Toronto Press, 1987; $95).

■ *Historical Atlas of the Outlaw West* (Boulder, Colo.: Johnson Books, 1984; $15.95).

■ *Historical Atlas of Political Parties in the United States Congress: 1789-1989* (New York: Macmillan, 1989; $190).

■ *Historical Atlas of the United States* (Washington, D.C.: National Geographic Society, 1988, $59.95).

■ *Historical Atlas of the United States, Deluxe Edition* (Washington, D.C., 1988, $74.95).

■ *Historical Atlas of the World* (Chicago: Rand McNally, 1965; $6.95).

■ *Peoples and Places of the Past: The National Geographic Illustrated Cultural Atlas of the Ancient World* (Washington, D.C.: National Geography Society, 1983, $69.95).

■ *Poland: A Historical Atlas* (New York: Hippocrene Books, 1986; $22.50).

■ *The Times Atlas of World History* (Maplewood, N.J.: Hammond Inc., 1989; $85).

■ *The Times Concise Atlas of World History* (Maplewood, N.J.: Hammond Inc., 1988; $24.95).

■ *The Times Past Worlds Atlas of Archaeology* (Maplewood, N.J.: Hammond Inc., 1988; $85).

POLITICAL, CULTURAL, AND SOCIOLOGICAL ATLASES

■ *A Comparative Atlas of America's Great Cities* (Minneapolis: Association of American Geographers/University of Minnesota Press, 1976; $95).

■ *Atlas of African Affairs* (New York: Methuen Inc., 1984˙ $14.95).

■ *Atlas of American Women* (New York: Macmillan, 1986; $90).

■ *Atlas of Disease Distribution* (New York: Basil Blackwell Inc., 1988; $150).

■ *Atlas of Environmental Issues* (New York: Facts On File, 1989; $16.95).

■ *Atlas of Global Strategy* (New York: Facts On File, 1985; $24.95).

■ *Atlas of Human Evolution* (New York: Holt, Rinehart & Winston Inc., 1979; $26.50).

■ *Atlas of North American Affairs* (New York: Methuen Inc., 1979; $11.95).

■ *Atlas of the North American Indian* (New York: Facts on File, 1989; $16.95).

■ *Atlas of Nuclear Energy* (Atlanta: Georgia State University, Dept. of Geography, 1984; $6).

■ *Atlas of the Arab World* (New York: Facts On File, 1983; $14.95).

■ *Atlas of the North American Indian* (New York: Facts On File, 1989, $16.95).

■ *Atlas of World Affairs* (New York: Methuen Inc., 1987; $12.95 soft, $39.95 hard).

■ *Atlas of World Cultures* (Pittsburgh, Pa.: University of Pittsburgh Press, 1981; $14.95).

■ *Atlas of the World Today* (New York: Harper & Row/Perennial, 1987; $10.95).

■ *Atlas of World Cultures* (Pittsburgh, Pa.: University of Pittsburgh Press, 1981; $14.95).

■ *Atlas of World Issues* (New York: Facts On File, 1989; $16.95).

■ *The Cambridge Atlas of the Middle East & North Africa* (New York: Cambridge University Press, 1987; $75).

■ *The Coin Atlas* (New York: Facts On File, 1989; $40).

■ *National Population Atlas of the People's Republic of China* (New York: Oxford University Press, 1987; $250).

■ *New State of the World Atlas* (New York: Simon & Schuster Inc., 1984; $13.95).

■ *The Nuclear War Atlas* (New York: Basil Blackwell Inc., 1988; $14.95).

■ *Oxford Economic Atlas of the World* (New York: Oxford University Press Inc., 1972; $16.95).

■ *Oxford Regional Economic Atlases: The United States & Canada* (New York: Oxford University Press Inc., 1975; $19.95).

■ *Strategic Atlas: A Comparative Geopolitics of the World's Powers* (New York: Harper & Row/Perennial, 1985; $14.95).

■ *The Stamp Atlas* (New York: Facts On File, 1986; $29.95).

■ *We The People: An Atlas of America's Ethnic Diversity* (New York: Macmillan, 1987; $125).

■ *Wine Atlas of the World* (New York: Simon & Schuster, 1985; $45).

■ *The Women's Atlas of the United States* (New York: Facts On File, 1986; $45).

■ *Women in the World: An International Atlas* (New York: Simon & Schuster, 1986; $12.95).

■ *Women in the World: An International Atlas* (New York: Simon & Schuster/Touchstone, 1986; $12.95).

■ *The World Bank Atlas* (Washington, D.C.: IBRD/ The World Bank, 1989; $5.95), an annual.

MILITARY AND WAR ATLASES

■ *The Month By Month Atlas of World War II* (New York: Simon & Schuster, Summit Books, 1989; $35).

■ *State of War Atlas: Armed Conflict-Armed Peace* (New York: Simon & Schuster, 1983; $9.95).

■ *World Atlas of Military History, 1945-84* (New York: Hippocrene Books, 1985; $24.95).

ATLASES OF RELIGION

■ *Atlas of the Bible Lands* (Nashville, Tenn.: Broadman Press, 1979; $4.95).

■ *Atlas of the Bible Lands* (Maplewood, N.J.: Hammond Inc., 1989; $10.95).

■ *Atlas of the Bible* (New York: Facts On File, 1985; $40).

■ *Atlas of the Christian Church* (New York: Facts On File, 1987; $40).

■ *Atlas of the Islamic World Since 1500* (New York: Facts on File, 1982; $40).

■ *Atlas of the Jewish World* (New York: Facts On File, 1984; $40).

■ *Maps of the Holy Land* (New York: Abbeville Press, 1985; $59.95).

■ *Oxford Bible Atlas* (New York: Oxford University Press, 1985; $14.95 soft, $24.95 hard).

■ *Reader's Digest Atlas of the Bible* (New York: Random House, 1982; $22.95).

CELESTIAL ATLASES

■ *Atlas Australis* (Belmont, Mass.: Sky Publishing Corp., 1964; $49.95).

■ *Atlas Borealis* (Belmont, Mass.: Sky Publishing Corp., 1964; $49.95).

■ *Atlas of Deep Sky Splendors* (Belmont, Mass.: Sky Publishing Corp., 1983; $39.95).

■ *Atlas of Deep Sky Splendors* (New York: Cambridge University Press, 1984; $54.50).

■ *Atlas of Selected Areas* (Belmont, Mass.: Sky Publishing Corp.; $33).

■ *The Cambridge Photographic Atlas of the Planets* (Belmont, Mass.: Sky Publishing Corp., 1982; ($24.95).

■ *The Cambridge Atlas of Astronomy* (Belmont, Mass.: Sky Publishing Corp., 1985; $90).

■ *Maps of the Heavens* (New York: Abbeville Press, 1989; $59.95).

■ *Norton's Star Atlas and Reference Handbook* (Belmont, Mass.: Sky Publishing Corp., 1986; $29.95).

■ *Sky Atlas* (Belmont, Mass.: Sky Publishing Corp., not dated; $15.95).

NATURE ATLASES

■ *Atlas of Distribution of Freshwater Fish Families of the World* (Lincoln, Neb.: University of Nebraska Press, 1981; $26.50 hard, $12.50 soft).

■ *Atlas of Living Resources of the Seas* (New York: Food and Agricultural Organization/Unipub, 1981; $90).

■ *Atlas of Natural Wonders* (New York: Facts On File, 1988; $35).
■ *Atlas of Wintering North American Birds* (Chicago: University of Chicago Press, 1989; $60).
■ *A Biogeographical Atlas of Corals, Seagrasses and Mangroves* (New York: Basil Blackwell Inc., 1989; $150).
■ *Climatic Atlas of the United States* (Asheville, N.C.: National Climatic Data Center, 1969; $17).
■ *California Walking Atlas* (New York: McGraw-Hill, 1989; $11.95).
■ *Mid-Atlantic Walking Atlas* (New York: McGraw-Hill, 1989; $11.95).
■ *New England Walking Atlas* (New York: McGraw-Hill, 1989; $11.95).
■ *The Walking Atlas of America* (New York: McGraw-Hill, 1989; $11.95).

OCEAN ATLASES

■ *Oceanographic Atlas of the Bering Sea Basin* (Seattle, Wash.: University of Washington Press, 1980; $35).
■ *Oceanographic Atlas of the Pacific Ocean* (Honolulu, Haw.: University of Hawaii Press, 1969; $25).
■ *Sea Ice Atlas of Arctic Canada, 1968-1974* (Ottawa: Canada Map Office, 1974; $20 Canada, $24 elsewhere).
■ *Sea Ice Atlas of Arctic Canada, 1975-1978* (Ottawa: Canada Map Office, 1978; $30 Canada, $36 elsewhere).
■ *Sea Ice Climatic Atlas* (Asheville, N.C.: National Climatic Data Center, 1989; free), an annual in three volumes.

RAILWAY AND SHIPPING ATLASES

■ *Trucker's Atlas for Professional Drivers* (Wood Dale, Ill.: Creative Sales Corp., 1989; $10.95), an annual.
■ *Rand McNally Atlas of the Oceans* (Chicago: Rand McNally, 1979; $40).
■ *Rand McNally Handy Railroad Atlas of the United States* (Chicago: Rand McNally, 1988; $14.95).

ATLASES FOR YOUNG PEOPLE

■ *Atlas of Ancient Egypt* (New York: Facts On File, 1990; $17.95).
■ *Atlas of Ancient Greece* [Cultural Atlas for Young People Series] (New York: Facts On File, 1989; $17.95).
■ *Atlas of Ancient Rome* [Cultural Atlas for Young People Series] (New York: Facts On File, 1989; $17.95).
■ *Atlas of Today* (New York: Franklin Watts Inc., 1987; $15.90).
■ *Atlas of the Middle Ages* (New York: Facts On File, 1990; $17.95).
■ *Atlas of the Presidents* (Maplewood, N.J.: Hammond Inc., 1989; $9.95).
■ *Bible Atlas* (Maspeth, N.Y.: American Map Corp., 1985; $2.95), available in English or Spanish.
■ *Facts On File Children's Atlas* (New York: Facts On File, 1987; $14.95).
■ *Rand McNally Student's World Atlas* (Chicago: Rand McNally, 1988; $5.95).
■ *Rand McNally Children's Atlas of World History* (Chicago: Rand McNally, 1989; $12.95).
■ *Rand McNally Children's World Atlas* (Chicago: Rand McNally, 1989; $12.95).
■ *Rand McNally Children's Atlas of the United States* (Chicago: Rand McNally, 1989; $12.95).
■ *Rand McNally Student's World Atlas* (Chicago: Rand McNally, 1988; $5.95).
■ *Scholastic World Atlas* (Maspeth, N.Y.: American Map Corp., 1989; $3.25).
■ *Students Atlas of the World* (Maspeth, N.Y.: American Map Corp., 1989; $2.50), available in English or Spanish.
■ *World Atlas* (Maspeth, N.Y.: American Map Corp., 1989; $1.75).
■ *World Atlas: A Resource for Students* (Chicago: Nystrom, 1989; $3.75).
■ *World Atlas for Young People* (Philadelphia, Pa.: Running Press/Courage Books, 1989; $12.98).

■ **See also: "Geography Education Materials."**

Globes

The globe—the most accurate distortion-free representation of the Earth's surface—has the greatest utility of any visual aid in geography. It is the only map of the world that shows the true size and shape of all land and water areas, as well as the true distances and geographical relationships between them. As a model of the rotating Earth, the globe can be used to illustrate illumination from the sun, the relationship of time and longitude, and related concepts.

Although mounting a world map onto a sphere does have special advantages, it also creates some unique problems that usually make flat maps more practical. Only part of the world can be seen at one time, for example. And whatever its size, a globe takes up more space than a flat map of equal scale, is not easily portable, and usually contains insufficient detail.

Perhaps the greatest disadvantage of globes is their greater cost compared to maps of similar scale. Like any material printed on a flat surface, maps can be produced cheaply in large quantities. Unfortunately, a satisfactory spherical printing press has yet to be invented, so globe maps are also printed on a flat surface and then transformed into a hemisphere by one of several manufacturing processes.

How globes are made. For centuries, globes were made by printing the world map onto 30-degree-wide *gores*, which were cut and pasted onto a wooden sphere by hand. In the Replogle and Cram globe factories, paper gores are still applied by hand on some models, but the wooden sphere has been replaced by pressboard or plastic.

This handmade process proved too slow to satisfy the public's growing demand for globes. By the middle of the 20th century, machine-made globes were being mass-produced by a method in which northern and southern polar projections were printed directly onto sheet aluminum and stamped into hemispheres with a hydraulic press. Susceptibility to dents and map stretching during the hemisphere-stamping process increases with globe size, however, and these globes could not be larger than 12 inches in diameter.

In the early 1960s, a new type of globe made from chipboard was introduced and has since become the most common type in current use. The maps are printed as interrupted polar projections resembling a propeller with 12 blades, referred to in the business as "rosettes." Each map is glued to chipboard, die-cut into the rosette shape, then glued to another rosette-shaped piece of chipboard, with seams overlapping, before the hydraulic press forms it all into a hemisphere.

One recent development in globe manufacture was the introduction of plastic, raised-relief globes. Using a procedure similar to that of making metal globes, the polar projections are printed onto sheet plastic, then formed using a vacuum process into their hemispheric shape. This process creates more accurate relief than is possible on chipboard globes. As with metal globes, the maps printed on plastic are drawn with some deliberate distortion, which occurs when the hemispheres are formed. Place names near the equator still appear vertically stretched, however. The northern and southern hemispheres of plastic, chipboard, and metal globes are matched at the International Date Line or the South Pacific island of Celebes when they are joined. A narrow strip of tape is applied over the equator to cover any gaps and to give a better appearance before the mounting and base are attached.

Types of globes. Terrestrial globes can be classified into three basic groups by type of coloring and information contained: political globes, physical globes, and political-physical combination globes.

The political globe is characterized by multicolor political units and solid-blue oceans. There is quite a variety in the amount of political information different manufacturers include on their globes. Some political globes show national units only. Other globe-makers add multicolor U.S. state units, and on some models Canada's provinces are also shown in different colors. Still other models show only provincial boundaries of non-North American countries. Names stretched across the length of bodies of water, deserts, and mountain ranges are typically the only evidence of the physical Earth on political globes, although some models also show ocean currents, steamship routes, and one type of relief portrayal.

The physical globe is primarily a representation of the world's relief features. The traditional method showing relief is by tints, with green for lowlands and lighter green, yellow, and brown for higher elevations. The use of hysometric tints has recently been subject to the criticism that it does not give a realistic image of the planet's surface. Perhaps as a response to this, some manufacturers have produced "eco-region" globes that use various shades of green, yellow, and brown to portray forests, grasslands, deserts, and other natural regions on land, and shades of blue to portray the ocean floor. Raised and shaped relief are used extensively on physical globes. Some even feature the ocean floor in indented relief. Almost all physical globes show place names of physical features and selected cities, and some also include spot heights, ocean currents, and political unit boundaries and names.

The third type of globe is the political-physical combination. It is similar to the political globe in that political units or boundaries are multicolor, but it also contains features found on a physical globe, such as raised and shaded relief, bathymetry of the ocean floor, and spot heights.

"Two-way illuminated" political-physical globes, available from some manufacturers, feature political cartography when not illuminated. When the light is turned on, the ocean floor is revealed in shaded relief, for example, or eco-region cartography. These globes can be considered two globes in one, unlike older illuminated globes, in which light is used for merely decorative purposes.

Special and extraterrestrial globes. Other types of globes include:

■ Outline or project globes, showing only the shapes or outlines of continents on a smooth plastic or slated surface. These are designed to be marked with pens or chalk to show various global concepts.
■ Political globes with either French or Spanish text. Globes with text in other foreign languages can be obtained from European globe companies.
■ Globes of celestial bodies other than the Earth, including the moon, Mars, and the solar system. Most are produced from NASA photographs. Celestial globes are available in two versions. One type has yellow stars and planets printed on an opaque blue sphere with lines connecting stars in constellations or the outlines of the characters these constellations represent. The other is a clear plastic sphere with the stars and planets printed on the inside surface and a small Earth model mounted in the center.

COMMERCIAL SOURCES

While thousands of globes exist, only a handful are readily available from globe manufacturers. Most map libraries have several terrestrial—and sometimes extraterrestrial—globes for study; the Library of Congress's map division has 400 globes on hand.

Globe prices range from less than $10 to well over $1,000, depending on size, quality of artwork, and such extras as floor stands, illumination, and raised relief.

George F. Cram Co. (P.O. Box 426, Indianapolis, IN 46206; 317-635-5564) manufactures 30 different globes, from a series of nine-inch "gift" globes ($19) to 16-inch deluxe illuminated models ($230 to $345).

A view of the 30" relief globe from Panoramic Studios.

cartoon-illustrated guide that discusses climate and fauna during each time period.

The Home Library (160 S. University Dr., Plantation, FL 33324; 305-579-2116; 800-367-7708) distributes Replogle, Scan, and various other globes at a discount. Their selection ranges from "The Lenox," a 12-inch raised-relief globe with antique styling ($47) to the 20-inch illuminated "Bainbridge" ($1,200). A popular product is the 20-inch "Bar Globe" ($399), hand-crafted in Italy, representing the world as it was known in the 16th century; the globe opens to reveal an ice container, as well as ample space for liqueurs and glasses.

Champion (200 Fentress Blvd., Daytona Beach, FL 32014; 904-258-1270; 800-874-7010; 800-342-1072 in Fla.) distributes more than 25 Replogle globe models, ranging from the nine-inch "World Scholar" ($12.95) to "The Statesman," a 20-inch illuminated globe on a floor stand ($900).

Geoscience Resources (2990 Anthony Rd., P.O. Box 2096, Burlington, NC 27216; 919-227-8300; 800-742-2677) produces the "Drift Globe" ($149.50), a 12-inch globe created to aid in understanding continental drift. The globe features nine continents that may be positioned anywhere on the globe with Velcro. Orientation marks on each continent match crosses on the globe, which represent 50-million-year intervals. The globe comes with a

Modern School (524 E. Jackson St., Goshen, IN 46526; 219-533-3111; 800-431-5929) produces seven globe models for classroom use, ranging from a 12-inch Student Political Globe ($37) to $161 for any of five 16-inch models on floor stands and casters.

National Geographic Society (17th & M Sts. NW, Washington, DC 20036; 202-921-1200) sells two physical and four political 12-inch and 16-inch globes, some illuminated, ranging from $49 to $119.

Nystrom (3333 Elston Ave., Chicago, IL 60618; 312-463-1144; 800-621-8086) produces 12-inch and 16-inch raised-relief globes with a variety of mountings. Made of vinyl, they are designed to be touched and are advertised as "guaranteed unbreakable." Models range from $73 to $227. Another Nystrom product is "Form-a-Globe," designed for classroom use, a build-it-yourself paper globe demonstrating how maps and globes differ from and are related to each other. A set of 30 "Form-a-Globes" is $18.75.

Panoramic Studios (1104 Churchill Rd., Wyndmoor, PA 19118; 215-233-4235) has manufactured large, intricate globes used by many news organizations, as well as in class-rooms and boardrooms. Currently, it produces only one globe, the 30-inch relief globe ($2,000), which is a popular geographic tool for teachers of the blind.

Rand McNally (P.O. Box 7600, Chicago, IL 60680; 312-673-9100) produces about 15 globes designed for educational use, ranging in price from $40 to $695. The terrestrial globes are divided into three series: Political, featuring country boundaries in contrasting colors; International, showing political boundaries in ribbons of blue; and World Portrait, showing the Earth as seen from space, with natural vegetation and underwater topography. Rand McNally also makes the Geography-Physical Globe, the largest model of the earth manufactured today, in 6-foot and 3-foot diameters. In addition, the company produces Geography-Physical plaques, each depicting a single continent, with diameters ranging from two feet to 60 inches.

Replogle Globes Inc. (2801 S. 25th St., Broadview, IL 60153; 312-343-0900) produces a variety of 12-, 16-, 20-, and 32-inch globes in both blue-ocean and antique styles, including several illuminated models. Their newest globe, the "President Antique" ($2,200), is a 20-inch illuminated globe with "touch-on" light control; three levels of brightness can be obtained by touching the meridian. The 9-inch "Weather

Watch" ($125) has a specially designed base that can aid in tracking global weather. Other popular models include the 16-inch nonilluminated "Commander," an antique-style floor model that retails for $80, and the 16-inch "Lafayette," an illuminated globe that comes with a hardwood base and is priced at $500.

Replogle also offers a 12-inch moon globe and a 12-inch celestial globe. Developed under the auspices of NASA, the moon globe ($28.95; 382455) has a three-dimensional look that reveals craters, seas, and mountains. The celestial globe ($32.95; 38848) locates 1,200 stars from 1st to 4th magnitude and shows constellations, star clusters, new stars, and double stars.

Spherical Concepts (2 Davis Ave., Frazer, PA 19355; 215-296-4119) produces silk-screened clear-acrylic globes, available in various colors. Models include "The Continental," in 12-, 16-, or 30-inch diameters, and "The International," a 12-inch political globe. Most models come with a wax marking pencil to use on the globe or an instruction booklet. Prices range from $29 for the 5-inch "Earthball," to $850 for the 30-inch "Continental" globe. Several celestial globes are also available (see "Astronomical Charts and Maps").

Trippensee Planetarium Co. (301 Cass St., Saginaw, MI 48602; 517-799-8102) produces hand-crafted clear-acrylic globes that are fashioned of individually blown hemispheres. The graphics are silk-screened on the interior surface, permitting the globe to be marked with a china marker, which is easily wiped off. Globes available include the "Harmony Globe," four-color 12- and 16-inch globes ($110 to $255) in three different color schemes; the "Flair Globe" ($27 to $49; 6 inches), in three color schemes; and the "Spectrum Globe" ($130 to $275; 12 and 16 inches), a nine-color globe depicting countries, oceans, major cities, and states and provinces.

■ **See also: "Astronomical Charts and Maps" and "World Maps."**

ET CETERA

Geography Education Materials

If you read the news, you've heard the stories: Today's young people—and Americans in particular—are geography illiterates. The tales would be funny if they weren't so serious. For example:

■ In one survey of the University of Miami, 30 percent of the students could not locate the Pacific Ocean on a world map.
■ Nearly half of the college students in a California poll could not identify the location of Japan.
■ In Baltimore, 45 percent of those tested could not respond correctly to the instruction: "On the attached map, shade in the area where the United States is located."
■ Twenty-five percent of the students tested in Dallas could not identify the country that borders the United States on the south.
■ In Boston, 39 percent of the students could not name the six New England states.
■ When a television station in Washington, D.C., asked 500 high school students to name the large country north of the United States, 14 percent guessed incorrectly. One said Delaware.

The statistics go on and on. According to a 1988 Gallup poll, Americans ranked among the bottom third in an international test of geographic knowledge, and those aged 18 to 24 came in last. Even less encouraging is that in this day of global communications, test scores are getting worse. Compared to surveys taken in the 1950s, American students know even less about the world than ever.

In response to this crying need, the 1980s saw the rebirth of geography education, in the form of "Geography Awareness Weeks," "Geography Bees," and dozens of new products designed specifically to help fill this void.

A few states, including South Dakota and California, have implemented geography requirements for graduation. From members of Congress to local schoolteachers to map dealers to the National Geographic Society—which budgeted some $4 million to improve geography education—everyone seems to have gotten into the act. Geography is at last finding its way back onto the map.

While most of the hundreds of maps, globes, atlases, software, and related products are worthy of being enlisted in this campaign, here are some of the products specifically aimed at the education market:

GAMES AND PUZZLES

Aristoplay (P.O. Box 7028, Ann Arbor, MI 48107; 313-995-4353; 800-634-7738) sells "Where In The World?" ($35), a colorful multi-use game designed to enhance world awareness for children six years old to adults. Included are "Crazy Countries" card games acquainting players with the countries of each continent, and "Statesman, Diplomat and Ambassador" board games introducing players to geographical locations and cultural and economic aspects of countries of the world.

Austin Peirce (19 Prairie Dunes, Hutchinson, KS 67502; 316-663-4192) produces the colorful, information-packed "Puzzlin' State" series of jigsaw puzzles. The puzzles feature county boundaries, county seats, rivers, lakes, points of interest, and historical trivia, as well as state flags, seals, animals, birds, rocks, plants, and trees. Puzzles for more than 20 states ($9.95 each), as well as one of the U.S. ($12.95), are also available through **Globe Pequot Press**, 10 Denlar Dr., Box Q, Chester, CT 06412; 203-526-9571; 800-243-0495.

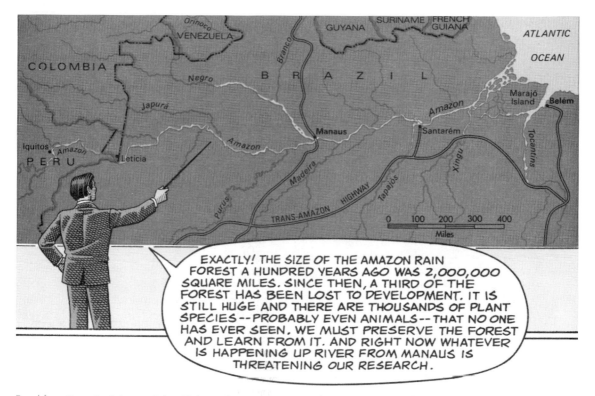

Panel from Captain Atlas and the Globe Riders®, Mystery of the Amazon, *part of a series of educational comic books from Hammond designed to teach geography and other skills.* © Hammond, Inc., Maplewood, New Jersey.

Educational Insights (19560 S. Rancho Way, Dominguez Hills, CA 90220; 213-637-2131; 800-367-5713) produces various fun games focusing on geography education. One example is "Geo-Safari" ($99.95), an electronic map game suitable for classroom or individual use that can both teach and quiz users. The game comes with 18 maps; a set of 44 additional maps is $19.95. "3-D Landforms" ($14.95) is a tactile teaching tool to help students visualize such geographic features as mesas, straits, reservoirs, and isthmuses. The kit includes reproducible maps and a teacher's guide. "Project: Earth" ($29.95) is a make-your-own-globe kit, consisting of a 14-inch styrofoam sphere that you cover with papier-mache, on which you then trace the outlines of the continents, add physical features, paint the countries, then plant flags. The company's catalog features additional games and map-skill products.

GeoLearning Corp. (555 Absaraka St., Sheridan, WY 82801; 307-674-6436) produces many fun and interesting map games, available in stores or directly from GeoLearning:

■ "The Spilhaus GeoGlyph" ($24.95), a puzzle showing the continuous surface of the world. More than 100 correct maps can be created from the pieces.
■ "GeOdyssey—The World Game" ($14.95), a game where players compete to build a world map, tile-by-tile, testing their geography knowledge by identifying continents, coastlines, and other geographical features.
■ "GeOdyssey Deluxe" ($24.95), a version of GeOdyssey featuring larger nine-color playing tiles and labeled oceans.
■ "GeoCards" ($7.95), a colorful game featuring a deck of 48 cards, with each card representing a different geographical location.
■ "TectoniCube" ($4.50), a fascinating seven-

The National Geography Bee

In which two hemispheres is North America located? Which continent contains no countries? The geographic center of North America is located in which state? Name the four oceans.

These are among the kinds of questions asked each year of thousands of fourth- through eighth-graders around the United States. Those who give the most correct answers qualify to participate in the National Geography Bee and vie for a $25,000 college scholarship and other prizes.

The annual event was established in 1989 by the National Geographic Society, part of the organization's multimillion-dollar campaign to improve geography education in the United States. Much like the age-old

spelling bees, the geography bee is conducted in an oral format, with questions addressing the broad range of geography topics, from the Earth itself to its inhabitants and cultures. Competitions begin at the school level, then proceed to the state level and culminate with the National Championships, held in the spring in Washington, D.C.

For more information on the National Geography Bee, contact the **National Geographic Society**, 17th & M Sts. NW, Washington, DC 20036; 202-921-1200.

(By the way, the answers to the above questions are: northern and western; Antarctica; North Dakota; and Atlantic, Artic, Indian, and Pacific.)

---

sided geometric form showing how the Earth's tectonic plates fit together. The cube illustrates the plates' various ridges, boundaries, zones, and faults.

■ "Flight Lines" ($5.95), a game featuring "Geography-Triangles," such as Los Angeles to Sydney, Paris to Rio, Tokyo to Chicago, and many more.

Geoscience Resources (2990 Anthony Rd., P.O. Box 2096, Burlington, NC 27216; 919-227-8300; 800-742-2677) produces the "Drift Globe" ($149.50), a 12-inch globe created to aid in understanding continental drift. The globe features nine continents, which may be positioned anywhere on the globe with Velcro. Orientation marks on each continent match crosses on the globe, which represent 50-million-year intervals. The globe comes with a cartoon-illustrated guide that discusses climate and fauna during each time period. Another product is "Puzzle of the Plates" ($14), released by the American Geophysical Union, a puzzle that features accurate reconstructions of the Earth's tectonic plates.

Hammond Inc. (515 Valley St., Maplewood, NJ 07040; 201-763-6000) publishes *Captain Atlas and the Globe Riders*, a colorful 48-page comic book that includes a 31″ X 21″ terrain and vegetation map of the Amazon basin.

National Geographic Society (17th & M Sts. NW, Washington, DC 20036; 202-921-1200) sells "National Geographic Global Pursuit" ($19.95), with nearly 1,000 geography trivia questions to test the global knowledge of students and adults alike. Includes a world map. A set of 324 additional trivia cards is $7.95.

The Nature Co. (P.O. Box 2310, Berkeley, CA 94702; 415-524-9052, 800-227-1114, plus 32 retail stores) offers "Continental Drift" ($19.95), a magnetic map with movable continents that encourages you to imagine the alignment of the continents at any time over the past 220 million years. When you tire of that, the continents double as refrigerator magnets.

Nystrom (3333 Elston Ave., Chicago, IL 60618; 312-463-1144; 800-621-8086) produces 12- and

16-inch raised-relief globes with a variety of mountings. Made of vinyl, they are designed to be touched and are advertised as "guaranteed unbreakable." Models range from $73 to $227. Another Nystrom product is "Form-a-Globe," designed for classroom use, a build-it-yourself paper globe demonstrating how maps and globes differ from and are related to each other. A set of 30 "Form-a-Globes" is $18.75.

Pacific Puzzle Co. (378 Guemes Island Rd., Anacortes, WA 98221; 206-293-7034) offers colorful hardwood puzzles, including three maps of the U.S. ($23 to $60), a political world map ($16), a physical world map ($16), two Dymaxion world maps ($22 and $40), and five continent puzzles ($16 to $19 each). Shipping is extra; write for catalog.

Parkwest Publications (238 W. 72nd St., New York, NY 10023; 212-877-1040) distributes the British-made "The Tarquin Globe" ($6.95), a paper build-it-yourself globe in full color, featuring all countries and some 1,200 major cities, seas, rivers, and other features. The kit is accompanied by a 24-page booklet full of facts and ideas. A similar product, "The Tarquin Star-Globe" ($6.95), is for constructing a three-dimensional globe of the night sky.

Pendergrass Publishing Co. (P.O. Box 66, Phoenix, NY 13135; 315-695-7261) publishes *Our United States* ($7.95; teacher's key free), a 256-page book at reading level 4-5 that introduces students to American history and geography with full-page graphically illustrated maps, as well as text.

Map Skills

George F. Cram Co. Inc. (P.O. Box 426, Indianapolis, IN 46206; 317-635-5564; 800-227-4199) offers the "Discovery Series," three sets of map-skills teaching units, and a 16-inch "Discovery Globe" in two mountings ($107 and $124). The classroom units ($115 to $185) include teaching legends, maps, inset maps, a teacher's guide, and hands-on activity books.

Cram also publishes a "Map Mastery Program" for all grade levels that includes wall maps and workbook exercises. Eighteen titles are available, covering the U.S., the world, continents, and geographical terms; $7.95 each.

Hammond Inc. (515 Maple St., Maplewood, NJ 07040; 201-763-6000; 800-526-4953) publishes "Captain Atlas and the Globe Riders," an innovative series of action comics for ages 8-12 that teach geography and map skills. Each 48-page book includes a folded 31″ X 21″ full-color wall chart or map that reinforces the subject matter and skills learned in each issue. Titles available include "Journey to the North Pole," "Mystery of the Amazon," "Return of the Dinosaurs," and "Mystery of the Tasmania Triangle." Price of the books is $6.95 each.

Modern School (524 E. Jackson St., Goshen, IN 46526; 219-533-3111, 800-431-5929) publishes *Map & Globe Skills* I, a 10-course unit centered on 14 large color charts, 24″ X 17″ and 17″ X 12″. The units cover map definition, symbols, direction, distance, scale, earth-globe relationship, projections, latitude-longitude, types of maps, and "our map address." Activities and tests are reproducible by photocopier for classroom use. The $42 set is intended for basic-beginners. (The company also produces a computer version; see "Software," below.) Another product, "Geography Terms" ($30), is a colorful 44″ X 32″ poster that identifies more than 50 geographical terms. A smaller 8-1/2″ X 11″ reproduction is priced at $13 for a set of eight.

National Geographic Society (17th & M Sts. NW, Washington, DC 20036; 202-921-1200) offers a filmstrip, "Maps and What We Learn From Them" ($90), intended for grades 3 through 6. The filmstrip (running time: 12-15 minutes) and accompanying cassettes reveal what maps reflect about the face of the Earth, and discuss the various types of maps.

Nystrom (3333 Elston Ave., Chicago, IL 60618; 312-463-1144; 800-621-8086) offers map and

A *sampling of geography education materials from GeoLearning Corp.*

90232; 213-839-2436; 800-421-4246) distributes a wide variety of map- and geography-skill products from a variety of publishers. Included are maps, globes, games, books, atlases, posters, reproducible masters, charts, photo aids, software, and a "Newsmap of U.S. History," three sets of double-sided charts that trace America's development from the arrival of the early pre-Columbian explorers to the conclusion of the war with Mexico in 1848 ($170). A catalog is available.

globe skills for several levels. Its "Primary Social Studies Skills" comes in three levels, designed to teach key concepts to early elementary school students ($210 to $235 for a class of 30). For older students, Nystrom offers six "Map and Globe Skills Programs" that include student books, maps, marking pens, and teachers' guides ($110 to $345 for a class of 30).

Pendergrass Publishing Co. (P.O. Box 66, Phoenix, NY 13135; 315-695-7261) publishes *Learning to Use Maps* ($4.25; teacher's key $1), an 80-page, low-reading-level book demonstrating basic map directions, map legends, abbreviations, scales, symbols, topography, and other features. Another title, *Finding Ourselves, or Where in the World Am I?* ($3.75; teacher's key $1.25), is a 44-page book teaching students to locate themselves, by map, in town, city, state, and country.

Social Studies School Service (10200 Jefferson Blvd., Rm. 1, P.O. Box 802, Culver City, CA

J. Weston Walch (321 Valley St., P.O. Box 658, Portland, ME 04104; 207-772-2846, 800-341-6094) offers several items, including: "Crossword Puzzle Maps in U.S. History" (grade 7 and up), offering a comprehensive review of U.S. history ($19.95 for 50 masters); "Current World Geography Problems" (grade 9 and up), teaching geography as a vital factor in world affairs ($11.95 for 26 masters); *Where On Earth: Understanding Latitude and Longitude* (grade 7 and up), a 100-page reproducible book ($15); "Mastering Geography" (grades 7-10), "clever puzzles and word games that add spark to your daily routine" ($11.95 for 26 masters); "Geography Skills Activity Pack" (grade 6 and up), a complete teaching unit including 64 masters and 18 colorful posters ($39.95); *Discovering America Through Map Activities* (grade 8 and up), a 76-page reproducible book plus 10 105-page student books ($81.95); and *Geography Teacher's Success Kit* (grades 7-12), a 148-page reproduc-

Some of the "Puzzlin' State" jigsaw puzzles from Austin Peirce.

ible book that contains "everything you need for success in teaching world geography" ($15).

SOFTWARE

Broderbund Software Inc. (17 Paul Dr., San Rafael, CA 94903; 415-492-3200) publishes *Where in the World Is Carmen Sandiego?*, *Where in the USA Is Carmen Sandiego?*, and *Where in Europe Is Carmen Sandiego?*, a series of secret-agent adventure games that teach geography, research, and reasoning skills. In the programs, recommended for grades 4 to 9, Carmen and her gang of thieves steal the world's priceless treasures and landmarks. The player's task is to capture her and her crooked cronies. To track them down, players begin at the scene of the crime and follow endless clues, including topography, cities, flags, currency, languages, historical events, and other facts. Programs are available for Apple II/IIe/IIc/IIgs, IBM PC and compatibles, Tandy and compatibles, Macin-

tosh, Amiga 500/1000/2000, and Commodore 64/128 computers. Prices range from $34.95 to $44.95 depending on the edition. School editions, with teacher's guide and additional disks, are $44.95 to $54.95.

CompuTeach (78 Olive St., New Haven, CT 06511; 203-777-7738; 800-448-3224) publishes *See the U.S.A.* ($49.95), which lets users take any of a series of imaginary drives across America. On the trip, users learn about states, capitals, famous places, and state facts, such as flowers, birds, and mottos. Intended for ages 8 and up, it operates on the Apple II or IBM PC and compatible computers.

Great Wave Software (5353 Scotts Valley Dr., Scotts Valley, CA 95066; 408-438-1990) publishes *American Discovery* ($69.95), a Macintosh geography game. The program starts with Find the States. On a map of the U.S., you click on the appropriate state or states in response to a question like, "Which state is Kansas?" When you answer correctly, a display of a horse race moves toward the finish line. Then you graduate to Click the States, Spell the States, Find the Capitals, and so on, including Rivers and Lakes. The program's customization feature allows teachers or parents to create interactive American geography and history lessons.

Maxis Software (953 Mountain View Dr., Ste. 113, Lafayette, CA 94549; 415-376-6434) publishes *SimCity*, an urban simulation game created for young people but one that has captured the eye of planning professionals and architects. The program—intended for those aged 12 and up—offers an interactive graphic representation of the dynamics of urban life. By clicking on icons, users can—budget permitting—build roads and transit lines, power plants (coal or nuclear), parks, police stations, airports, and many other urban features. As the city grows, the program calculates the impact of each new development on population, pollution, traffic, crime, and other variables. To liven things up, random-occurring events—nuclear meltdowns and airline crashes, for example—emulate real life (or some dramatic variation thereof). Prices are $49.95 for the Macintosh version, $44.95 for the Amiga version, and $29.95 for the Commodore 64/128 version.

Mindscape, Inc. (Educational Div., Dept. D, 3444 Dundee Rd., Northbrook, IL 60062; 312-480-1948; 800-221-9884), as part of its *Social Studies Explorer Set*, produces a four-disk "World Geography Set." The disks cover "Asia and Australia," "Central United States," "Eastern United States," and "Western Europe." The program allows students to identify the state or country they're "visiting" and answer questions about the locale. Price is $39.95 per disk. Another program, *America Coast to Coast* ($49.95), is described as a "high-resolution

color, animated map" that illustrates the relative sizes and locations of the 50 states. Students select a state, then learn its location, capital, shape, motto, and other facts. Both programs are available for Apple IIe/IIc, Apple II+, and IBM-compatible computers.

Modern School (524 E. Jackson St., Goshen, IN 46526; 219-533-3111, 800-431-5929) publishes *Computer Map Skills* for all Apple II computer models. The colorful screen images teach skills similar to *Map & Globe Skills* I (see "Map Skills," below) and comes on two disks: "Beginner (Readiness)" and "Basic." Price is $44 each or $79 for both.

Soft Horizon (P.O. Box 2115, Harker Heights, TX 76543; 817-699-0493) publishes *Know Your World*, which teaches countries, capitals, and major cities of the world using maps and challenging learning activities. The program ($39.95), for Macintosh and IBM PC and compatibles, teaches and then quizzes users, scoring on both correct answer and correct spelling.

Springboard Software Inc. (7808 Creekridge Cir., Minneapolis, MN 55435; 612-944-3915) publishes *Atlas Explorer*, a world geography program that displays maps of states and countries along with general facts about the locations. Moreover, the program tests users on any location—a continent, country, state, province, or city—scoring on both accuracy and speed of answers. The program ($49.95) is intended for users aged 8 to 18 and runs on Apple IIe/IIc/IIgs, IBM PC and compatibles, Tandy and compatibles, and Macintosh Plus/SE/II computers. Also available is an educator's Lab Pack ($100) consisting of five disks and a manual, for Apple and IBM versions only.

■ **See also: "Map Software" and "Map Stuff."**

Map Accessories

Having a map is a good first step toward finding your way, but sometimes even the best map isn't quite enough. Here are some accessories that may help you plot your route or reach your destination. Many of these (or similar) products are readily available in map, travel, and camping-supply stores; or you may contact the manufacturers directly. Prices shown are suggested retail prices; many retailers and mail-order houses discount these items. Be aware that there may be shipping and handling charges added for mail orders.

MAGNIFIERS

These come in several shapes, sizes, and materials, depending on whether they will be used in a library or must be sufficiently compact to fit in a pocket or knapsack.

■ **Tasco Sales Inc.** (7600 NW 26th St., Miami, FL 33122; 305-591-3670) carries several sizes of pocket loupes, compact powerful fold-out magnifiers designed to slip into a pocket or purse. Price range is $4.95 to $29.95.
■ **Bausch & Lomb** (P.O. Box 22810, Rochester, NY 14692; 716-338-6000; 800-452-6789) manufactures a wide array of magnifiers, some with built-in lights, called "minilights." Prices range from about $4 to $35.

MAP CASES

A good map gets a lot of use—and abuse. It gets folded and refolded, usually not back to its original form. Used on a hike or a boat, it gets dirty or wet. Map cases provide protection, letting you study the map while it remains safe and dry. There are several varieties of map cases, not all are watertight, for example:

■ **Omniseal** (83 S. King St., Seattle, WA 98104; 206-587-3755) produces "Omniseal Waterproof Pouches" of varying sizes ($1.45 to $5.95),

some designed specifically for maps. Also available are "Omniseal Courier" tubes, ideal for carrying rolled maps ($2.35 to $4.15).
■ **The Outsiders** (Box 626, North Salt Lake, UT 84054; 801-292-7354) manufactures clear "Show-It" map cases, available in many stores for about $4.

MAP PRESERVERS

One alternative to a map case is a map preserver—basically, a liquid or spray applied to maps that renders them waterproof and increases their durability.

■ **Adventure 16** (4620 Alvarado Canyon Rd., San Diego, CA 92120; 619-283-2374; 800-854-0222, in Calif.) distributes two products: "Map Life," a map-treatment liquid that also permits writing on a map with a grease pencil, and a similar product, "Map Shield." Both retail for around $4.
■ **Martensen Co.** (P.O. Box 261, Williamsburg, VA 23187; 804-565-1760) manufactures "Stormproof," a clear penetrating liquid treatment that impregnates paper maps. An eight-ounce can will treat about a dozen topographic maps or four large nautical charts. Available in many map and camping stores in eight-ounce ($5.50) and 16-ounce ($8.50) sizes.
■ **Tamerica Products Inc.** (20722 Currier Rd., City of Industry, CA 91789; 714-594-3888; 800-822-6555) produces but does not retail "Print Protectors," plastic sleeves that can be used in place of lamination, glass or acrylic framing. They are available in a variety of sizes, including 30″ X 31″ and 42″ X 41″. Custom-order sizes are also available.

MAP MEASURERS

These help you obtain accurate measurements of distance from maps, whether measuring hikes, highways, or the high seas. Most models

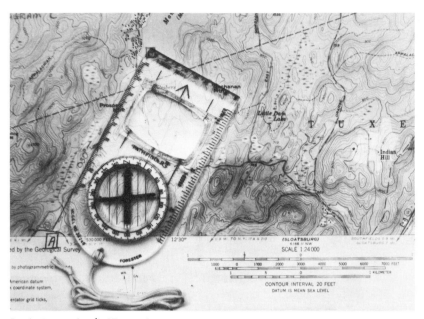

Precise International's "Forester" compass.

measure both miles and kilometers; some also measure nautical miles.

■ **Precise International** (3 Chestnut St., Suffern, NY 10901; 914-357-6200; 800-431-2996) sells "Pathfinder," a topographic map measurer that converts maps into miles using 1:24,000, 1:31,680, and 1:62,500 scales; and "Roadrunner," similar to "Pathfinder," which is better suited for road trips, and which also measures in nautical miles. Both are $9.95. Also available are compasses and altimeters.

■ **Johnson Camping** (P.O. Box 2050, Binghampton, NY 13902; 607-779-2200) sells "Silva Map Measure Type 40," calibrated to read topo maps in scales of 1:24,000, 1:62,500, and 1:250,000 ($19.35).

MAP DISPLAYS
■ **Versa-Lite Systems, Inc.** (5834 Kirby Rd., Clinton, MD 20735; 301-297-7710; 800-638-0982) produces "Versa-Lite Display System," display boards with thumbtack-like lights that automatically go on or blink when placed anywhere on the board. Prices range from

$395 to $3,000, depending on board size and other features.

COMPASSES
When hiking, a compass can help you orient yourself with landmarks on a map, keeping you on track. Compasses are sold by dozens of companies, including several listed above. While there are a seemingly endless number of compass "bells and whistles"— sizes, mountings, magnifiers, wrist attachments, luminescence, mirrors, and on and on—almost any compass will do. To function properly, it need only be capable of pointing toward magnetic North and have some direction and degree markings around its perimeter.

■ **Precise International** (3 Chestnut St., Suffern, NY 10901; 914-357-6200; 800-431-2996) produces "Pathfinder" compasses, available in assorted styles.

■ **Suunto Oy USA** (2151 Las Palmas Dr., Carlsbad, CA 92009; 619-931-6788) distributes a variety of Suunto compasses, including the "A-Series" compasses, which includes the competitive orienteering range and can fit into the hand of a child; a "Geologist's Compass," complete with a protractor, level, and clinometer; the "M-3" series, designed for scouting and orienteering under the most difficult conditions; and "thumb compasses," small compasses with a strap designed to fit on the thumb. A number of marine compasses are also available. Suunto also offers compass accessories, such as compass cases, pouches, and rubber covers. A catalog is available upon request.

Map Organizations

American Association of Petroleum Geologists (AAPG, P.O. Box 979, Tulsa, OK 74101; 918-584-2555) is a membership organization of geoscientists dedicated to encouraging scientific research and advancing the science of geology, particularly as it relates to petroleum and energy minerals. AAPG publishes and distributes to the public many geologic and energy-related maps. AAPG holds an annual meeting and regional meetings throughout the year. Annual membership dues vary from $10 to $48, and entitle members to receive AAPG's three publications: *The Explorer*, a monthly magazine, *The Bulletin*, a monthly technical journal, and *Geobyte*, a bimonthly magazine addressing computer usage in geology.

The American Congress on Surveying and Mapping (ACSM, 210 Little Falls St., Falls Church, VA 22046; 703-241-2446) is a large (more than 11,000 members) organization devoted to advancing and promoting the surveying and mapping profession. There are various branches of ACSM, including "Government Affairs," active in the federal legislative and regulatory arenas; the "State Legislative Clearinghouse Program," which monitors major legislation affecting the mapping and surveying profession and issues monthly updates; the "ACSM/NSPS Political Action Committee," established to enable members to contribute to congressional campaigns; "Education," offering short courses in hydrographic and technician certification, three annual conventions, and student fellowships and scholarships; "GPS Education," offering a series of short courses on global positioning systems; and "Field Services Program," serving as a liaison between ACSM, state societies, and sections. ACSM produces a number of books, videos, and technical papers; a catalog is available upon request.

Membership is based on concurrent memberships in ACSM and a member organization; dues vary. Membership benefits include a subscription to ACSM *Bulletin*, a bimonthly magazine; subscriptions to ACSM's professional periodicals; discounts on ACSM publications; participation in ACSM's standing and special committees; and a MasterCard program. ACSM is the umbrella organization for the American Association for Geodetic Surveyors, the American Cartographic Association, and the National Society for Professional Surveyors; information on these groups can be obtained from ACSM.

The American Geographical Society (AGS, 156 Fifth Ave., Ste. 600, New York, NY 10010; 212-242-0214) is an independent, nonprofit corporation founded in 1851 to gather and disseminate geographical information. AGS sponsors expeditions, lectures, conferences, and symposiums; conducts research on geographical topics; and publishes educational and scientific periodicals, books, maps, and atlases. Individual membership ($20 annually) includes receipt of the AGS *Newsletter* and discounts on the *Geographical Review*, *Focus*, and other AGS publications and activities. Corporate and student membership rates are also available. AGS publications are available to nonmembers; a list is furnished upon request.

American Library Association, Map and Geography Roundtable (50 E. Huron St., Chicago, IL 60611; 312-944-6780) is committed to the exchange of ideas between persons working with or interested in map and geography collections, to improving communication and cooperation between map and geography libraries, and encouraging the improvement of education and training of map and geography information specialists. It meets twice a year, in connection with American Library Association conferences, and holds business meetings and workshops on various topics throughout the year. Map and Geography Roundtable publications include *base line*, a bimonthly newsletter; *Meridian*, a semi-annual journal, presenting articles on maps; "Open File Reports," working documents for cartographic professionals; and "Occasional Papers," a series of monographs of interest to a wider audience. The organization is open to anyone; annual dues are $10.

American Society for Photogrammetry and Remote Sensing (ASPRS, 210 Little Falls St., Falls Church, VA 22046; 703-534-6617) is a nonprofit scientific and technical association founded in 1934 and dedicated to advancing knowledge in and encouraging use of photogrammetry, geographic information systems, and remote sensing. ASPRS membership includes more than 7,500 people involved in or studying photogrammetry, remote sensing, geographic information systems, and related sciences. ASPRS sponsors annual conventions and workshops and maintains an extensive publications program. Membership ($18 to $125) includes a subscription to ASPRS's monthly journal, *Photogrammetric Engineering and Remote Sensing*; invitations to local regional meetings, seminars, field trips, continuing-education workshops and symposiums; and special rates for life and accident insurance.

Association of American Geographers (AAG, 1710 16th St. NW, Washington, DC 20009; 202-234-1450) is a nonprofit group founded in 1904 with the goals of advancing professional studies in geography and encouraging the application of geographic research in education, government, and business. AAG boasts a membership of 5,900 students and professionals from around the world. The organization's activities include sponsoring research projects that apply a geographic perspective to national issues; furthering geographic education through Geography Education National Implementation Project; supporting symposiums; and producing a number of publications. Through its 40 "Specialty Groups," which range from "Biogeography" to "Microcomputers" to "Socialist Geography," members can meet and

work with others who share their interests. Membership ($35 to $1,400) includes receipt of the AAG *newsletter*, and entitles members to receive two quarterly scholarly journals, *Annals of the Association of American Geographers* (featuring scholarly articles, commentaries, and book reviews) and *The Professional Geographer* (including research articles, technical reports, and book and software reviews), as well as many other benefits.

Association of Map Memorabilia Collectors (AMMC, c/o Siegfried Feller, 8 Amherst Rd., Pelham, MA 01002; 413-253-3115) is a small association founded to provide and exchange information and news between collectors of maps and maps appearing in other forms (postcards, stamps, envelopes, etc.). The primary focus of AMMC is the publication of the quarterly *Cartomania*, a chatty newsletter full of information, illustrations, and member news. Membership, open to anyone, is $10 annually in the U.S. and Canada, $13 elsewhere. Single issues of *Cartomania* are also available.

Carto-Philatelists (c/o Thomas H. Sutter, Secty., 303 S. Memorial Dr., Appleton, WI 54911) is a society of map-stamp enthusiasts, founded in 1955, which promotes cartophilately. Through the society, members are able to exchange information, trade ideas on collecting and mounting their miniature maps, and, occasionally, meet one another. *The Carto-Philatelist*, an illustrated quarterly publication, features various background articles, listings of new issues, and member news. Membership is open to anyone; dues are $8.50 in the U.S., $9.50 in Canada and Mexico, and $10 elsewhere.

Chicago Map Society (c/o Secty./Treas., 60 W. Walton St., Chicago, IL 60610; 312-943-9090), claiming to be the oldest map society in North America, was founded in 1976 to bring together people interested in all aspects of maps and mapping. The society holds monthly meetings (September through May) at Chicago's Newberry Library, featuring speakers, field trips, and other events. The society publishes a quarterly newsletter, *Mapline*, as well as the *World Directory of Dealers in Antiquarian Maps* (compiled by George Ritzlin). Membership is open to anyone; annual dues are $15 and include a subscription to *Mapline*. The society welcomes guests at its meetings.

Comité Européen des Responsables de la Cartographie Officielle (CERCO) (CERCO, Institut Géographique National, Abbaye de la Cambre 13, B-1050 Brussels, Belgium) is an organization promoting mutual information, consultation, and cooperation in cartography among the official mapping agencies of European countries. The committee meets once a year to discuss developments in cartography and related fields and to share ideas.

Geography Education National Implementation Project (GENIP, 1710 16th St. NW, Washington, DC 20009; 202-234-1450) is a joint project of The National Council for Geographic Education, the Association of American Geographers, the American Geographical Society, and the National Geographic Society. GENIP was formed in 1984 to improve the status and quality of geographic education in grades K-12 in the United States. GENIP produces various publications designed to aid geography educators and is active in many aspects of geography education, including teacher training, developing new learning materials, providing curriculum development guidelines, developing state and national networks to support geographic education, overseeing the definition of teacher certification and competency standards, and establishing links for funding to public and private agencies. GENIP publishes a monthly newsletter, GENIP NEWS, issued to all members of the GENIP Network. Membership is free.

Geological Society of America (GSA, P.O. Box 9140, Boulder, CO 80301; 303-447-2020), a membership organization of 16,000 geoscientists worldwide, was founded in 1888 to publish scientific literature and organize scientific meetings. It holds an annual fall meeting; regional sections hold spring meetings. GSA members receive *Bulletin*, a monthly scientific journal; *Geology*, a monthly journal featuring book reviews and editorials; and GSA *News & Information*, the group's newsletter. Members are entitled to discounts on GSA publications and maps, and to participate in GSA's specialized divisions and receive their newsletters; insurance discounts and discounts on subscriptions to geologic magazines also come with membership. Annual member dues vary.

International Cartographic Association (ICA, c/o Donald Pierce, Secty. General, 24 Strickland Rd., Mt. Pleasant, W. Australia 6153; (09) 316-1876) is an international society aimed at advancing the study of cartographic problems; the coordination of cartographic research between different nations; and the organization of conferences, meetings, and exhibitions. Members of ICA form various commissions and working groups on such subjects as "Advanced Technology," "Population Cartography," and "The Marketing of Spatial Information." Many of these commissions and working groups publish the results of their research and seminar papers; these publications are available through Geo Books (Regency House, 34 Duke St., Norwich, England NR3 3AP). ICA is also pledged to assist with cartographic growth in developing countries and to promote cooperation with sister organizations, such as the ISPRS. Membership in ICA is limited to one organization for each nation; the ACA is the official U.S. representative to ICA.

International Map Dealers Association (IMDA, P.O. Box 1789, Kankakee, IL 60901; 815-939-4627) is a membership organization founded in 1982 to bring together map sellers, publishers, and others related to the field of map publishing and retailing. It holds an annual conference and trade show where its 200-plus members share ideas and attend workshops and meetings. Membership ($100 annually) includes subscription to *The Map Report*, a monthly newsletter containing industry and member news and feature articles.

International Society for Photogrammetry and Remote Sensing (ISPRS, c/o Chunji Murai, ISPRS Secretary General, Institute of Industrial Science, University of Tokyo, 7-22 Roppongi, Minato-ku, Tokyo, Japan; 011-813-402-6231 Ext. 2560) is an international association founded in 1907 to encourage cooperation and exchange of information among the various national societies of photogrammetry. ISPRS holds quadrennial congresses, where members report, review, and discuss developments in photogrammetry and remote sensing, as well as attending social events, technical and scientific tours, and excursions. Membership is available only to national societies; ASPRS is the U.S. representative to ISPRS.

Michigan Map Society (Clements Library, University of Michigan, Ann Arbor, MI 48109; 313-764-2347) is an 80-member society addressing the history of cartography, mapping, and mapping technology. Monthly meetings are held at the library September through May. Annual dues of $15 entitle members to attend meetings and receive the Chicago Map Society quarterly newsletter, *Mapline*.

National Council for Geographic Education (NCGE, Leonard 16A, Indiana Univ. of Pennsylvania, Indiana, PA 15705; 412-357-6290) was chartered in 1915 to promote geographic education at all levels of instruction. NCGE has directed its efforts to encouraging teacher training, providing active leadership in formulating geographic-education policies, developing effective geographic-education programs in schools and colleges, stimulating the production and use of geographic teaching materials and media, and enhancing public awareness of and appreciation for geography and geographers. NCGE members receive subscriptions to the following publications: *The Journal of Geography*, a bimonthly journal, and *Perspective*, a

newsletter appearing five times yearly. It also holds annual meetings, which provide members an opportunity to meet, and hosts a number of other activities. Membership is divided into eight categories, with fees corresponding.

National Council for the Social Studies (NCSS, 3501 Newark St. NW, Washington, DC 20016; 202-966-7840) is the largest association in the U.S. devoted solely to social studies. NCSS membership includes 25,000 members in all 50 states and in 69 foreign countries. NCSS is formed of three subgroups: The Council of State Social Studies Specialists (CS4); the Social Studies Supervisors Association (SSSA); and the College and University Faculty Assembly (CUFA). Membership dues vary.

National Geographic Society (NGS, 17th & M Sts. NW, Washington, DC 20036; 202-857-7000), one of the best-known geographic organizations, was founded in 1888 as a nonprofit scientific and educational organization "for the increase and diffusion of geographic knowledge." The society supports exploration and research projects and publishes *National Geographic*, an illustrated monthly magazine featuring articles on geographic and cultural topics. Membership dues are $18 per year, most of it designated for subscription to *National Geographic*. NGS is a major sponsor of "Geography Awareness Week" activities (usually held in November).

In 1985, realizing the lack of geography education in American schools, NGS created the "Geography Education Program," dedicated to encouraging geography education in school curriculums and to promoting public awareness of the need for geography education. The program sponsors the annual "National Geography Bee" for students (see "Geography Education Materials") and also coordinates state "alliances," funded by the NGS Education Foundation. The alliances, usually based in state universities, conduct "Summer Institutes" for teachers to learn how to teach geography; a more in-depth four-week national summer institute is held annually in Washington, D.C. The Geography Education Program produces *Update*, a twice-yearly newsletter featuring geography news and lesson plans. Subscription is free and available to anyone.

New York Map Society (c/o Map Div., New York Public Library, Fifth Ave. and 42nd St., New York, NY 10018; 212-930-0587) was founded in 1978 to support and encourage the study and preservation of maps and related materials, especially antique maps. The society holds monthly meetings (September through May) at the Museum of Natural History in New York City, featuring guest speakers and field trips. Membership ($20 annually) includes receipt of *Rhumb line*, the society's monthly newsletter providing a calendar of events and an "Exhibits of Note" section.

North American Cartographic Information Society (NACIS, 6010 Executive Blvd., Ste. 100, Rockville, MD 20852; 301-443-8075) is a professional society founded in 1980 to facilitate the exchange of information in the map community. NACIS membership consists of map-makers, collectors, libraries, users, educators, and distributors. Annual dues range from $5 to $35, and entitle members to receive *Cartographic Perspective*, a quarterly journal. NACIS holds an annual meeting, where members attend seminars and workshops, as well as smaller technical meetings throughout the year.

Special Libraries Association, Geography and Map Division (1700 18th St. NW, Washington, DC 20009; 202-234-4700) is the oldest geography and map librarians' association in the U.S. The group promotes the exchange of information in geography, map librarianship, and cartography. Division membership entitles one to attend and participate in annual conferences and quarterly meetings, as well as receive a subscription to the quarterly journal, *Bulletin*. The Division produces various publications, including *Map Collections in the United States and Canada: A Directory*. Annual dues are based on five classes, ranging from Student/Retired ($15) to Sustaining ($300).

Map Software

It was inevitable: the marriage of cartography and computers. Maps, after all, represent an efficient organization of information, a method of bringing together sometimes widely disparate data about a geographical entity. Computers, for their part, are the world's information machines. They enable us to organize, store, and manipulate vast quantities of data in ways never previously imagined. And so it would follow that computers would become an efficient means for organizing, storing, and manipulating the information contained on maps. It was a match made in heaven.

As computer graphics have made great strides, computerized mapping has become one of the fastest-growing areas of computer software. Coupled with the growth of "personal" desktop computers, it has created the ability for just about anyone to engage in what has come to be called "desktop mapping." There are now dozens of computer programs related to maps and mappings, some containing just a few simple black-and-white outline maps that you can insert into other documents, others containing highly sophisticated geographical data bases that, combined with cartographic software, enable you to produce sophisticated full-color maps in a few hours or even a few minutes. While such programs are not about to put government or commercial map-makers out of business—there will always be the need for the cartographer's keen eye and a geographer's intelligence and intuition—the technology is enabling more and more people to become instant map-makers, for business or pleasure.

The more sophisticated programs are finding their way into a variety of businesses, in both the public and private sectors. In Syracuse, New York, for example, the police department has used a relatively inexpensive program called *MapInfo* to track all "part one" crimes—homicide, rape, robbery, larceny, car theft, and arson. The city's Crime Analysis Unit enters information on victims and suspects into a computer and produces weekly and monthly maps summarizing crime patterns. The maps are distributed to the police on the street, allowing them to stake out areas of heavy activity. The department has estimated that the system has solved or stopped dozens of crimes.

GEOGRAPHIC INFORMATION SYSTEMS

At the more sophisticated end of the computerized-mapping spectrum is a geographic information system, or GIS. This is a data base, a complex set of information about a geographic unit that can be expressed digitally in graphic terms—a computer-generated map. A GIS might contain coordinates and other attributes of all the streets in a city, including the locations of signals, fire hydrants, buildings, and utilities lines. Starting with the map of the entire city, a user could click on a certain intersection, which would "explode" the view to show locations of water, sewerage, and electrical lines at that intersection. Geographic information systems have long been used on large mainframe computers, but only recently have they been accessible to those with IBM-compatible, Apple Macintosh, and other personal computers. Even the most sophisticated and expensive GIS have plummeted in price. One of the oldest, the 15-year-old *Ultimap* (**Ultimap Corp.**, 2901 Metro Dr., Ste. 312, Minneapolis, MN 55425; 612-854-2382) has dropped in price from $300,000 to a mere $50,000, a price that includes an Apollo minicomputer. But even at its original price, the program may be a bargain: Minneapolis used *Ultimap* to reduce its staff of street designers from 90 to 30, saving an estimated $25 million.

The greatest stimulus for geographic data bases is expected to come with publication of the 1990 U.S. census. The Census Bureau has created computer-readable maps known as TIGER (for "topologically integrated geographic encoding and referencing") files. The entire set

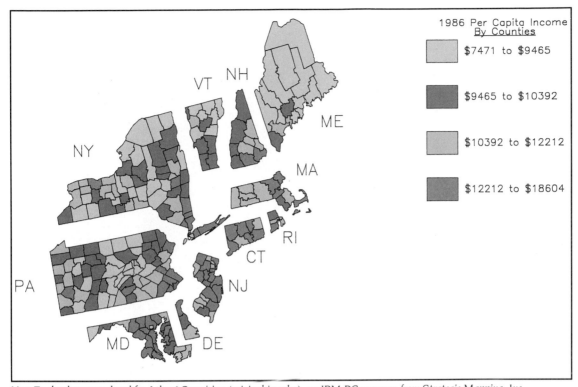

New England map produced by Atlas*Graphics *(original in color), an* IBM-PC *program from* Strategic Mapping, Inc.

of TIGER files (with data from the 1980 census and its updates) available for $56,025, provides a wealth of information about communities and their residents that can be incorporated into computerized maps. The 1990 version promises additional features and will be eagerly awaited by marketing firms, retailers, insurance companies, and other industries that rely on demographics. It is predicted that a flurry of new GIS products will directly result from the release of the 1990 TIGER files.

CD-ROMs

The ability to use GIS on low-cost personal computer systems has been enhanced by the growth of compact disks, known as CD-ROMs, which can hold incredible amounts of data on a thin removable platter identical to those that offer superior audio reproductions of music. A single compact disk can hold the equivalent of 250,000 printed pages of information; previously, such data would require more than

1,600 floppy disks. The capacity and speed of CD-ROMs make them ideal for holding geographic information. The potential created by CDs is truly astounding. Already, there are entire atlases on a single compact disk. What used to take several pounds of paper can now fit on a single ounce of plastic. (It is this same CD-ROM technology that will bring sophisticated navigational systems to the dashboards of our cars. See "Mapping the Future" for more on this coming technology.)

As most audiophiles already know, compact disks can also hold high-quality sound. Indeed, the new generation of compact disks—known as CDI, for "compact disk interactive"—will feature both pictures and sound. Already, Rand McNally is planning a CDI edition of its *Rand McNally Family Atlas*, which will feature the sounds of Old Faithful gushing and Kentucky Bluegrass bands strumming, among many other aural sensations.

In addition to the programs described below, which are intended for use with IBM-compatible and Apple Macintosh computers, there are hundreds of additional programs, routines, and subroutines. A comprehensive listing of more than 1,000 such products is contained in *Sources for Software for Computer Mapping and Related Disciplines*, published by the **U.S. Geological Survey's Earth Science Information Center.** Examples of programs described are *Orser*, produced by Pennsylvania State University's Office for Remote Sensing of Earth Resources, which "is used primarily to process Landsat data but also accommodates other remote-sensing inputs including Skylab and aircraft data"; *Automated Hydrographic Survey System*, from the U.S. Army Corps of Engineers District in Wilmington, Delaware, "a series of programs developed to take position coordinates and soundings acquired through use of an automated hydrographic survey system"; and "General Map Plot Library," from Arizona's Department of Transportation, consisting of "31 Fortran subroutines for drafting cadastral-planimetric road and highway features according to specified measurements and positions." The publication is available for $22 from Reston ESIC, 507 National Center, Reston, VA 22092; 703-860-6045.

Here is a summary of popular commercial software that contains map-making capabilities. As you will see, these range widely in price and capabilities, from less than $100 well into the thousands of dollars. (Map-related programs intended primarily for teaching geography skills can be found in "Geography Education Materials.") Some of these programs require that you use other "shell" programs. Keep in mind that the publishers of these programs frequently issue updated versions, often containing impressive enhancements. Because it isn't possible to fully list these programs' capabilities (or limitations) in the space below, you are encouraged to contact the publishers directly to obtain their current literature, sample disks, sample output, and price information. Also be aware that some of these programs are available through discount mail-order retailers, often at substantial discounts.

PROGRAMS FOR IBM PCs AND COMPATIBLES

Atlas*Graphics (Strategic Mapping Inc.., 4030 Moorpark Ave., Ste. 123, San Jose, CA 95117; 408-985-7400), intended primarily for business use, is designed to create presentation-quality maps in color. You may enter data to be mapped—sales figures, for example—in the program's spreadsheet or import data from other popular programs, including 1-2-3 and *dBase*, as well as ASCII and DIF files. The program includes base maps for all 50 states and some metropolitan areas, and additional files are available for zip code groups, telephone area codes, census tracts, and other demographic configurations, at prices ranging from $195 to $7,500. The program itself retails for $450 and requires 512K of memory, a graphics display, and a hard disk.

Desktop Maps (CGMpix, 41 Sutter St., Ste. 1850, San Francisco, CA 94104; 800-626-9541 ext. 1130 in California, 800-452-4445 ext. 1130 elsewhere) is a clip-art package containing more than 200 maps and related images in computer-graphics metafile (CGM) format. Maps included are states (in three versions: three-dimensional, line, and with neighboring states), regional, U.S., continents, and world. There are three versions, for use in *Ventura Publisher* ($149.95), *PageMaker* ($149.95), and *WordPerfect* ($99.95).

Electromap World Atlas (Electromap, Inc., P.O. Box 1153, Fayetteville, AR 72702; 501-442-2309) is a full-color atlas, available either on three diskettes or a CD-ROM. The program combines text and graphics to create a combination atlas, almanac, and world fact book. The package includes 238 colorful country, regional, topographic, and statistical maps, which can be accessed through an index map, as well as an on-screen alphabetical index. The program uses drop-down menus, making its maps accessible even to novice users. The $159 program requires 640K of memory, EGA or VGA graphics, and a color monitor; the program supports use of a Microsoft mouse, although it works with a keyboard as well. The program does not offer printing capabilities.

MapInfo (MapInfo Corp., 200 Broadway, Troy, NY 12180; 518-274-8673, 800-327-8726) combines mapping and database management capabilities. Combined with files (purchased separately) containing boundaries and locations for cities, county zip codes, census tracts, and other demographic divisions, the program allows you to pinpoint addresses on maps or to combine other data (such as sales figures) with the maps. The graphics of this program, while easily manipulated, aren't as flashy as on other similar programs, but its capabilities are powerful. The program can produce maps that show, among other things, store locations, sales territories, or the number of residences on each street in a neighborhood. It has been described as a "PC-based road map." The basic program ($750) includes a sample map—of Troy, New York—and a map containing all zip code centroids. Additional maps must be purchased separately, starting from $250. Requires 640K memory, hard disk, and graphics capability.

Map-Master (Ashton-Tate, 6411 Guadalupe Mines Rd., San Jose, CA 95120; 408-268-2300) is designed to create presentation-quality maps showing both data and locations. The powerful package includes boundary files for every country and every state, with population, age, income statistics, and retail sales. A long list of additional boundary files is available separately. *Map-Master* allows you to enter data or import statistics files from popular spreadsheet and database programs. The program, which supports most printers and color plotters, requires 384K memory and a graphics adapter. Price is $395.

PC Globe (PC Globe Inc., 4700 S. McClintock Dr., Ste. 150, Tempe, AZ 85282; 602-730-9000 is an inexpensive ($69.95) and relatively powerful program that provides a comprehensive database for 177 countries—including population statistics, political information, demographic information, and currency conversion —and the ability to generate what the publisher calls "geographic graphics" for slide presentations or desktop publishing. While more an electronic atlas than a mapping program, *PC Globe* does allow you to add data. The graphics and text from the program can be exported to other PC-based programs, including *WordPerfect*, *PageMaker*, and *Lotus 1-2-3*. The program requires 384K memory and a

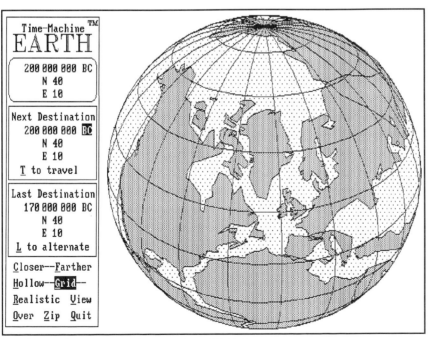

Sample screen from Time-Machine Earth, *an IBM PC program that demonstrates continental drift from more than 500 million years in the past to far in the future, from Sageware Corp.*

Hercules monochrome or EGA or VGA color-graphics board.

PC USA (PC Globe Inc., 2100 S. Rural Rd., Tempe, AZ 85282; 602-894-6866) is a single-country version of PC *Globe* (see above), offering more in-depth information on the country and its individual cities and states. The $69.95 program requires 384K memory and a Hercules monochrome or EGA or VGA color-graphics board.

StreetSmart (Street Map Software, 1014 Boston Cir., Schaumburg, IL 60193; 312-529-4044) is designed to create street maps and generate detailed directions to specific locations. Intended for those who must navigate their way around cities and counties—fire departments, police, delivery companies, and salespeople, for example—it allows you to design maps one intersection at a time, using on-screen commands. Also available are separate files containing computerized map files created by the Census Bureau in preparation for the 1990 census. Those files, which cost $600 per county, cover virtually all streets in the U.S. With *StreetSmart*, users can edit the files to generate custom maps. The program itself is $349 and requires 640K of memory, a hard disk, and a graphics display.

Time-Machine Earth (Sageware Corp., 1282 Garner Ave., Schenectady, NY 12309; 518-377-1052) is a fascinating program that allows you to view the globe as it would appear at virtually any point in human history, from 20,000,000 B.C. to 20,000,000 A.D. By entering the specific year, and by "spinning" the globe to show a specific viewing point, you can watch continental shift take place. Any view may be printed on a dot-matrix or laser printer. The "Personal Edition" of the program is $99; *Time-Machine Earth Professional* with higher-resolution graphics and color, is $245. Requires an IBM PC-compatible computer with a graphics card.

U.S. *MapMaker* (Software Publishing Corp., 1901 Landings Dr., Mountain View, CA 94039; 415-

962-8910) is designed for use with the company's *Harvard Graphics* and requires that "shell" program to operate. U.S. *MapMaker* provides a database of U.S. state outlines, city coordinates, and data to generate color-coded maps. Data may be entered directly or imported as an ASCII file from popular spreadsheet and database programs. The program requires 512K memory and a graphics adapter. Price is $149; *Harvard Graphics* is $495.

PROGRAMS FOR MACINTOSH

Business Class (Activision Inc., 3885 Bohannon Dr., Menlo Park, CA 94025; 415-329-0800) is a *HyperCard* stack designed for use as a "desktop travel-planning" program. In addition to world, regional, and country maps, the program offers a treasure trove of information about each country, from customs and health regulations to currency rates and social customs. With the use of *HyperCard's* telephone-dialing capabilities, *Business Class* can automatically dial reservations for international airlines, hotels, car-rental companies, tourist and business bureaus, embassies, and other important numbers. Requires a Macintosh Plus with 1 MB memory and *HyperCard*. Price is $49.95.

Geoquery (Odesta Corp., 4084 Commercial Ave., Northbrook, IL 60062; 312-498-5615; 800-323-5423) allows you to import information from a database file and display it on a map, from a single county to a series of states to the entire country. The program uses the "push pin" analogy: It allows you to place computerized "pins" to mark and annotate specific points on a map. It is a relatively simple and inexpensive program, with limited capabilities, but a good basic geographic front end to any database program. The program ($349) requires 1 MB of memory and two 800K disk drives.

HyperAtlas (Micromaps Software Inc., Box 757, Lambertville, NJ 08530; 609-397-1611; 800-334-4291) is a *Hypercard* stack containing information about countries, states, and U.S. cities. The program includes maps and information

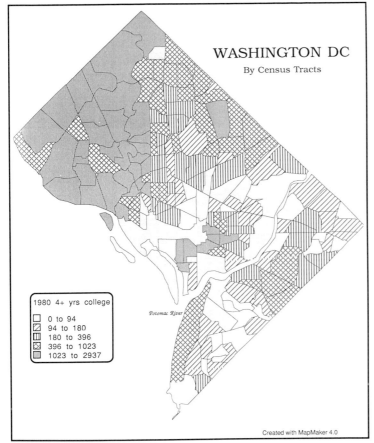

Census tract map of Washington, D.C., created by MapMaker, a Macintosh program from Strategic Mapping, Inc.

stacks that are linked. Users can add new cards, stacks, or buttons. Price is $99 and requires *Hypercard*.

MacAtlas (Micromaps Software Inc., Box 757, Lambertville, NJ 08530; 609-397-1611; 800-334-4291) includes a set of customizable maps. The program includes maps of the world, world regions, the U.S., and the 50 states. They can be used as they appear on disk without modification, or they can be enhanced with fill patterns, shadows, and colors. The program comes in several formats: PICT ($199; requires *MacDraw*, *MacDraft*, or *Canvas*), Encapsulated PostScript ($199; requires Adobe *Illustrator* or Aldus *Freehand*), and Paint ($99; requires *MacPaint*, *FullPaint*, or *SuperPaint*).

Map II (John Wiley & Sons, Inc., 605 Third Ave., New York, NY 10158; 212-850-6222) calls itself a "map processor." In reality it is a geographical information system designed for map making, viewing, marking, measuring, and transforming. Maps can be created using graphic techniques analogous to working with a map base on a light table, or data can be imported in a variety of formats. Cartographic marking, measuring, detecting, interpreting, analysis, and modeling can be accomplished by screen tools as well as by specifying text-based transformations. Finished maps may be printed in black and white or color and can be exported to other applications. The program ($99.95) requires 2 MB of memory and a hard disk.

MapGrafix (ComGrafix Inc., 620 E St., Clearwater, FL 34616; 813-443-6807, 800-448-6277) is a geographical information system used primarily to analyze fire and emergency services, transportation systems, land-use forecasting, site planning, zoning, and the like. The program lets you create, store, edit, merge, and exchange maps and map data with databases, both in Macintosh and IBM-compatible computers. *MapGrafix* allows you to link every feature on a map—whether graphic or text—to a relational database, providing powerful capabilities to generate and analyze both graphics and text. The program retails for a hefty $8,500. A demonstration disk is $100 and can be credited against the purchase price. Requires SE (with 1 MB memory) or Mac II (with 2 MB), hard disk, printer or plotter, digitizer, and relational database program. Another ComGrafix program, *MapStar*, is for vehicle tracking applications.

MapMaker (Strategic Mapping Inc.., 4030 Moorpark Ave., Ste. 123, San Jose, CA 95117;

408-985-7400) is a map-making and data-analysis program that can create maps of the U.S. by county, individual states, groups of states, or the entire country, as well as continents of the world. The program includes a palatte of *MacDraw*-like object-oriented color drawing tools that allows you to create or customize maps without going to another program. The program also features a data base, into which you can import data from Macintosh spreadsheets or enter data directly. Maps as large as 6 feet by 8 feet can be created, built from multiple tiled pages. Maps can be saved as *MacPaint* or *MacDraw* (PICT) files for import into other applications. The program ($349) comes with 58 U.S. census demographic and population statistics for each state, 13 for each county in the U.S. and 20 for each of 176 foreign countries. Other map boundary files can be purchased separately at prices ranging from $50 per state for more detailed county maps to $395 for a U.S. map divided into 435 congressional districts. Requires 512K memory.

QuickMap (Micromaps Software Inc., Box 757, Lambertville, NJ 08530; 609-397-1611; 800-334-4291) is a geographic analysis tool that allows you to represent data on a map. You may import data sets from spreadsheets and databases, or enter data directly into the program. Data can be analyzed by calculating ratios, sums, differences, and percentages. Completed maps can be pasted into other programs, including desktop publishing applications. Price is $99 and requires *Hypercard*.

Sun Clock (MLT Software, P.O. Box 98041, 6325 SW Capitol Hwy., Portland, OR 97201; 503-245-7093) isn't really a map-making program. Instead it is an inexpensive desk accessory that graphically displays where the sun is shining (and where it isn't) on a map of the world. *Sun Clock* displays the areas of day and night for the current date and time. The shape dividing day from night changes with the seasons. Viewing *Sun Clock* you can actually see the days lengthen and shorten as the year progresses. Price is $15 plus $2 shipping.

OTHER SYSTEMS

MicroStation GIS (Intergraph, One Madison Industrial Park, Huntsville, AL; 205-772-2700, 800-826-3515) is a high-end Unix-based software family used on a hardware system provided by Intergraph. The system includes five modules intended for map publishing, spatial analysis, modeling, image processing, and projection manipulation. The base system, which includes a workstation, software, and Hewlett-Packard plotter, costs $38,000; modules sell for $3,000 to $5,000 each. An even higher-level system, TIGRIS, sells for $60,000 for the base system, plus $5,000 to $20,000 for the modules. The system is intended for professional map-makers and publishers. Among the company's clients are Hammond, Rand McNally, and the U.S. Geological Survey.

CD-ROM PACKAGES

These require your computer to be equipped with a compact disk reader, which runs from about $600 on up, along with appropriate cables, circuit boards, and software to connect it with your computer. Most of these packages work on either IBM-compatible or Macintosh computers but not both. Some of the above-mentioned programs also are sold on CD-ROM.

The Electronic Map Cabinet (Highlighted Data, P.O. Box 17229, Washington Dulles International Airport, Washington, DC 20041; 703-533-1939) is a Macintosh-based package that includes data for most of the U.S. The program does not store maps, but generates them on request from a geographical database containing more than 400 million bits of information. The maps range from the complete continental U.S. down to areas encompassing only a few square blocks. Maps can show political boundaries, coastlines, bodies of water, public facilities, and, for the 300 largest metropolitan areas, highways down to the local street level. Maps can be annotated with text and enhanced with standard drawing tools. Completed images can be imported as PICT files into other programs. The disk sells for $199.95.

Window On the World (Geovision Inc., 270 Scientific Dr., Ste. 1, Norcross, GA 30092; 404-448-8224) is an IBM-compatible geographic database system of CD-ROM-based disks. Each disk has a capacity of 550 megabytes of information containing highways, roads, boundaries, bodies of water, topographic detail, land use, and landmarks, among other things. Users can create, display, and modify overlays and store composites of graphics, text, and symbols on a hard disk. The program includes a built-in "surveyor" that provides distance, bearing, and position information. There are four map series available: National Series (scale 1:2,000,000), Regional Series (six regions; scale 1:250,000), State Series (50 states; scale 1:100,000), and Metro Series (approximately 60; scale 1:24,000). Price of the basic program is $495, plus $495 for the National Series database, "Geodisk U.S. Atlas"; purchased together, the price is $595. Prices for the other disks were not established as this book went to press.

CHARTING AND NAVIGATION SYSTEMS

Navigate! (Fair Tide Technologies Inc., 18 Ray Ave., Burlington, MA 01803; 617-229-6409, 800-332-3426) is a digital-chart navigation system for either Macintosh or IBM-compatible computers. The program provides full-color nautical charts, including depth-contour lines to warn sailors of shallow waters and other hazards. You can plot a course by pointing at various locations along the route and clicking on the mouse. The program automatically calculates the latitude and longitude of each point and its range and bearing from the previous point. It then draws a course line connecting them. An adjunct product, *Chart Catalog!* is a Hypercard front-end program. By clicking on a point on a map—at San Francisco Bay, for example—*Chart Catalog!* automatically loads *Navigate!* along with the correct chart. There are two versions, one in black and white ($395 for Macintosh, $595 for IBM) and one in color ($695 for Macintosh, $895 for IBM). Each comes with one free chart of your choice;

additional charts are $95. There are nearly 100 charts available (examples: "Cape Canaveral to Key West" or "Monterey Bay to Coos Bay"). Also available are packages of five to eight maps of a given region (e.g., northern New England, from Boston to the Canadian border) for $395.

NAVplus (Ocean Products Corp., 8 Bayberry Ln., New Fairfield, CT 06812; 203-746-1175) is a charting and navigation system for Macintosh computers that allows sailors to plan a course by a number of methods. Images representing reefs, rocks, shipwrecks, and other features—including fishing holes—appear on the screen. The basic program sells for $599.95 and includes one free chart; additional charts are $199.95 each.

GEOGRAPHIC DATA BASES

In addition to the databases available with the products described above, the U.S. Geological Survey's Earth Science Information Center offers inventories of digital spatial databases—collections of computerized data that have some coordinate system that can be used by cartographic software. The publication *Sources for Digital Spatial Data* describes more than 500 such data sets containing spatially referenced base or thematic categories of data, from federal, state, and local government agencies and the private sector. Examples include "Natural Resource Digital Data Base for Death Valley National Monument and Surrounding Region," produced by the National Park Service, "Sanitary Sewers Data Base," produced by the City of San Jose (California) Planning Department, and "Larimer County Area Soil Survey, Colorado," produced by the U.S. Soil Conservation Service. The complete survey, which is updated and printed on a weekly basis, is $22 from Reston ESIC, 507 National Center, Reston, VA 22092; 703-860-6045.

■ **See also: "Geography Education Materials."**

Map Stuff

It used to be that maps appeared only on flat or folded sheets of paper and on globes. But no more. Today, you can find maps printed on just about everything, from scratch pads to shower curtains, wastebaskets to basketballs. Not all of these are suitable for getting from here to there, or even for locating the world's capital cities. But no matter: These maps are more for entertainment than enlightenment.

The proliferation of map images on everyday products isn't surprising. After all, it's a big, beautiful world—why not display it proudly? Thanks to a growing collection of map "stuff," you can.

Here is a representative sampling of what's out there:

CALENDARS

Prices indicated are for 1990 calendars. You are advised to inquire about price and availability of calendars for subsequent years.

Graphic Learning International (1123 Spruce St., Boulder, CO 80302; 303-440-7620) publishes "Earth 1990, The World in Maps," a calendar filled with environmental maps and 86 smaller maps illustrating climate, soils, ocean currents, annual rainfall, time zones, treaty and trade cartels, endangered animal species, and physical, political, and population division ($15.95); and "Earth 1990, The International Date Book," a 160-page engagement calendar with environmental and political maps, with a 47-page statistical section ($14.95).

Pomegranate Publications (Box 808022, Petaluma, CA 94975; 707-765-2005; 800-227-1428) publishes a colorful 16" X 28" "Masters of Cartography" calendar ($12.95), featuring maps of 15th-, 16th-, and 17th-century European cartographers. A "Masters of Cartography" address book ($14.95) is also available,

with colorful illustrated maps from the Huntington Library reproduced.

CLOCKS

Geochron Enterprises Inc. (899 Arguello St., Redwood City, CA 94063; 415-361-1771) manufactures the Geochron World Time Indicator, an attractive chalkboard-size electronic map that continually shows the exact time of day and the amount of sunlight anywhere in the world (basic unit is $1,200; deluxe models up to $2,000).

Wuersch Time Inc. (1273 Robeson St., Fall River, MA 02720; 617-672-8018) produces a variety of clocks with topographic maps and nautical charts on the face. Clocks available range from a 6" X 8" nautical chart wall clock with a teak frame ($45) to a 12 1/2-inch topographic map wall clock framed in brass ($70). Call or write for brochure.

CLOTHING

The Cockpit (33-00 47th Ave., Long Island City, NY 11101; 718-482-1860; 800-354-5514) sells several items, including an "Allied Pilot's Cloth Invasion/Escape Map," an 18-inch-square reproduction ($14); "14th Air Force 'Briefing Map' Underwear" ($14.50), boxer shorts made from reproductions of Flying Tigers World War II maps; "Original World War II Cloth Escape Maps" ($50), pilot's AAF cloth charts printed by the Army Map Service in May 1945 for the Allied invasion of the Japanese mainland. One side shows the East China Sea and Japanese mainland; the other portrays Japan and the South China seas.

Buckminster Fuller Institute (1743 S. La Cienega Blvd., Los Angeles, CA 90035; 213-837-7710) sells the "Dymaxion Map T-Shirt"

A *selection of "Map Stuff" from several companies.*

($9.95), a 100-percent cotton T-shirt with Buckminster Fuller's Dymaxion world map printed on the front and the caption "A One World Island in a One World Ocean."

Interarts, Ltd. (15 Mt. Auburn St., Cambridge, MA 02138; 617-354-4655) sells unique and colorful "Wearin' the World" map jackets made of lightweight Tyvek, featuring Sweden's attractive Esselte map graphics ($49.95). Also available is a laminated world map desk blotter for $14.95.

Michael Poisson Associates (P.O. Box 84, Boulder Creek, CA 95006; 408-338-2984) produces "Visual Dialogue Cloth Maps," topographic maps reproduced on bandanas, T-shirts, and sweatshirts. "Bandanamaps" ($7.95 each), full-color 22" X 22" cloth maps, are available for more than 30 areas, including Gold Country, California, Lake Tahoe, San Francisco, Smoky Mountain National Park, and Yosemite National Park. All-cotton "Teemaps" ($9.95 for T-shirts, $17.95 for sweatshirts), are available for more than 20 areas, including Mt.

Lassen, California; Crater Lake, Oregon; and the Smoky Mountains.

PUZZLES AND GAMES

Austin Peirce (19 Prairie Dunes, Hutchinson, KS 67502; 316-663-4192) produces the colorful, information-packed "Puzzlin' State" series of jigsaw puzzles. The puzzles feature county boundaries, county seats, rivers, lakes, points of interest, and historical trivia, as well as state flags, seals, animals, birds, rocks, plants and trees. Puzzles for more than 20 states, as well as one for the U.S., are available from Austin Peirce or through **Globe Pequot Press**, 10 Denlar Dr., Chester, CT 06412; 203-526-9571.

Better Boating Association (295 Reservoir St., Needham, MA 02194) produces "Chart Kit Puzzles" ($8.95 each), full-color, 550-piece puzzles of nautical charts. The 18" X 22" puzzles are available for Annapolis/St. Michaels, Maryland; Fisher's Island Sound, Connecticut; Martha's Vineyard, Massachusetts; Mt. Desert/Frenchman's Bay, Maine;

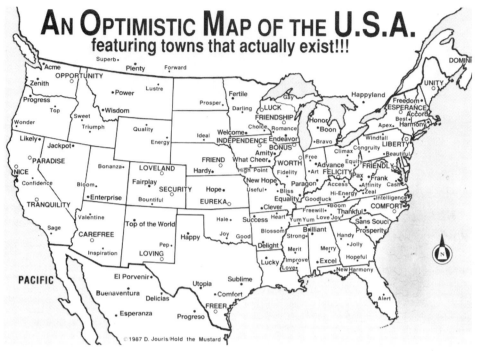

"An Optimistic Map of the U.S.A.," ©1987 David Jouris/
Hold the Mustard Productions.

837-7710)
sells copies
of the
Dymaxion
map in
several
forms:
"Dymaxion
Sky-Ocean
World Map"
($5), a 22" X
14 1/2" map
on heavy
stock that
folds into a 5
1/2-inch
icosahedron;
and post-
cards ($1.95
each).

Newport, Rhode Island; and Sanibel/Captiva,
Florida.

Coburn Designs (P.O. Box 331, Nashua, NH
03061; 603-889-2070) sells jigsaw puzzles of
U.S. National Parks ($12 each). The full-color,
500-piece puzzles, reproduced from U.S.
Geological Survey maps, show scenic routes,
trails, elevations, contour lines, and points of
interest. Puzzles are available for Acadia
National Park, Grand Canyon National Park,
Mount McKinley/Denali National Park, White
Mountain National Forest, Yellowstone National Park, and Yosemite National Park.

David Fox (P.O. Box 533, Narberth, PA 19072;
215-667-2136) produces jigsaw puzzles of its
axionometric city maps. The 1,200-piece map
puzzles are available for Boston, Chicago,
lower Manhattan, New Orleans, Philadelphia,
and Washington, D.C. for $35 each.

Buckminster Fuller Institute (1743 S. La
Cienega Blvd., Los Angeles, CA 90035; 213-

**GeoLearning
Corp.** (555
Absaraka St., Sheridan, WY 82801; 307-674-
6436) produces many fun and interesting
things, among them:

■ "The Spilhaus GeoGlyph" ($24.95), a puzzle
showing the continuous surface of the world.
More than 100 correct maps can be created
from the pieces.
■ "GeOdyssey—The World Game" ($14.95), in
which players compete to build a world map,
tile-by-tile, testing their geography knowledge
by identifying continents, coastlines, and other
geographical features.
■ "GeOdyssey Deluxe" ($24.95), a version of
GeOdyssey featuring larger nine-color playing
tiles and labeled oceans.
■ "GeoCards" ($7.95), a colorful game featur-
ing a deck of 48 cards, with each card representing a different geographical location.
■ "Tectonic Cube" ($4.50), a fascinating seven-
sided geometric form showing how the Earth's
tectonic plates fit together. The cube illustrates
the plates' various ridges, boundaries, zones,
and faults.
■ "Tectonocycle" ($3), which the company de-

scribes as "a type of hexaflexagon or kaleido-
cycle." By assembling and rotating this ingen-
ious cardboard model, the Tectonocycle demonstrates the breakup of the continent Pangea
in four stages.

■ "Flight Lines" ($5.95), an educational puzzle
game featuring "Geography-Triangles," such as
Los Angeles to Sydney, Paris to Rio, Tokyo to
Chicago, and many more.

Geoscience Resources (2990 Anthony Rd.,
P.O. Box 2096, Burlington, NC 27216; 919-227-
8300; 800-742-2677) distributes national-park
jigsaw puzzles, made from USGS topographic
maps ($10.95 each), available for six national
parks; and a "Landforms Puzzle" ($17.95), a 36″
X 24″ puzzle of more than 1,000 pieces, showing geological landforms.

National Geographic Society (17th & M Sts.
NW, Washington, DC 20036; 202-921-1200)
sells "National Geographic Global Pursuit"
($19.95), with nearly 1,000 geography ques-
tions to test the global knowledge of students
and adults alike. Comes with a world map.

Pacific Puzzle Co. (378 Guemes Island Rd.,
Anacortes, WA 98221; 206-293-7034) sells a
variety of hand-cut birchwood puzzles, including three maps of the U.S. ($23 to $60), a
political world map ($16), a physical world map
($16), two Dymaxion world maps ($22 and
$40), and five continent puzzles ($16 to $19
each). Shipping is extra; write for catalog.

Optimago (43 Perrymead St., London SW6
3SN, England; 01-736-2380) produces wooden
jigsaw puzzles of several interesting maps.
Among puzzles available are the 1750 "Nouvelle Mappe Monde" by Bailleul le Jeune, a
1574 map of London, a celestial chart from the
Folio Atlas of the Heavens by Andreas Cellarius in 1600, a map of Japan by Adrian Reland,
and the "Leo Hollandicus" by Nicolas Visscher.

World Impressions Inc. (1493 Beach Park
Blvd., Ste. 312, Foster City, CA 94404; 415-571-
8859) produces colorfully illustrated sports and

"Wearin' the World" jacket, distributed by Interarts, Ltd.

travel theme maps of the U.S., many available
as jigsaw puzzles ($12.95 each), postcards (50
cents each, minimum of 10), and "Coloring
ArtMaps" ($10.95), black-and-white line art with
pens to color in the map. Jigsaw puzzles and
postcards include "The Official Major League
Baseball ArtMap," "The Official National Football League ArtMap," "Skiing of North America,"
and "San Francisco." Coloring ArtMaps are
available for the baseball and football ArtMaps,
"California Adventure," "Oregon Adventure,"
and "Yosemite Adventure."

Et Cetera

Alan Spigelman/Wings Over the World (15 W.
27th St., New York, NY 10001; 212-727-1818)
produces but does not retail several fun map
products, including a colorful "World Shower
Curtain" ($30 to $35); "Black Star Gazer Shower
Curtain" ($30 to $35), showing the night sky; a
handy-size "World Umbrella" ($16 to $18);
inflatable political globes, available in blue (four
sizes—9-, 12-, 16-, and 27-inch), a 12-inch clear
inflatable globe ($3 each in 12-pack), and an
inflatable "Star Gazer" ball, showing the night

sky ($2.25 each in six-pack, 12-inch ball; $2.25 each in 12-pack, 27-inch ball). Also available is the Earth Zone Yo-Yo ($3), a globe on a string.

Baden Sports Inc. (1120 SW 16th, Renton, WA 98055; 206-235-1830) produces "World Ball," a basketball with a seven-color world map printed on it. The ball is available in official National Basketball Association size (29 1/2 inches), $21.95; and in a "mini-star" size (22 1/2 inches), for $15.95.

Fantasy Productions (2124 Westlake Ave., Seattle, WA 98121; 206-343-0903; 800-678-8697) produces but does not retail inflatable globes. The "Dinosaur Globe" features major fossil finds marked with descriptions of prehistoric creatures; the "Star System Globe" illustrates the constellations and planets; a "Blue World Globe"; and a "Clear World Globe."

Gabelli U.S. Inc. (11500 W. Olympic Blvd., #475, Los Angeles, CA 90064; 213-312-4546) distributes brightly colored map clipboards, briefcases, duffel bags, pencil cases, rucksacks, trashcans, and wallets.

Hold the Mustard Productions (P.O. Box 822, Berkeley, CA 94701; 415-549-9555) produces eight humorous postcards featuring real names of cities and towns across the United States. Titles include: "An Optimistic Map of the U.S.A.," "A Pessimistic Map of the U.S.A.," "An Exotic Map of the U.S.A.," "An Animal Map of the U.S.A.," "A Confusing Map of the U.S.A.," "A Lovers Map of the U.S.A.," "An Eccentric Map of the U.S.A.," and "An Artists Map of the U.S.A." Poster-sized versions ($5.50 each; 16″ X 20″) of the "Optimistic" and "Pessimistic" maps are available by mail from **Sonoma Portal**, 605 Broadway, Sonoma, CA 95476; 707-996-2225.

Maptec International Ltd. (5 S. Leinster St., Dublin 2, Ireland; 353-1-766266) produces posters, mini-guides, T-shirts, and jigsaw puzzles from its satellite images of U.S. and foreign metropolitan areas.

Mino Publications (9009 Paddock Ln., Potomac, MD 20854; 301-294-9514) produces a four-color laminated placemat map of downtown Washington, D.C., with subway and Capital Beltway maps on the reverse ($2.95).

New England Cartographics (P.O. Box 369, Amherst, MA 01004; 413-253-7415) produces "Topolopes," envelopes made from U.S. government-surplus topographic maps. These novel envelopes require only an address label to meet U.S. postal regulations. They cost $.10 to $.20 each, and are available in map stores.

Pacific Drum Co. (P.O. Box 4226, Bellingham, WA 98227; 206-671-8108) produces the "Wholearth" ball, an inflatable 16-inch globe depicting the Earth as seen from space. The colorful ball is made from NASA photographs, weather-satellite pictures, and geophysical maps, and comes with a "Global Handbook" of interesting facts and ideas ($7.95).

Parkwest Publications Inc. (238 W. 72nd St., Ste. 2F, New York, NY 10023; 212-222-6100) distributes "The Tarquin Globe" ($6.95), a full-color paper build-it-yourself globe, illustrating all countries and more than 1,000 major cities, seas, rivers, and other features. The kit comes with a 24-page booklet full of facts and ideas. A similar product, "The Tarquin Star Globe" ($6.95), is for constructing a three-dimensional globe of the night sky.

XTC Products Inc. (247 Rockingstone Ave., Larchmont, NY 10538; 914-833-0200) sells "Hugg-A-Planet Earth," colorful soft-sculpture globes illustrating continents, oceans, and countries. Three sizes are available: 6″ "Baby Hugg-A-Planet" ($10); 12″ "Hugg-A-Planet" ($17); and 24″ "Super Hugg-A-Planet" ($110). Also available is a 12″ "Geo Hugg-A-Planet" ($17), showing the Earth's natural features, including mountain ranges, lakes, deserts, oceanic trenches, and population densities.

■ **See also: "Geography Education Materials."**

APPENDIXES

APPENDIX A

State Map Agencies

TRANSPORTATION/HIGHWAY OFFICES

Alabama
State of Alabama Highway Dept.
Bureau of State Planning
11 S. Union St., Rm. 313
Montgomery, AL 36130
205-832-6128

Alaska
Div. of Design & Construction
Alaska Dept. of Transportation & Public Facilities
Pouch 6900, 411 Aviation Dr.
Anchorage, AK 99502
907-266-1500

Bridge Design
Alaska Dept. of Transportation & Public Facilities
P.O. Box 1467
Juneau, AK 99802
907-789-0841

Arizona
Arizona Highways
2039 W. Lewis Ave.
Phoenix, AZ 85009
602-258-6641

Arizona Dept. of Transportation
206 S. 17th Ave., Rm. 181A
Phoenix, AZ 85007
602-261-7325

Arkansas
Arkansas State Highway & Transportation Dept.
Map Sales, Rm. 203
9500 New Benton Hwy.
P.O. Box 2261
Little Rock, AR 72203
501-569-2444

California
California Dept. of Transportation
1120 N St., P.O. Box 1499
Sacramento, CA 95807
916-324-6732

Colorado
Report Preparation Section
Div. of Transportation Planning
Colorado Dept. of Highways
4201 W. Arkansas Ave.
Denver, CO 80222
303-757-9523

Connecticut
Connecticut Dept. of Transportation
24 Wolcott Hill Rd.
P.O. Drawer "A"
Wethersfield, CT 06109
203-566-2410

Delaware
Office of Planning & Programming
Delaware Dept. of Transportation
P.O. Box 778
Highways Admin. Bldg.
Dover, DE 19903
302-736-4346

Florida
Dept. of Transportation
Hayden Burns Bldg.
605 Suwannee St.
Tallahassee, FL 32301
904-488-6721

Georgia
Map Sales Unit
Georgia Dept. of Transportation
2 Capitol Square
Atlanta, GA 30334
404-656-5336

Planning Data Services
Cartographic Branch
Georgia Dept. of Transportation
5025 New Peachtree Rd. NE
Chamblee, GA 30341
404-393-7370

Hawaii
Hawaii Dept. of Transportation
869 Punchbowl St.
Honolulu, HI 96813
808-548-4719

Idaho
Management Services Section
Idaho Transportation Dept.
P.O. Box 7129
Boise, ID 83707
208-334-2569

Illinois
Illinois Dept. of Transportation
2300 S. Dirksen Pkwy.
Springfield, IL 62764
217-782-7820

Indiana
Public Information Office
Indiana Dept. of Highways
1106 State Office Bldg., 100 N. Senate Ave.
Indianapolis, IN 46204
317-232-5115

Iowa
Planning & Research Div.
Iowa Dept. of Transportation
800 Lincoln Way
Ames, IA 50010
515-239-1661

Kansas
Bureau of Transportation Planning
Kansas Dept. of Transportation
State Office Bldg.
Topeka, KS 66612
913-296-3841

Kentucky
Division of Mass Transportation
7th Fl., State Office Bldg.
Frankfort, KY 40622
502-564-7433

Louisiana
Office of Aviation & Public Transportation
Louisiana Dept. of Transportation & Development
P.O. Box 44245, Capitol Station
Baton Rouge, LA 70804
504-925-7742

Dept. of Transportation & Development
P.O. Box 44155, Capitol Station
Baton Rouge, LA 70804
504-379-1100

Maryland
Cartographic Section
Maryland State Highway Admin.
2323 W. Joppa Rd.
Brooklandville, MD 21022
301-321-3518

Massachusetts
Executive Office of Transportation & Construction
10 Park Plaza
Boston, MA 02116
617-973-7000

Massachusetts Turnpike Authority
State Transportation Bldg.
10 Park Plaza, Ste. 5170
Boston, MA 02116
617-973-7300

Michigan
Michigan Dept. of Transportation
425 W. Ottawa, P.O. Box 30050
Lansing, MI 48909
517-373-2090

Minnesota
Minnesota Dept. of Transportation
Transportation Bldg., Rm. 809
St. Paul, MN 55155
612-296-1680

Mississippi
Mississippi State Highway Dept.
P.O. Box 1850
Jackson, MS 39205
601-354-7176

Montana
Montana Dept. of Highways
2701 Prospect Ave.
Helena, MT 59620
406-444-6115

Nebraska
Nebraska Dept. of Roads, Planning Div.
P.O. Box 94759
Lincoln, NE 68509
402-473-4519

Nevada
Nevada Dept. of Transportation
1263 S. Stewart St.
Carson City, NV 89710
702-885-4322

New Hampshire
Planning & Economics Div.
Dept. of Public Works & Highways
P.O. Box 483
Concord, NH 03301
603-228-1382

New Jersey
New Jersey Dept. of Transportation
1035 Parkway Ave., CN600
Trenton, NJ 08625
609-292-3105

New Mexico
New Mexico State Highway Dept.
P.O. Box 1149
Santa Fe, NM 87503
505-983-0668

New York
Map Information Unit
Dept. of Transportation
State Campus, Bldg. 4, Rm. 105
Albany, NY 12232
518-457-3555

North Carolina
North Carolina Dept. of Transportation
P.O. Box 25201
Raleigh, NC 27611
919-733-2520

Office of Public Affairs
North Carolina Dept. of Transportation
P.O. Box 25201
Raleigh, NC 27611
919-733-3463

Location & Survey
North Carolina Dept. of Transportation
P.O. Box 25201
Raleigh, NC 27611
919-733-7241

Ohio
Map Sales
Ohio Dept. of Transportation
25 S. Front St., Rm. B100
Columbus, OH 43215
614-446-4430

Public Information Office
Ohio Dept. of Transportation
25 S. Front St., Rm. 308
Columbus, OH 43215
614-466-7170

Oklahoma
Reproduction Branch
Oklahoma Dept. of Transportation
200 NE 21st St.
Oklahoma City, OK 73105
405-521-2586

Pennsylvania
Bureau of Strategic Planning, Rm. 912
Pennsylvania Dept. of Transportation
Transportation & Safety Bldg.
Harrisburg, PA 17120
717-787-4247

Puerto Rico
Photogrammetry & Geodesy Section
Puerto Rico Dept. of Transportation & Public Works
Minillas Government Center
G.P.O. Box 8218
San Juan, PR 00910
809-722-2929

Rhode Island
Planning Div.
Rhode Island Dept. of Transportation
368 State Office Bldg., Smith St.
Providence, RI 02903
401-277-2694

South Carolina
Map Sales, Traffic Engineering Div.
Dept. of Highways & Public Transportation
P.O. Box 191
Columbia, SC 29202
803-758-3001

South Dakota
South Dakota Dept. of Transportation
State Highway Bldg.
Pierre, SD 57501
605-773-3267

Tennessee
Information Office
Tennessee Dept. of Transportation
700 James K. Polk Bldg., 505 Degderick St.
Nashville, TN 37219
615-741-2331

Texas
Texas Travel & Information Div.
State Dept. of Highways & Public Transportation
11th & Brazos Sts.
Austin, TX 78701
512-475-2877

Utah
Office of Community Relations
Utah Dept. of Transportation
4501 South, 2700 West
Salt Lake City, UT 84119
801-965-4101

Vermont
Planning Div.
Agency of Transportation
133 State St.
Montpelier, VT 05602
802-828-2671

Vermont Agency of Transportation
State Admin. Bldg.
133 State St.
Montpelier, VT 05602
802-828-2828

Virginia
Dept. of Highways & Transportation
1221 E. Broad St.
Richmond, VA 23219
804-786-2838

Washington
Washington Dept. of Transportation
Transportation Bldg., KF-01
Olympia, WA 98504
206-753-2150

West Virginia
Systems Planning Div.
Corridor & Project Planning Section
West Virginia Dept. of Highways
1900 Washington St. E.
Charleston, WV 25305
304-348-2764

Wisconsin
Wisconsin Dept. of Transportation
Special Services Section
P.O. Box 7916
Madison, WI 53707
608-266-0309

Wyoming
Document Sales—Maps
Wisconsin Dept. of Transportation
3617 Pierstorff St.
P.O. Box 7713
Madison, WI 53707
608-266-8921

TOURISM OFFICES

Alabama
Alabama Bureau of Publicity & Information
532 S. Perry St.
Montgomery, AL 36130
205-261-4169; 800-ALABAMA

Alaska
Div. of Tourism
Alaska Dept. of Commerce & Economic Development
Pouch E
Juneau, AK 99811
907-465-2010

Arizona
Arizona State Office of Tourism
3507 N. Central Ave., Ste. 506
Phoenix, AZ 85012
602-255-3618

California
California Office of Tourism
P.O. Box 189
Sacramento, CA 95812
916-322-1397; 800-862-2543

Colorado
Colorado Tourism Board
1625 Broadway, Ste. 1700
Denver, CO 80202
303-592-5410; 800-433-2656

Connecticut
Div. of Tourism
Connecticut Dept. of Economic Development
210 Washington St.
Hartford, CT 06106
203-566-3385

Delaware
Delaware Development Office
P.O. Box 1401
99 Kings Hwy.
Dover, DE 19903
302-736-4271

District of Columbia
D.C. Convention and Visitor Association
1212 New York Ave. NW
Washington, DC 20005
202-789-7000

Florida
Div. of Tourism
Florida Dept. of Commerce
107 W. Gaines St., Rm. 505
Tallahassee, FL 32301
904-487-1462

Guam
Guam Visitors Bureau
1220 Pale San Vitores Rd.
P.O. Box 3520
Tamuning, Guam U.S.A. 96911
671-646-5278

Hawaii
Hawaii Visitors Bureau
2270 Kalakaua Ave., Ste. 801
Honolulu, HI 96815
808-923-1811

Idaho
Department of Commerce
700 State Street
Boise, ID 83720
208-334-2470

Illinois
Illinois Office of Tourism
Dept. of Commerce & Community Affairs
2209 W. Main St.
Marion, IL 62959
618-997-4371

Illinois Office of Tourism
Dept. of Commerce & Community Affairs
620 E. Adams St.
Springfield, IL 62701
217-782-7139

Illinois Tourist Information Center
310 S. Michigan Ave., Ste. 108
Chicago, IL 60604
312-793-2094

Indiana
Tourism Div.
Indiana Dept. of Commerce
1 N. Capitol, Ste. 700
Indianapolis, IN 46204
317-233-8870

Iowa
Tourism, Iowa Development Commission
600 E. Court Ave.
Des Moines, IA 50306
515-281-3100

Kansas
Travel & Tourism Div.
Kansas Dept. of Economic Development
503 Kansas Ave., 6th Fl.
Topeka, KS 66603
913-296-2009

Kentucky
Kentucky Dept. of Travel Development
Capitol Plaza Tower, 22nd Fl.
Frankfort, KY 40601
502-564-4930

Louisiana
Louisiana Office of Tourism
P.O. Box 44291
Capitol Station
Baton Rouge, LA 70804
504-925-3860

Maryland
Office of Tourist Development
Dept. of Economic & Community Development
1748 Forest Dr.
Annapolis, MD 21401
301-269-3517

Massachusetts
Citizen Information Center
Secretary of the Commonwealth
1 Ashburton Pl., 16th Fl.
Boston, MA 02108
617-727-7030

Michigan
Travel Bureau
Michigan Dept. of Commerce
P.O. Box 30226
Lansing, MI 48909
517-373-1195; 800-292-2520 (in Mich.); 800-248-5700
    (out of state)

Minnesota
Office of Tourism
Dept. of Energy & Economic Development
240 Bremer Bldg., 419 N. Robert St.
St. Paul, MN 55101
612-296-5029; 800-652-9747 (in Minn.); 800-328-1461
    (out of state)

Mississippi
Mississippi Dept. of Economic Development
P.O. Box 849
Jackson, MS 39205
601-354-6715; 800-647-2290

Missouri
Missouri Div. of Tourism
Truman State Office Bldg.
301 W. High St., P.O. Box 1055
Jefferson City, MO 65101
314-751-4133

Montana
Travel Promotion Div.
Montana Dept. of Commerce
1424 9th Ave.
Helena, MT 59620
406-444-2654

Nebraska
Travel & Tourism Div.
Nebraska Dept. of Economic Development
P.O. Box 94666
Lincoln, NE 68509
402-471-3111; 800-742-7595 (in Neb.); 800-228-4307
    (out of state)

Nevada
Nevada Commission on Tourism
Capitol Complex
Carson City, NV 89710
702-885-4322

New Hampshire
Office of Vacation Travel
Dept. of Resources & Economic Development
P.O. Box 856
Concord, NH 03301
603-271-2665

New Jersey
New Jersey Dept. of Transportation
1035 Parkway Ave., CN600
Trenton, NJ 08625
609-292-3105

New Mexico.
New Mexico Economic Development & Tourism Dept.
Bataan Memorial Bldg.
Santa Fe, NM 87503
505-827-6230; 800-545-2040

New York
Div. of Tourism
New York State Dept. of Commerce
1 Commerce Plaza
Albany, NY 12245
518-474-4116; 800-225-5697 (Northeast)

North Dakota
North Dakota Tourism Promotion
Capital Grounds
Bismarck, ND 58505
701-224-2525; 800-472-2100 (in N.D.); 800-437-2077
(out of state)

Ohio
Office of Travel & Tourism
Dept. of Development
P.O. Box 1001
Columbus, OH 43216
614-466-8844

Oklahoma
Marketing Services Div.
Oklahoma Tourism & Recreation Dept.
500 Will Rogers Bldg.
Oklahoma City, OK 73105
405-521-2406

Oregon
Tourism Div.
Economic Development Dept.
595 Cottage St. NE
Salem, OR 97310
503-373-1200; 800-547-7842 (out of state)

Pennsylvania
Bureau of Travel Development
Pennsylvania Dept. of Commerce
416 Forum Bldg.
Harrisburg, PA 17120
717-787-5453

Rhode Island
Tourist Promotion Div.
Dept. of Economic Development
7 Jackson Walkway
Providence, RI 02903
401-277-2601; 800-556-2484 (out of state)

South Carolina
Div. of Tourism
Recreation & Tourism, Ste. 110
1205 Pendleton St.
Columbia, SC 29201
803-758-2536

Tennessee
Tennessee Dept. of Tourist Development
P.O. Box 23170
Nashville, TN 37202
615-741-2158

Utah
Utah Travel Council
Council Hall, Capitol Hill
Salt Lake City, UT 84114
801-533-5681

Vermont
Vermont Travel Div.
Agency of Development & Community Affairs
134 State St.
Montpelier, VT 05602
802-828-3236

Virgin Islands
Div. of Tourism
Box 6400 - Charlotte Amalie
St. Thomas, U.S. Virgin Islands 00801
809-774-8784

Virginia
Virginia State Travel Service
202 N. 9th St., Ste. 500
Richmond, VA 23219
804-786-2051

West Virginia
Travel Div.
Office of Economic & Community Development
State Capitol Bldg.
Charleston, WV 25305
304-348-2286

Wisconsin
Div. of Tourism
P.O. Box 7606
Madison, WI 53707
608-266-2161; 800-ESCAPES in-state and in Midwest

Wyoming
Wyoming Travel Commission
Cheyenne, WY 82002
307-777-7777.

NATURAL RESOURCES OFFICES

Alabama
Div. of State Parks
Alabama Dept. of Conservation & Natural Resources
64 N. Union St.
Montgomery, AL 36130
205-832-6323

Alaska
Div. of Geological & Geophysical Surveys
Dept. of Natural Resources
Pouch 7-028
Anchorage, AK 99510
907-276-2653

Mines Information Office
794 University Ave., Basement
Fairbanks, AK 99701
907-474-7062

Div. of Technical Services
Dept. of Natural Resources
Pouch 7035
Anchorage, AK 99510
907-786-2291

Div. of Land & Water Management
Dept. of Natural Resources, Pouch 7-005
Anchorage, AK 99510
907-265-4355

Arizona
Arizona Bureau of Geology & Mineral Technology
845 N. Park Ave.
Tucson, AZ 85719
602-626-2733

Information Resources Div.
Arizona State Land Dept.
1624 W. Adams, Rm. 302
Phoenix, AZ 85007
902-255-4061

California
Div. of Mines & Geology
California Dept. of Conservation
1416 9th St., Rm. 1341
Sacramento, CA 95814
916-445-0514

Map Information Office
California Dept. of Water Resources
P.O. Box 388
Sacramento, CA 95802
916-445-9259

Colorado
Colorado Geological Survey
Dept. of Natural Resources
1313 Sherman St., Rm. 715
Denver, CO 80203
303-866-2611

Connecticut
Natural Resources Center
Dept. of Environmental Protection
State Office Bldg., Rm. 553
Hartford, CT 06106
203-566-3540

Connecticut Geological & Natural History Survey
Dept. of Environmental Protection
State Office Bldg., Rm. 561
Hartford, CT 06106
203-566-3540

Florida
Bureau of Geology
903 W. Tennessee St.
Tallahassee, FL 32301
904-488-4191

Div. of Recreation & Parks
3900 Commonwealth Blvd.
Tallahassee, FL 32303
904-488-6131

Hawaii
Div. of Water & Land Development
Dept. of Land & Natural Resources
P.O. Box 373
Honolulu, HI 96809
808-548-7539

Idaho
Idaho Bureau of Mines & Geology
Morrill Hall, Rm. 332
University of Idaho
Moscow, ID 83843
208-885-7991

Idaho Dept. of Fish & Game
P.O. Box 25
Boise, ID 83707
208-334-3700

Illinois
Illinois Dept. of Conservation
160 N. LaSalle St.
Chicago, IL 60601
312-793-2070

Indiana
Div. of State Parks
616 State Office Bldg.
Indianapolis, IN 46204
317-232-4124

Div. of Water
Indiana Dept. of Natural Resources
605 State Office Bldg.
Indianapolis, IN 46204
317-232-4180

Iowa
Iowa Conservation Commission
Wallace State Office Bldg.
Des Moines, IA 50319
515-281-5145

Kansas
Kansas Fish & Game Commission
Box 54A, RR 2
Pratt, KS 67124
316-672-5911

Maryland
Maryland Forest & Park Services
Dept. of Natural Resources
Tawes State Office Bldg.
Annapolis, MD 21401
301-269-3761

Massachusetts
Dept. of Fisheries & Wildlife
Route 135
Westboro, MA 01581
617-366-4479

Waterways Div., State Geologist
Dept. of Environmental Quality Engineering
1 Winter St., 7th Fl.
Boston, MA 02108
617-292-5690

Div. of Forests & Parks
Dept. of Environmental Management
100 Cambridge St.
Boston, MA 02202
617-727-3180

Michigan
Div. of Land Resource Programs
Michigan Dept. of Natural Resources
Mason Bldg., 7th Fl., P.O. Box 30028
Lansing, MI 48909
517-373-3328

Minnesota
Div. of Parks & Recreation
Minnesota Dept. of Natural Resources
Box 39, Centennial Bldg.
St. Paul, MN 55155
612-296-4776

Mississippi
Bureau of Geology
Mississippi Dept. of Natural Resources
P.O. Box 5348
Jackson, MS 39216
601-359-1270

Missouri
Div. of Geology & Land Survey
Missouri Dept. of Natural Resources
P.O. Box 250
Rolla, MO 65401
314-364-1752

Montana
Montana Bureau of Mines & Geology
Montana College of Mineral Science & Technology
Main Hall, Rm. 200
Butte, MT 59701
406-496-4167

Nebraska
Information & Education Div.
State Game & Parks Commission
P.O. Box 30370
Lincoln, NE 68503
402-464-0641

Conservation & Survey Div.
113 Nebraska Hall
University of Nebraska
Lincoln, NE 68588
402-472-3471

Nevada
Geologic Information Specialist
Nevada Bureau of Mines & Geology
University of Nevada
Reno, NV 89557
702-784-6691

New Hampshire
Div. of Forests & Lands
105 Loudon Rd., Prescott Park
P.O. Box 856
Concord, NH 03301
603-271-3456

New Hampshire Div. of Parks & Recreation
Prescott Park, Bldg. 2
105 Loudon Rd., Box 856
Concord, NH 03301
603-271-3254

New Mexico
New Mexico Bureau of Mines & Mineral Resources
Campus Station
Socorro, NM 87801
505-835-5420

New York
Dept. of Environmental Conservation
50 Wolf Rd.
Albany, NY 12205

New York State Office of Parks, Recreation & Historic
    Preservation
Agency Bldg. 1
Empire State Plaza
Albany, NY 12238
518-474-0456

North Carolina
Div. of Land Resources
Geological Survey Section
Dept. of Natural Resources & Community Development
P.O. Box 27687
Raleigh, NC 27611
919-733-2423

North Dakota
North Dakota State Water Commission
State Office Bldg.
209 E. Boulevard Ave.
Bismarck, ND 58501
701-224-2750

Ohio
Div. of Geological Survey
Ohio Dept. of Natural Resources
Fountain Square, Bldg. B
Columbus, OH 43224
614-265-6605

Publications Center
Ohio Dept. of Natural Resources
Fountain Square, Bldg. B
Columbus, OH 43224
614-265-6608

Div. of Parks & Recreation
Ohio Dept. of Natural Resources
Fountain Square
Columbus, OH 43224
614-265-7000

Oregon
Oregon Dept. of Geology & Mineral Industries
1005 State Office Bldg.
Portland, OR 97201
503-229-5580

Oregon State Parks & Recreation Div.
525 Trade St., SE
Salem, OR 97310
503-378-6305

Pennsylvania
Bureau of Topographic & Geologic Survey
Pennsylvania Dept. of Environmental Resources
P.O. Box 2357
Harrisburg, PA 17120
717-787-2169

South Carolina
Div. of State Parks
Edgar A. Brown Bldg., Ste. 110, 1205 Pendleton St.
Columbia, SC 29201
803-758-3622

South Carolina Land Resources
Conservation Commission
2221 Devine St., Ste. 222
Columbia, SC 29205
803-758-2823

South Dakota
South Dakota Dept. of Game, Fish, & Parks
Anderson Bldg.
Pierre, SD 57501
605-224-3485

Tennessee
Div. of Geology
Tennessee Dept. of Conservation
G-5 State Office Bldg.
Nashville, TN 37219
615-741-2726

Div. of State Parks
Dept. of Conservation
701 Broadway
Nashville, TN 37203
615-742-6667

Texas
Texas Natural Resources Information System
P.O. Box 13087
Austin, TX 78711
512-475-3321

Utah
Utah Geological & Mineral Survey
Dept. of Natural Resources
606 Black Hawk Way
Salt Lake City, UT 84108
801-581-6831

Div. of Parks & Recreation
1596 W. North Temple
Salt Lake City, UT 84116
801-533-6012

Vermont
Agency of Environmental Conservation
Dept. of Forests, Parks & Recreation
324 Main St.
Barre, VT 05641
802-479-3621

Virginia
Dept. of Conservation & Economic Development
Natural Resource Bldg.
P.O. Box 3667
Charlottesville, VA 22903
804-293-5121

Div. of State Parks
1201 Washington Bldg.
Richmond, VA 23219
804-786-6140

Washington
State of Washington Water Research Center
Washington State University
Pullman, WA 99164
509-335-5531

Washington State Parks & Recreation Commission
7150 Clearwater Ln., KY-11
Olympia, WA 98504
206-753-5755

Dept. of Natural Resources
Photos, Maps & Reports Section QW-21
Public Lands Bldg.
Olympia, WA 98504
206-753-5338

Bureau of Surveys & Maps
Dept. of Natural Resources
1102 S. Quince EV-11
Olympia, WA 98504
206-753-5337

West Virginia
Div. of Parks & Recreation
1800 Washington St. E., Rm. 311
Charleston, WV 25305
304-348-2764

AVIATION/AERONAUTICAL OFFICES

Alabama
Alabama Dept. of Aeronautics
11 S. Union
State Highway Bldg.
Montgomery, AL 36130
205-832-6290

Arizona
Aeronautics Div.
Arizona Dept. of Transportation
1801 W. Jefferson, Rm. 426M
Phoenix, AZ 85007
602-261-7691

Arkansas
Arkansas Div. of Aeronautics
Adams Field - Old Terminal Bldg.
Little Rock, AR 72202
501-376-6781

California
Div. of Aeronautics
California Dept. of Transportation
1120 N St.
Sacramento, CA 95814
916-322-3090

Connecticut
Bureau of Aeronautics
Connecticut Dept. of Transportation
24 Wolcott Hill Rd.
Wethersfield, CT 06109
203-566-3076

Illinois
Illinois Div. of Aeronautics
Capital Airport, 1 Langhorne Bond Dr.
Springfield, IL 62706
217-753-4400

Indiana
Aeronautics Commission of Indiana
100 N. Senate Ave.
Indianapolis, IN 46204
317-232-3794

Iowa
Iowa Aeronautics Div.
Des Moines Municipal Airport
Des Moines, IA 50321
515-281-4280

Kansas
Kansas Aviation Div.
7th Fl., State Office Bldg.
Topeka, KS 66612
913-296-2553

Louisiana
Office of Aviation & Public Transportation
P.O. Box 44245, Capitol Station
Baton Rouge, LA 70804
504-925-7742

Maine
Maine Div. of Aeronautics
State Airport
Augusta, ME 04330
207-289-3185

Maryland
Office of General Aviation Service
Maryland State Aviation Admin.
BWI Airport, MD 21240
301-859-7064

Massachusetts
Massachusetts Aeronautics Commission
Logan Airport
East Boston, MA 02128
617-727-5350

Michigan
Aviation Information
Capital City Airport
Lansing, MI 48906
517-373-2146

Minnesota
Aeronautics Div.
Transportation Bldg., Rm. 417, John Ireland Blvd.
St. Paul, MN 55155
612-296-2216

Mississippi
Mississippi Aeronautics Commission
400 Robert E. Lee Bldg., P.O. Box 5
Jackson, MS 39205
601-359-1270

Missouri
Aviation Div.
Missouri Highway & Transportation Dept.
P.O. Box 270
Jefferson City, MO 65102
314-634-6469

Montana
Montana Aeronautics Div.
P.O. Box 5178
Helena, MT 59601
406-444-2506

Nebraska
Nebraska Dept. of Aeronautics
P.O. Box 82088
Lincoln, NE 68501
402-471-2371

New Hampshire
New Hampshire Aeronautics Commission
Municipal Airport
Concord, NH 03301
603-271-2551

New Mexico
Aviation Div.
New Mexico Transportation Dept.
P.O. Box 579
Santa Fe, NM 87504
505-827-4590

North Carolina
North Carolina Div. of Aviation
P.O. Box 25201
Raleigh, NC 27611

Oklahoma
Oklahoma Aeronautics Commission
200 NE 21st St.
Oklahoma City, OK 73105
405-521-2377

Pennsylvania
Pennsylvania Bureau of Aviation
Capital City Airport
New Cumberland, PA 17070
717-783-2280

South Carolina
South Carolina Aeronautics Commission
P.O. Drawer 1987
Columbia, SC 29202
803-758-2766

Utah
Div. of Aeronautical Operations
Dept. of Transportation
135 N. 2400 West
Salt Lake City, UT 84116
801-533-5057

Virginia
Dept. of Aviation
Commonwealth of Virginia
4508 S. Laburnum Ave., P.O. Box 7716
Richmond, VA 23231
804-786-3685

Wyoming
Aeronautics Commission
State of Wyoming
Cheyenne, WY 82002
307-777-7481

GEOLOGICAL SURVEYS

Alabama
Geological Survey of Alabama
P.O. Drawer O
University, AL 83486
205-349-2852

Arkansas
Arkansas Geological Commission
3815 W. Roosevelt Rd.
Little Rock, AR 72204
501-371-1488

Delaware
Delaware Geological Survey
University of Delaware
101 Penny Hall
Newark, DE 19716
302-451-2833

Illinois
Illinois State Geological Survey
Natural Resources Bldg.
615 E. Peabody Dr.
Champaign, IL 61820
217-344-1481

Indiana
Indiana Geological Survey
611 N. Walnut Grove
Bloomington, IN 47405
812-335-7636

Iowa
Iowa Geological Survey
123 N. Capitol St.
Iowa City, IA 52242
319-338-1173

Kansas
Kansas Geological Survey
Raymond C. Moore Hall
University of Kansas
Lawrence, KS 66044
913-864-3965

Kentucky
Kentucky Geological Survey
228 Mining and Resources Bldg.
University of Kentucky
Lexington, KY 40506
606-257-5863

Louisiana
Louisiana Geological Survey
Box G, University Station
Baton Rouge, LA 70893
504-342-6754

Maine
Maine Geological Survey
Ray Bldg., State House, Station 22
Augusta, ME 04333
207-289-2801

Maryland
Maryland Geological Survey
711 W. 40th St., Ste. 440
Baltimore, MD 21211
301-338-7066

Minnesota
Minnesota Geological Survey
2642 University Ave.
St. Paul, MN 55114
612-373-3372

New Jersey
New Jersey Geological Survey
1474 Prospect St. CN029
Trenton, NJ 08625
609-292-2576

North Dakota
North Dakota Geological Survey
University Station
Grand Forks, NJ 58202
701-777-2231

Oklahoma
Oklahoma Geological Survey
University of Oklahoma
Van Vleet Oval, Rm. 163
Norman, OK 73019
405-325-3031

South Dakota
South Dakota Geological Survey
Science Center
University of South Dakota
Vermillion, SD 57069
605-624-4471

Tennessee
Tennessee Div. of Geology
701 Broadway
Nashville, TN 37203
615-742-6696

Texas
Texas Bureau of Economic Geology
University of Texas at Austin
University Station, Box X
Austin, TX 78712
512-471-1534

West Virginia
West Virginia Geological & Economic Survey
West Virginia Cartographic Center
P.O. Box 879
Morgantown, WV 26507
304-594-2331

Wisconsin
Wisconsin Geological Survey
University of Wisconsin
1815 University Ave.
Madison, WI 53705
608-263-7389

Wyoming
Geological Survey of Wyoming
P.O. Box 3008
University Station
Laramie, WY 82071
307-742-2954

APPENDIX B
Federal Map Agencies

AGRICULTURAL STABILIZATION AND CONSERVATION SERVICE

2222 West 2300 South, P.O. Box 30010, Salt Lake City, UT 84130; 801-524-5856.

The U.S. Department of Agriculture's ASCS distributes aerial photographs produced by it and other branches of the USDA. Payment should be in the form of a money order or check, drawn from a U.S. or Canadian bank in U.S. dollars, made payable to ASCS.

ASCS's Aerial Photography Field Office will provide free "photoindexes" of specific areas in aerial photographic form. Photoindexes of Forest Service and Soil Conservation Service aerial photographs, as well as those by the ASCS, are available. A price list, order form, and descriptive pamphlet, "ASCS Aerial Photography," will be sent upon request.

ARMY CORPS OF ENGINEERS

Public Affairs, Army Corps of Engineers, Dept. of the Army, Dept. of Defense, 20 Massachusetts Ave. NW, Rm. 1101, Washington, DC 20314; 202-272-0011.

The Army Corps of Engineers produces and distributes inland waterway charts and maps of federal water-recreation lands within its regional districts. Write to the district nearest the areas for which you need maps. Some maps are free, while others are offered for a minimal fee. For a listing of Corps of Engineers districts and publications, write the Office of Public Affairs at the above address.

BUREAU OF THE CENSUS

Data User Services Div., Customer Services (Publications), Bureau of the Census, Washington, DC 20233; 301-763-4100.

The Customer Services Branch will accept Visa, MasterCard, personal checks, and money orders made payable to the "Superintendent of Documents" for census publications. "Statistical Maps of

the United States, GE-50 Series," "Congressional District Atlas," and other order forms describing available maps or series are free upon request.

Certain census map products are available from the Government Printing Office (Superintendent of Documents, U.S. Government Printing Office, Washington, DC 20402; 202-783-3238), rather than from the Bureau of the Census. These include the state and county subdivision maps, the *Congressional District Atlas* and "1982 Census of Agriculture, Graphic Summary." The free GPO subject index "Maps and Atlases" describes Census Bureau products available from GPO.

BUREAU OF LAND MANAGEMENT

USGS is the distributor for BLM Surface Management and Minerals Management maps, published as 30-minute X 60-minute maps at a scale of 1:100,000 ($4). Map coverage is shown on the USGS "Index to Intermediate Scale Mapping," available from Map Distribution Centers in Denver, Colorado, and Fairbanks, Alaska, as well as from ESIC offices. Copies can be secured by mail from the Distribution Centers.

Plats of townships and mineral surveys are available for public distribution, as follows:

■ For Illinois, Indiana, Iowa, Kansas, Missouri, and Ohio, plats may be ordered from:

National Archives and Record Administration
Cartographic Archives Div.
Pennsylvania Ave. at 8th St. NW
Washington, DC 20408

■ For all other public-land states, plats may be ordered from:

Bureau of Land Management
Eastern States Office
350 Pickett St.
Alexandria, VA 22304

CENTRAL INTELLIGENCE AGENCY

CIA maps are available from several sources. The

National Technical Information Service distributes most CIA maps published since 1980. Orders should be directed to:

National Technical Information Service (NTIS)
U.S. Dept. of Commerce
5285 Port Royal Rd.
Springfield, VA 22161
703-487-4650

Use the NTIS document number (PB number) when ordering. NTIS accepts personal checks and money orders drawn on U.S. or Canadian banks and made payable to NTIS.

Maps and other CIA publications produced before 1980 are available from:

Library of Congress
Photoduplication Service
Washington, DC 20504
202-287-5650

Staff in the Library's Geography and Map Division can help locate CIA maps in its collection.

A free catalog listing CIA maps and where to obtain them, "CIA Maps and Publications Released to the Public," is available from:

Central Intelligence Agency
Public Affairs Office
Washington, DC 20505
703-351-2053.

DEFENSE MAPPING AGENCY

DMA Combat Support Center, ATTN: PMSR, Washington, DC 20315; 202-227-2495.

The Defense Mapping Agency is the central map-producing branch of the Department of Defense. Products may be purchased with a check or money order in U.S. dollars, drawn from a U.S. or Canadian bank and made payable to "Treasurer of the United States."

Catalogs of DMA products are available from the Combat Support Center or authorized DMA map dealers for $2.50 each. The "Department of Defense/ Defense Mapping Agency Catalog of Maps, Charts and Related Products" is divided into the following public sales catalogs: *Aerospace Products* (free; CATP6V03); *Hydrographic Products* (10 volumes, $2.50 each; CATP2V01 through CATP2V10); and *Topo-*

graphic Products (free; CATP6V01). Product prices and listings of authorized DMA map dealers are listed in the catalogs.

EARTH SCIENCE INFORMATION OFFICE

The Earth Science Information Office operates a nationwide information and sales service for the results of Earth science research, maps, and related products and publications. A network of Earth Science Information Centers (ESIC) provides information on geologic, hydrologic, topographic, and land-use maps, books, and reports; aerial, satellite, and radar images and related products; Earth science and map data in digital form and related applications software; and geodetic data. ESIC offices can take orders for such customized products as digital cartographic data, and geographic names gazeteers.

ESIC OFFICES

Alaska
Anchorage-ESIC
4230 University Dr., Rm. 101
Anchorage, AK 99508
907-561-5555

Anchorage/Courthouse-ESIC
U.S. Courthouse, Rm. 113
222 W. 7th Ave., #53
Anchorage, AK 99513
907-271-4307

California
Los Angeles-ESIC
Federal Bldg., Rm. 7638
300 N. Los Angeles St.
Los Angeles, CA 90012
213-894-2850

Menlo Park-ESIC
Bldg. 3, MS 532
345 Middlefield Rd.
Menlo Park, CA 94025
415-329-4309

San Francisco-ESIC
504 Custom House, 555 Battery St.
San Francisco, CA 94111
415-556-5627

Colorado
Denver-ESIC
169 Federal Bldg., 1961 Stout St.
Denver, CO 80294
303-844-4169

Lakewood-ESIC
Box 25046
Federal Center, MS 504
Denver, CO 80225
303-236-5829

District of Columbia
Washington, DC-ESIC
Dept. of the Interior Bldg.
18th & C Sts. NW, Rm. 2650
Washington, DC 20240
202-343-8073

Mississippi
Stennis Space Center-ESIC
Bldg. 3101
Stennis Space Center, MS 39529
601-688-3544

Missouri
Rolla-ESIC
1400 Independence Rd., MS 231
Rolla, MO 65401
314-341-0851

Utah
Salt Lake City-ESIC
8105 Federal Bldg., 125 S. State St.
Salt Lake City, UT 84138
801-524-5652

Virginia
Reston-ESIC
507 National Center
Reston, VA 22092
703-860-6045

Washington
Spokane-ESIC
678 U.S. Courthouse, W. 920 Riverside Ave.
Spokane, WA 99201
509-353-2524

ESIC State Affiliates

Alabama
Geological Survey of Alabama
420 Hackberry Ln.
P.O. Box O, Univ. Station
University, AL 35486
205-349-2852

Alaska
Geophysical Institute
Univ. of Alaska-Fairbanks
Fairbanks, AK 99775
907-474-7487

Arizona
Arizona State Land Dept., Resource Analysis Div.
1616 W. Adams
Phoenix, AZ 85007
602-255-4061

Arkansas
Arkansas Geological Commission
3815 W. Roosevelt Rd.
Little Rock, AR 72204
501-371-1488

California
San Diego State University
Library, Map Collection
San Diego, CA 92812
619-594-5650

Map and Imagery Laboratory Library
Library, University of California
Santa Barbara, CA 93106
805-961-2779

Connecticut
Natural Resources Center
Dept. of Environmental Protection
165 Capitol Ave., Rm. 553
Hartford, CT 06106
203-566-3540

University of Connecticut
Map Library, Level 4
Storrs, CT 06268
203-486-4589

Delaware
Delaware Geological Survey
Cartographic Information Center
University of Delaware, Geological Survey Bldg.
Newark, DE 19716
302-451-8262

Florida
Resources and Environmental Analysis Center
361 Bellamy Bldg.
Florida State University
Tallahassee, FL 32306
904-644-2883

Georgia
Office of Research and Information
1200 Equitable Bldg., 100 Peachtree St.
Atlanta, GA 30303
404-656-5526

Hawaii
Dept. of Business and Economic Development
Kamamalu Bldg., P.O. Box 2359
Honolulu, HI 96804
808-548-3047

University of Hawaii at Manoa
Thomas Hale Hamilton Library
2550 The Mall
Honolulu, HI 96822
808-948-6199

Idaho
Idaho State Historical Library
610 N. Julia Davis Dr.
Boise, ID 83702
208-334-3356

University of Idaho Library
Map Collection
Moscow, ID 83843
208-885-6344

Illinois
University of Illinois, Urbana-Champaign
Map and Geography Library
1408 W. Gregory Dr.
Urbana, IL 61801
217-333-0827

Illinois State Geological Survey
615 W. Peabody Dr.
Champaign, IL 61820
217-244-0933

Indiana
Purdue University
214 Entomology Hall
West Lafayette, IN 47907
317-494-6305

Iowa
Iowa Geological Survey
123 N. Capitol St.
Iowa City, IA 52242
319-335-1575

Kansas
Kansas Geological Survey
University of Kansas
1930 Constant Ave., Campus West
Lawrence, KS 66046
913-864-3965

Kansas Applied Remote Sensing Program
Space Technology Center
University of Kansas
2291 Irving Hill Dr.
Lawrence, KS 66045
913-864-7720

Kentucky
Kentucky Geological Survey
228 Mining & Mineral Resources Bldg.
University of Kentucky
Lexington, KY 40506
606-257-5500

Louisiana
Office of Public Works
Dept. of Transportation and Development
P.O. Box 94245, Capitol Station
Baton Rouge, LA 70804
504-379-1473

Maine
Maine Geological Survey
State House Station, #22
Augusta, ME 04333
207-289-2801

University of Maine
College of Forest Resources
208 Nutting Hall
Orono, ME 04469
207-581-6277

Maryland
Maryland Geological Survey
2300 St. Paul St.
Baltimore, MD 21218
301-554-5524

Massachusetts
University of Massachusetts
Cartographic Information Research Services
102D Hasbrouck Laboratory
Amherst, MA 01003
413-545-0359

Michigan
Land and Water Management
Michigan Dept. of Natural Resources
Steven T. Mason Bldg., Box 30028
Lansing, MI 48909
517-373-9123

Minnesota
Minnesota State Planning Agency
Land Management Information Center
300 Centennial Office Bldg., 658 Cedar St.
St. Paul, MN 55155
612-296-1201

University of Minnesota
S76 Wilson Library
Map Library
Minneapolis, MN 55455
612-624-4549

Mississippi
Geographic Information Systems Div.
Mississippi Research and Development Center
3825 Ridgewood Rd.
Jackson, MS 39211
601-982-6354

Missouri
Missouri Dept. of Natural Resources
Div. of Geology and Land Survey
P.O. Box 250
Rolla, MO 65401
314-364-1752

Montana
Montana Bureau of Mines and Geology
Montana Tech
Main Hall, Rm. 200
Butte, MT 59701
406-496-4167

Nebraska
Conservation and Survey Div.
University of Nebraska, Lincoln
901 N. 17th St.
Lincoln, NE 68508
402-472-2567

Nevada
Univ of Nevada, Las Vegas—Library
Government Documents Dept.
4505 Maryland Pkwy.
Las Vegas, NV 89154
702-739-3409

Nevada Bureau of Mines and Geology
University of Nevada, Reno
Reno, NV 89557
702-784-6691

New Hampshire
Documents Dept.
Dimond Library
University of New Hampshire
Durham, NH 03824
603-862-1777

New Jersey
Dept. of Environmental Protection
New Jersey Geological Survey
CN-029
Trenton, NJ 08625
609-292-2576

New Mexico
University of New Mexico
Technology Applications Center
2808 Central Ave. SE
Albuquerque, NM 87131
505-277-3622

New York
Map Information Unit
New York Dept. of Transportation
Albany, NY 12232
518-457-3555

North Carolina
North Carolina Survey
ESIC Affiliate Coordinator
P.O. Box 27687
Raleigh, NC 27611
919-733-2423

North Carolina Geological Survey
ESIC Affiliate Coordinator
59 Woodfin Pl.
Asheville, NC 28801
704-251-6208

North Dakota
North Dakota Geological Survey
University of North Dakota
University Station
Grand Forks, ND 58202
701-777-2231

North Dakota Water Commission
State Office Bldg.
900 East Blvd.
Bismark, ND 58505
701-224-2750

Ohio
Ohio Dept. of Natural Resources
Div. of Soil and Water Conservation
Remote Sensing Section - ESIC
Fountain Square, Bldg. E
Columbus, OH 43224
614-265-6770

Oklahoma
Oklahoma Geological Survey
University of Oklahoma
830 Van Vleet Oval
Norman, OK 73019
405-325-3031

Oregon
Oregon State Library
State Library Bldg.
Salem, OR 97301
503-378-4368

University of Oregon Library
Map Library
165 Condon Hall
Eugene, OR 97403
503-686-3051

Pennsylvania
Dept. of Environmental Resources
Bureau of Topographic and Geological Survey
P.O. Box 2357
Harrisburg, PA 17120
717-787-2169

Rhode Island
Rhode Island Cartographic Information Center
Pell Library
University of Rhode Island
Narragansett, RI 02882
401-792-6539

South Carolina
South Carolina Land Resources Conservation
  Commission
2221 Devine St., Ste. 222
Columbia, SC 29205
803-734-9100

South Dakota
South Dakota Geological Survey
Science Center
University of South Dakota
Vermillion, SD 57069
605-677-6146

Tennessee
Tennessee Valley Authority
Maps and Survey
101 Haney Bldg.
Chattanooga, TN 37402
615-751-6277

Texas
Texas Natural Resources Information System
P.O. Box 13231
Austin, TX 78711
512-463-8406

Utah
Utah Geological and Mineral Survey
606 Black Hawk Way
Research Park
Salt Lake City, UT 84108
801-581-6831

Vermont
University of Vermont
Documents/Map Dept.
Bailey/Howe Library
Burlington, VT 05405
802-656-2503

Virginia
Dept. of Mines, Minerals and Energy
Div. of Mineral Resources, Natural Resources Bldg.
Box 3667
Charlottesville, VA 22903
804-293-5121

Washington
Washington State Library
Information Services Div.
Olympia, WA 98504
206-753-4027

University of Washington
Map Collection & Cartographic Information Center
University of Washington Libraries
FM-25
Seattle, WA 98195
206-543-9392

West Virginia
West Virginia Geological and Economic Survey
West Virginia Cartographic Center
Box 879
Morgantown, WV 26507
304-594-2331

Wisconsin
State Cartographer's Office
155 Science Hall
550 N. Park St.
Madison, WI 53706
608-262-6850

Wyoming
State Engineer
Herschler Bldg.
Cheyenne, WY 82002
307-777-7354

EROS Data Center

User Services Section, EROS Data Center, U.S. Geological Survey, Sioux Falls, SD 57198; 605-594-6151.

EROS is the clearinghouse for aerial photographs and satellite and space imagery created by the federal government. EROS (like all USGS branches) accepts checks or money orders in U.S. dollars made payable to the Department of the Interior/USGS, for the exact amount of the order. Order forms for products can be obtained from either EROS or the Earth Science Information Center (507 National Center, 12201 Sunrise Valley Dr., Reston, VA 22092; 703-860-6045) and affiliated offices (see ESIC listing for addresses).

Order forms are available for "Aerial Mapping Photography," "NASA Aircraft Photography," "Inquiry for Geographic Search for Aircraft Data," "NHAP: National High Altitude Photography Program/NAPP: National Aerial Photography Program," "Land Satellite Images," "Manned Spacecraft Photographs & Major Metropolitan Area Photographs and Images," "Understanding Color-

Infrared Photographs and False Color Composites," and "Catalog of Cartographic Data." EROS will supply price lists upon request and USGS produces a series of free pamphlets that explain products and ordering procedures. Pamphlet titles include "How to Order Landsat Images," "Looking for an Old Aerial Photograph," "EROS: A Space Program for Earth Resources," "The Aerial Photography Summary Record System," "Manned Spacecraft Photographs & Major Metropolitan Area Photographs and Images," "Understanding Color-Infrared Photographs and False Color Composites," "Catalog of Cartographic Data," and "How to Obtain Aerial Photographs."

FEDERAL ENERGY REGULATORY COMMISSION

Div. of Public Affairs, Federal Energy Regulatory Commission, 825 Capitol St. NE., Washington, DC 20426; 202-357-8055.

FERC produces an energy map relating to interstate gas pipelines. The map is available from GPO and the FERC Public Reference Branch. The map is in the free index, "Maps and Atlases," available at GPO bookstores.

FISH AND WILDLIFE SERVICE

Div. of Realty, Fish and Wildlife Service, Washington, DC 20240; 703-358-1713.

The Fish and Wildlife Service provides free maps and brochures for its protected lands. There is no catalog, but the maps can be obtained either at the sites themselves or from the Division of Realty. Address lists of FWS hatcheries can be obtained free from the Division of Hatcheries, and lists of refuges can be obtained free from the Division of Refuges, both located at the above address. Information about FWS's "National Wetlands Inventory" can be obtained from FWS offices; maps produced as part of NWI are available through USGS ESIC offices.

GOVERNMENT PRINTING OFFICE

Superintendent of Documents, U.S. Government Printing Office, Washington, DC 20402; 202-783-3238.

GPO accepts Visa, MasterCard, and money orders or personal checks in U.S. dollars drawn on U.S. and

Canadian banks and made payable to the "Superintendent of Documents." There are more than 100 free subject indexes available at most GPO stores, listing products by category, including "Maps and Atlases"; other subject indexes cover areas such as "Farms and Farming," "Oceanography," and "Public and Private Utilities." A free catalog, "U.S. Government Books," lists numerous GPO books, including atlases and other map-filled volumes. Although few GPO bookstores carry the full catalog of available maps, most have certain Census Bureau and CIA maps in stock. There are GPO bookstores around the country:

Alabama

Birmingham GPO Bookstore
9220-B Parkway E.
Birmingham, AL 35206
205-731-1056

California

Los Angeles GPO Bookstore
ARCO Plaza, Level C
505 S. Flower St.
Los Angeles, CA 90071
213-894-5841

San Francisco GPO Bookstore
Rm. 1023, Federal Office Bldg.
450 Golden Gate Ave.
San Francisco, CA 94102
415-556-0643

Colorado

Denver GPO Bookstore
Rm. 117, Federal Bldg.
1961 Stout St.
Denver, CO 80294
303-844-3964

Pueblo GPO Bookstore
Majestic Bldg.
720 N. Main St.
Pueblo, CO 81003
303-544-3142

District of Columbia

Main GPO Bookstore
710 N. Capitol St.
Washington, DC 20402
202-275-2091

Commerce Dept. GPO Bookstore
14th & Pennsylvania NW
Washington, DC 20230
202-377-3527

Farragut West GPO Bookstore
1510 H St. NW
Washington, DC 20006
202-653-5075

Florida
Jacksonville GPO Bookstore
Rm. 158, Federal Bldg.
400 West Bay St., P.O. Box 35089
Jacksonville, FL 32202
904-791-3801

Georgia
Atlanta GPO Bookstore
Rm. 100, Federal Bldg.
275 Peachtree St. NE
Atlanta, GA 30303
404-331-6947

Illinois
Chicago GPO Bookstore
Rm. 1463, Everett McKinley Dirksen Bldg.
219 S. Dearborn St.
Chicago, IL 60604
312-353-5133

Maryland
Retail Sales Outlet
8660 Cherry Ln.
Laurel, MD 20810
301-953-7974

Massachusetts
Boston GPO Bookstore
Rm. G25, Kennedy Federal Bldg.
Sudbury St.
Boston, MA 02203
617-565-2488/9

Michigan
Detroit GPO Bookstore
McNamara Federal Bldg., Ste. 160
477 Michigan Ave.
Detroit, MI 48226
313-226-7816

Missouri
Kansas City GPO Bookstore
Rm. 144, Federal Office Bldg.
601 E. 12th St.
Kansas City, MO 64106
816-374-2160

New York
New York GPO Bookstore
Federal Bldg., Rm. 110
26 Federal Plaza
New York, NY 10278
212-264-3825

Ohio
Cleveland GPO Bookstore
Federal Office Bldg.
1240 E. 9th St.
Cleveland, OH 44199
216-522-4922

Columbus GPO Bookstore
Rm. 207, Federal Bldg.
200 N. High St.
Columbus, OH 43215
614-469-6956

Pennsylvania
Philadelphia GPO Bookstore
Morris Bldg., 100 N. 17th St.
Philadelphia, PA 19103
215-597-0677

Pittsburgh GPO Bookstore
Rm. 118, Federal Office Bldg.
1000 Liberty Ave.
Pittsburgh, PA 15222
412-644-2721

Texas
Dallas GPO Bookstore
Rm. 1C50, Federal Bldg.
1100 Commerce St.
Dallas, TX 75242
214-767-0076

Houston GPO Bookstore
9319 Gulf Fwy.
Houston, TX 77017
713-226-5453

Washington
Seattle GPO Bookstore
Rm. 194, Federal Office Bldg.
915 Second Ave.
Seattle, WA 98174
206-442-4270

Wisconsin
Milwaukee GPO Bookstore
Rm. 190, Federal Bldg.
517 E. Wisconsin Ave.
Milwaukee, WI 53202
414-291-1304

INTERNATIONAL BOUNDARY AND WATER COMMISSION, U.S. AND MEXICO

The Commons, Bldg. C, Ste. 310, 4171 N. Mesa, El Paso, TX 79902.

The International Boundary and Water Commission

produces several maps of the region in and around the United States-Mexico border. Map prices vary; a complete listing of maps available to the public will be sent upon request.

INTERNATIONAL BOUNDARY COMMISSION, U.S. AND CANADA

425 I St. NW, Rm. 150, Washington, DC 20001; 202-632-8058.

IBC sells maps of the U.S.-Canadian borders for $3 each. A listing that details available maps will be sent upon request.

LIBRARY OF CONGRESS

Geography and Map Div., Library of Congress, Washington, DC 20540; 202-707-4177.

The Geography and Map Division's Reading Room (located in the James Madison Memorial Building, Rm. LM B01, 101 Independence Ave. SE, Washington, DC 20540) is open to the general public. Maps may not be taken from the room, but reproduction services are available. Two free pamphlets, available in the Reading Room or through the mail, "Geography and Map Division, Library of Congress," and "Publications of the Geography and Map Division," provide information about the division, and its services and products.

Complete price and reproduction information can be obtained from the Library's Photoduplication Service (202-707-5650) at the above address.

NATIONAL AERONAUTIC AND SPACE ADMINISTRATION

NASA aircraft photography is available from the EROS Data Center (User Services Section, U.S. Geological Survey, Sioux Falls, SD 57198; 605-594-6511). NASA space imagery is distributed through Bara-King Photographic Inc., 4508 Frolich Ln., Hyattsville, MD 20781; 301-322-7900.

A complete price list and information on ordering available NASA space imagery can be obtained from Bara-King Photographic Inc. or NASA's Audiovisual Department (National Aeronautics & Space Admin., Rm. 6035, 400 Maryland Ave. SW, Washington, DC 20546; 202-453-1000).

NATIONAL ARCHIVES CARTOGRAPHIC AND ARCHITECTURAL BRANCH

Washington, DC 20408; 703-756-6700.

You must have a researcher's card to use the Archives, which can be obtained at the front desk of the Archives map headquarters (located at 841 S. Pickett St., Alexandria, VA 22304) to use the reading room of the Archives Cartographic and Architectural collection. A price list for reproduction services is available upon request, or can be quoted over the telephone. (Photocopy facilities are located on the premises; higher-quality reproduction work generally takes six to eight weeks.) It helps to know the file number of the maps you need reproduced, but the staff can sometimes help locate this information. Free inventories and catalogs of various parts of the Archives collection are listed in the free pamphlet "Cartographic and Architectural Branch," available both at the branch location and from the above mailing address. Among the inventories available are "Cartographic Records of the Bureau of the Census," "Cartographic Records of the Forest Service," and "Cartographic Records of the United States Marine Corps."

NATIONAL CLIMATIC DATA CENTER

Federal Bldg., Asheville, NC 28801; 704-259-0682.

The National Climatic Data Center, part of the National Oceanic and Atmospheric Administration, has a Historical Climatology Series that includes several long-term weather records, maps, and atlases. A free catalog of the series, as well as a price list, are available upon request. Payment can be made with American Express, Visa, and Master-Card, or check and money order drawn on U.S. funds and made payable to "Commerce-NOAA-NCDC."

NATIONAL GEOPHYSICAL DATA CENTER

Code E/GC4, Dept. ORD, 325 Broadway, Boulder, CO 80303; 303-497-6423.

The National Geophysical Data Center, part of the National Oceanic and Atmospheric Administration, compiles data on various seismic and natural resource subjects for domestic and foreign areas. Many maps are produced from this data; a publications catalog is available. Prepayment is required on all nonfederal orders; make checks and money orders payable to "Commerce/NOAA/NGDC." NGDC accepts payments made with American Express, MasterCard, and Visa. A $10 handling fee is charged for each order; there is an additional $10 charge for non-U.S. orders.

National Ocean Service

Distribution Branch, (N/CG33), National Ocean Service, Riverdale, MD 20737; 301-436-6990.

NOS, a branch of the National Oceanic and Atmospheric Administration, distributes nautical and aeronautical charts of the U.S. created by NOAA. The products may be purchased with Visa and MasterCard or with a check or money order made payable to "NOS, Department of Commerce, N/CG33." Five catalogs of available nautical charts and tables — "Atlantic and Gulf Coasts," "Pacific Coast," "Alaska," "Great Lakes and Adjacent Waterways," and "Bathymetric Maps and Special Purpose Charts" — are free upon request, as is the "Catalog of Aeronautical Charts and Related Publications."

Maps and charts can also be purchased from two NOS services offices:

National Ocean Service
Chart Sales Office
6501 Lafayette Ave.
Riverdale, MD 20737
301-436-6980

Chart Sales & Geodetic Control
Federal Bldg. and Courthouse
701 C St., Box 38
Anchorage, AK 99513
907-271-5040

National Ocean Service maps can also be obtained at hundreds of authorized dealers around the country (most are map, sports equipment, boating and tackle, and aviation equipment stores). The dealers are listed in the free catalogs.

Nautical and aeronautical maps of foreign countries are distributed through the Defense Mapping Agency.

National Park Service

Office of Public Inquiries, National Park Service, Rm. 1013, Washington, DC 20240; 202-343-4747.

The National Park Service supplies the visitor-information centers of its national parks, forests, seashores, and historical sites with maps and information folders, most of which are free to the public. The maps and folders can also be obtained through the Office of Public Inquiries. The Government Printing Office is the distributor for all Park Service publications, including booklets or studies of parks and historic sites, many of which also contain maps and guides.

Tennessee Valley Authority

Maps and Surveys Branch, 101 Haney Bldg., 311 Broad St., Chattanooga, TN 37402; 615-751-6277.

TVA produces a wide variety of maps related to its region. Maps may be purchased with a check or money order in U.S. dollars, drawn on a U.S. or Canadian account, and made payable to the Tennessee Valley Authority.

A "Catalog of Selected Maps and Data Available," as well as indexes of TVA nautical and topographic maps, will be sent free upon request. More information on TVA's mapping program can be obtained by writing to the above address.

U.S. Forest Service

Public Affairs Office, U.S. Forest Service, Rm. 3107, South Bldg., P.O. Box 96090, Washington, DC 20090; 202-447-3957.

There are maps available for each of the U.S. Forest Service's 156 National Forests. They are available either at each forest visitor center or from regional offices of the Forest Service. Most maps cost from $1.00 to $3.50 and may be purchased with checks or money orders drawn on U.S. dollars and made payable to USFS. A national map with an address list of National Forests will be sent free upon request.

U.S. Geological Survey

Map Distribution, Federal Center, Box 25286, Denver, CO 80225; 303-236-7477.

USGS accepts money orders and personal checks made payable to the "Department of the Interior/USGS." State-by-state indexes of topographic, geologic, and general maps, as well as listings of maps by category (land use, national atlas, etc.);

indexes by map scale; price lists; and order forms are available free upon request from either the map distribution center or from regional Earth Science Information Centers.

Alaska residents may purchase geologic, hydrologic, and topographic maps of Alaska from the Alaska Distribution Section, USGS, New Federal Bldg., Box 12, 101 12th Ave., Fairbanks, AK 99701; 907-456-7535.

EROS Data Center is the USGS distributor of images produced from Landsat imagery, aerial photographs and photomaps. Hundreds of map and travel stores, as well as state offices, are authorized dealers of USGS products. Names and addresses of USGS map dealers are listed on the state indexes already mentioned. The maps may also be purchased over the counter at the USGS's Reston, Va., location.

International Cartographic Agencies

Algeria
Institut des Sciences de la Terre
Universite des Sciences et de la Technologie Houari
    Boumediene
B.P. 9
Dar Al Beida—Alger

Institut National de Cartographie
123 rue de Tripoli
B.P. 32 Hussein-Dey
Algiers

Angola
Direccao de Servicos de Geologia e Minas
Caixa Postal 1260-C
Luanda

Argentina
Instituto Geografico Militar
Avenida Cabildo 381
1246 Buenos Aires
Tel.: 771-3031

Aruba
Dienst Landmeetkunde en Vastgoedregistratie
Yrausquinplein 10
Oranjestad
Aruba
Tel.: 297-8-22605/21311

Australia
Department of Primary Industries and Energy
G. P. O. Box 858
Canberra City, A.C.T. 260l
Tel.: 777-7520
Telex: AA62308

Austria
Federal Authority for Standardization and Geodesy
Gruppe Landesaufnahme
Schiffamtsgasse 1-3
A-1025 Vienna
Tel.: (0222) 35 76 11

Federal Institute for Pedological Research
Denisgasse 31
A-1200 Vienna
Tel.: (0222) 33 46 31-O DW

Bahamas
Director of Lands and Surveys
P.O. Box N-592
Nassau

Bangladesh
Directorate of Land Records and Survey
Tejgaon Industrial Area
Dhaka 1208
Tel.: 326380/327215

Survey of Bangladesh
Tejgaon Industrial Area
Dhaka 1208
Tel.: 326276

Belgium
Institut Geographique National
13 Abbaye de la Cambre
1050 Brussels

Belize
Belize Meteorologist/Hydrologist
Ladyville
Tel.: 501-025-2012
Telex: 251 BTL BUREAU DZ

Benin
Direction de la Topographique et du Cadastre
B.P. 360
Cotonou

Bhutan
Department of Geology and Mines
Department of Trade, Industry and Power
P.O. Box 173
Thimphu
Tel.: 2879

Bolivia
Servicio Geologico de Bolivia
Federico Zuazo, Esq. Reyes Ortiz
Casilla de Correos 2729
La Paz
Tel.: 326278

Instituto Geografico Militar
Avenida Saavedra, Cuartel Miraflores
La Paz
Tel.: 360513/378194

Botswana
Geological Survey Department
Ministry of Mineral Resources and Water Affairs
Private Bag 14
Lobatse
Tel.: 330428/330327
Telex: 2293 BD

Department of Surveys and Lands
Ministry of Local Government and Lands
Private Bag 37
Gaborone
Tel.: 353251
Telex: 2589 MLGL BD

Brazil
Instituto de Pesquisas Espaciasis
Caixa Postal 515
12201 Sao Jose dos Campos
Sao Paulo, SP
Tel.: (0123) 22-9509
Telex: (123) 3530 INPE BR

Diretoria do Servico Geografico do Exercito
SMU—QG Ex., Bloco F, 20 Pav.
70630 Brasilia, D.F.
Tel.: (061) 223-8529
Telex: (061) 1094

Diretoria de Geodesia e Cartografia
Instituto Brasileiro de Geografia e Estatistica
Av. Franklin Roosevelt, 166 100 Andar
20021 Rio de Janeiro, RJ
Tel.: (021) 220-6671/6821
Telex: (021) 30939

Projeto RADAM
CRS 509, Bloco A, Loja 1 a 5
70360 Brasilia, D.F.
Tel.: (061) 244-9432
Telex: (061) 2243

Brunei
Survey Department of Brunei
Ministry of Development
Bandar Seri Begawan 2070
Tel.: 43171
Telex: BU2228

Bulgaria
Central Laboratory for Higher Geodesy
Acad. G. Bonchev Street, Block 1, 4th Floor
Sofia 1113
Tel.: 24-64 72-08-41
Telex: 22424

Central Laboratory for Seismic Mechanics and Seismic
    Engineering
Acad. G. Bonchev Street, Block 3
Sofia 1113
Tel.: 33-41 70-31-07
Telex: 22424

Burkina Faso
Cadastral Direction
B.P. 7054
Ouagadougou

Burma (Union of Myanmar)
Central Research Organization
Kanbe Road, Yankin
Rangoon
Tel.: 50917/50544
Telex: 21503 KASALA BM

Survey Department
Ministry of Agriculture and Forestry
Thirimingala Lana
Kaba Aye Pagoda Road
Rangoon
Tel.: 62923/62933

Cameroon
Institute for Geological and Mining Research
B.P. 4110
Yaounde
Tel.: 22 00 08

Ministry of Higher Education, Computer Services
    and Scientific Research
B.P. 1457
Yaounde
Tel.: 23 16 50
Telex: 8418 KN

Denmark
Geodetic Institute
Rigsdagsgaardent
DIL-1218
Copenhagen

Egypt
Aerial Survey of Egypt
308 El-Haram Street
Giza, Cairo
Tel.: 852-950
Telex: 739794 WAZRA UN/93794 WAZRA UNH

Survey of Egypt
1 Abdel Salam Arif Street
Giza, Cairo
Tel.: 348-4904/348-4422
Telex: 739794 WAZRA UN/93794 WAZRA UNH

Egyptian Remote Sensing Center
101 Kasr El Aini Street
Cairo
Tel.: 354-0173/355-7110

El Salvador
Ministry of Public Works
La. Av. Sur No. 630
San Salvador
Tel.: 71-6026

Center for Geotechnical Research
Avenida Peralta, final, contiguo a Talleres El Coro
San Salvador
Tel.: 22-9011

National Geographic Institute
"Ingeniero Pablo Arnoldo Guzman"
Avenida Juan Bertis No. 79
San Salvador
Tel.: 25-5060

Weather Forecast and Hydrology Service
Renewable Natural Resources Center
Canton El Matazano, Soyapango
San Salvador
Tel.: 27-0484/27-0622

Ministry of Agriculture
Alameda Roosevelt 2823
San Salvador
Tel.: 23-24434/24-2944

General Directorate of Irrigation and Drainage
Canton El Matazano, Soyapango
San Salvador
Tel.: 77-0490

Ethiopia
Cartography Unit
UNECA
Africa Hall
P.O. Box 3001
Addis Ababa

Ethiopian Mapping Agency
P.O.Box 597
Addis Ababa

Fiji
Mineral Resources Division
Ministry of Energy and Mineral Resources
Private Mail Bag. P.O.
Suva
Tel.: 383-611
Telex: 2330 FJ

Lands and Survey Department
Ministry of Lands, Local Government and Housing
P.O. Box 2222, Government Building
Suva
Tel.: 211-516
Telex: 2167 FOSEC FJ

Finland
National Board of Survey
Opastinsilta 12
00520 Helsinki
Tel.: 1541
Telex: 125 254 MAP SE

Geodetic Institute
Ilmalankatu l A
00240 Helsinki
Tel.: 410 433

Cartographic Department
Opastinsilta 12
00520 Helsinki
Tel.: 1541
Telex: 125 254 MAP SF

France
SPOT Image
15, avenue Edouard Belin
BP 4359
Toulouse Cedex
Tel.: 61 53 99 76
Telex: 532 079 SPOTIM

Institut Geographique National
136, rue St. Grenelle
75700 Paris
Tel.: 45 50 34 95
Telex: 204 989 F

Gabon
Bureau de Recherches Geologiques et Minieres
B.P. 175
Libreville
Tel.: 76 06 09/764498
Telex: 5576 GO

Direction Generale de la Geologie
Ministere des Mines
B.P. 576
Libreville
Tel.: 76 35 56
Telex: 5352 GO

The Gambia
Survey Department
Half Die, Banjul

German Democratic Republic
Research Center for Soil Fertility
Department of Cartography
Wilhelm-Pieck-Strasse 72
Munchenberg DDR-1278
Tel.: 820
Telex: 016367

German Library
Map Collection
Deutscher Platz 1
DDR-7010 Leipzig
Tel.: 8812323
Telex: 51562 DBUECH DD

Germany, Federal Republic of
Federal Office for Geosciences and Raw Materials
Alfred-Bentz-Haus
Postfach 510153
Stilleweg 2
D-3000 Hannover 51
Tel.: 49-511-643-2243
Telex: 923730 BG-RHAD

Federal Minister for Defense
Office for Military Geosystems
Frauenbergerstrasse 250
5350 Euskirchen
Bonn 2
Tel.: 02251/7092
Telex: 8-882527

German Hydrographic Institute
Bernhard-Nocht-Strasse 78
2000 Hamburg 4
Tel.: 040/31901
Telex: 2-11138

Association of Geodetic Administrations of the Federal
   States
Warmbuechenkamp 2
3000 Hannover 1
Tel.: 0511/1201

Institute for Applied Geodesy
Richard-Strauss-Allee 11
6000 Frankfurt am Main 70
Tel.: 069-63331
Telex: 4-13592

Greece
Hellenic Army Geographic Service
3 Evelpidon Street
GR-113 62 Athens
Tel.: 8842811

Guatemala
Military Geographic Institute
Avenida de las Americas 5-76, Zona l3
Guatemala
Tel.: 363281 to 83

Guinea
Ministry of Mines and Geology
B.P. 295
Conakry
Tel.: 44-28-01

Guyana
Lands and Surveys Department
22 Upper Hadfield Street
D'urban Backlands
Greater Georgetown
Tel.: (02) 72582/60524-9

Haiti
Service de Geodesie et de Cartographie
Cite de l'Exposition/Boulevard Harry Truman
Port-au-Prince
Tel.: 509-1-23225

Honduras
Direccion General de Minas e Hidrocarbouros
Ministerio de Recursos Naturales
Boulevard Centro America
Tegucigalpa, D.C

Instituto Geografico Nacional
Edificio SECOPT
Barrio La Bolsa
Comayaguela, D.C.
Tel.: 33-7403

Oficina Nacional del Catastro
Edificio Didemo, 4 Piso
Tegucigalpa, D.C.
Tel.: 33-2081

Hong Kong
Survey Division
Lands Department
Murray Building, Second Floor
Garden Road
Kowloon

Hungary
Geodesy and Cartography Bureau
H-l442 Bosnyak ter 5
Budapest
Tel.: 635-260
Telex: 635-260

Iceland
Iceland Geodetic Survey
Laugavegi 178
105 Reykjavik
Tel.: 681611

India
National Remote Sensing Agency
Department of Space
Balanagar
Hyderabad 500037
Andhra Pradesh
Tel.: 263360
Telex: 0915-6522

Survey of India
Surveyor General Office
Hathi Bukla Estate
P.O. Box 37
Dehra Dun 24800l
Tel.: 23468/23467
Telex: 0585218 SURVEYS IN

Geodetic and Geophysical Surveys
Survey of India
Dehra Dun 24800l
Tel.: 24528
Telex: 0585210-G & RB IN

Indian Institute of Remote Sensing
4, Kalidas Road
P.B. No. 135
Dehra Dun, Uttar Pradesh
Tel.: 24583

Indonesia
Geologic and Mineral Resources
Department of Mines and Energy
Jalan Janderal Gatot Soebroto Kav. 49
Jakarta Selatan
Tel.: 510134
Telex: 62471 DIPU-IE

Directorate General of Mines
J1 Gatot Soebroto Kav 40
Jakarta Selatan
Tel.: 510447
Telex: 44030 PERTUM IA

Army Topographic Service
Jalan Gunung Sahari 90
Jakarta Pusat
Tel.: 341983

Directorate of Land Use
Ministry of Internal Affairs
Jalan Sisingamangaraja 2
Jakarta Selatan
Tel.: 798004

Armed Forces Surveys and Mapping Agency
Jalan Dr. Wahidin I/11
Jakarta Pusat
Tel.: 364474

National Coordinating Body for Survey and Mapping
    Board
Jalan Raya Jakarta km 46, Cibionong
Java Barat
Tel.: (99)82062-67

Aerial Mapping Service
Komplek Lapangan Udara Halim Perdanakusuma
Jakarta Timur
Tel.: 8091010

Iran
National Cartographic Center
Azadi Square—Merage Ave.
P.O. Box 11365-5167 Cab. Sanca.
Tehran
Tel.: 021-960031-37
Telex: 212701 NCC IR

National Geographic Organization
Imperial Iranian Army
Tehran

Iraq
State Establishment for Geological Survey and
    Investigation
Ministry of Industry and Military
Industrialization
P.O. Box 2330 and 2730
Baghdad
Tel.: 719-5123
Telex: 212293 SOM IK

Seismic Exploration Company
Ministry of Oil
P.O. Box 476
Khullani Square
Baghdad
Tel.: 887-1115
Telex: 212208/212204 INCO IK

State Commission for Dams
P.O. Box 5982
Baghdad
Tel.: 416-9141
Telex: 212107 DAMS IK

Water Wells Drilling Company
P.O. Box 3231—Waziriyah
Baghdad
Tel.: 422-2021
Telex: 212795 ARTESIAN IK

State Establishment for Surveying
Ministry of Agriculture and Irrigation
P.O. Box 5813
Gailani Square
Baghdad
Tel.: 888-6101
Telex: 212916 SURVEY IK

Ireland
Ordnance Survey
Phoenix Park
Dublin 8
Tel.: (01) 213 171
Telex: 30126

Israel
Department of Surveys
Ministry of Labor
2 Caplan Street
Hakirya
Jerusalem 91008 Aviv
Tel.: 02-694-222

Italy
Military Geographic Institute
Via Cesare Battisti 10
50122 Firenze
Tel.: 055 27751
Telex: 575597

Jamaica
Survey Department
P.O. Box 493
231/2 Charles Street
Kingston
Tel.: 809-922-6630/5

Japan
Geographical Survey Institute
Ministry of Construction
l, Kitasato, Tsukuba
Ibaraki 305
Tel.: 0298-64-1111

Jordan
Natural Resources Authority
P.O. Box 7, 39, or 2220
Amman

Department of Lands and Surveys
Ministry of Finance
P.O. Box 70
Amman

Jordan National Geographic Center
P.O. box 20214
Amman

Kenya
Survey of Kenya
P.O. Box 30046
Nairobi

Regional Center for Services in Surveying, Mapping
and Remote Sensing
P.O. Box 18332
Nairobi

Korea (North)
Geology and Geography Research Institute
Academy of Sciences
Mammoon-dong, Central District
P'yongyang

Korea (South)
Korea Institute of Energy and Resources
2l9-5, Karibong-dong
Kuro-Ku
Seoul
Tel.: 856-0041-7
Telex: K24337

Hydrographic Office
Ministry of Transportation
1-17ang-dong, 7-ka, Chung-ku
Inchon City
Tel.: 032-885-3821
Telex: K25134

Korean Society of Geodesy, Photogrammetry and
  Cartography
4th Fl., Kyohaksa Bldg
105-67, Kongdok-dong, Mapop-ku
Seoul
Tel.: 701-8383

Kuwait
Director of Building and Survey Department
Kuwait Municipality
P.O. Box 10, Safat
Kuwait

Laos
State Geographic Department
Office of the Council of Ministers
P.O. Box 2159
Vientiane, Lao P.D.R.
Tel.: 2661

Lebanon
Directorate of Geographic Affairs
Army H.Q. Al Yarze
Republic of Lebanon

Lesotho
Department of Lands, Surveys and Physical Planning
Ministry of Interior, Chieftainship Affairs and Social
  Welfare
P.O. Box 174
Maseru 100

Liberia
Ministry of Lands, Mines and Energy
Capitol Hill
P.O. Box 9024
Monrovia
Tel.: 221580

Liberian Cartographic Service
P.O. Box 9024
Monrovia
Tel.: 261842

Libya
Planning Division
Secretariat of Agrarian Reform and Land Reclamation
P.O. Box 190
Tripoli

Geological Department
Faculty of Science
Al-Fateh University
P.O. Box 398
Tripoli

Department of Surveying
Secretariat of Planning
P.O. Box 600
Tripoli

Liechtenstein
Landesbauamt des Furstentums
Liechtenstein
Stadtle 49
9490 Vaduz

Agricultural Office
Stadtle 49
9490 Vaduz

Luxembourg
Administration du Cadastre et de la Topographie
54 Avenue Gaston Diderich
L-1420 Luxembourg
Tel.: 44901-222

Madagascar
Ministere de l'Industrie, de l'Energie at de Mines
B.P. 527
101 Antananarivo
Tel.: 255-15
Telex: 22-540 MIEM MG

Institut National de Geodesie et de Cartographie
(Foibe-Taosanrintanin'i Matagascar-FTM)
3 Lalana J. J. Ravelomanantsoa
B.P. 323, Ambanidia
l0l Antananarivo
Tel.: 229-35

Malawi
Survey Department
Ministry of Natural Resources
P.O. Box 349
Blantyre
Tel.: 633 722

Malaysia
Directorate of National Mapping
Bangunan Ukor, 1st Floor
Jalan Gurney
50578 Kuala Lumpur
Tel.: 60-3-292-5311/ext. 291
Telex: MA 28148

Mali
Institut National de Topographie
B.P. 240
Bamako

Malta
Information Division
Auberge de Castille
Valletta
Tel.: 224901/225231
Telex: 1448 DOI MW

Mauritania
Service Topographique et Cartographique
B.P. 237
Nouakchott

Mauritius
Ministry of Housing, Lands, and Town and Country
    Planning
Survey Department
Port Louis

Mexico
Departmento Geografico Militar
Servicio Cartografico
Secretaria de la Defensa Nacional
Lomas de Sotelo
Mexico 10, D.F.

Micronesia, Federated States of
Department of Resources and Development
Box 490, Kolonia, Pohnpei
Federated States of Micronesia 96941
Tel.: (691) 320-2646
Telex: 729-6807 FSMGOV FM

Mongolia
Academy of Sciences
U1, Leniadom 2
Ulaanbaatar

Morocco
Ministere de L'Agriculture et de la Reforme Agraire
Division de la Cartographie
31, Avenue Moulay Hassan
Rabat
Tel.: 212-7-646-08
Telex: 31038M

Ministere de l'Agriculture et de la Reforme Agraire
Conservation Fonciere et des Travaux Topographiques
Avenue Moulay Youssef
Rabat
Tel.: 212-7-657-17
Telex: 31038M

Mozambique
National Directorate for Geography and Mapping
Avenida Josina Machel, 589/Maputo
P.O. Box 288
Maputo
Tel.: 22786
Telex: 6-500 DNHMC-MO

Namibia
The Surveyor General
P/B 13182 Kaiser Street
Windhoek

Nepal
Survey Department
Ministry of Land Reform
Dilli Bazar
Kathmandu
Tel.: 411897

Netherlands
Land Registry Service
Waltersingel 1
Postbus 9046
7300 GH Apeldoorn
Tel.: 31-55-28-51-11

Cartography Service
Ministry of Defense
Bendienplein 5
7815 SM Emmen
Tel.: 31-70-5910-96911

Netherlands Antilles
Dienst van het Kadaster (Office of the Land Registry)
President Romulo Betancourt Boulevard 4
Willemstad, Curacao
Tel.: 599-611188

New Caledonia
Service Topographique
Territorial Administration Center
Avenue Paul Doumer
Noumea

New Zealand
Survey and Land Information Department
Head Office
Private Bag, Charles Fergusson Building
Wellington
Tel.: (04) 735-022
Telex: (04) 722-244

Niger
Direction du Service Topographique et du Cadastre
Ministere des Finances
B.P. 250
Niamey

Nigeria
Cartographic Section
Geological Survey Department
P.M.B. 2007
Kaduna South, Kaduna State

Surveys Division
Federal Ministry of Works
Igbosere Road at Okesuna Street
Lagos

Norway
Norwegian Institute for Polar Affairs
Rolfstangveien 12
Postboks 158
1330 Oslo Lufthavn
Tel.: 47-2-123650
Telex: 74745

Oman
Ministry of Defence
P.O. Box 113
Muscat
Tel.: 701-109/702-195
Telex: 3228 DEFENCE ON

Pakistan
Survey of Pakistan
P.O. Box 1068
Murree Road
Rawalpindi
Tel: 842229

Panama
Instituto Geografico Nacional "Tommy Guardia"
Apartado Postal 5267, Zona 5
Panama
Tel.: 64-0444

Paraguay
Military Geodetic Service
Avenida Artigas y Via Ferrea
Asuncion
Tel.: 208858

Ciudad Universitaria
Km 10, San Lorenzo
Institute of Basic Sciences
Tel.: 501517

Peru
Geological Mining and Metallurgical Institute
Pablo Bermudez 211
Apartado 889
Lima
Tel.: 316233

National Office for Evaluation of Natural Resources
Calle 17 # 355, Urb. El Palomar
San Isidro, Lima
Tel.: 410425

General Bureau of Aerophotography
Base Aerea "Las Palmas"
Barranco, Lima
Tel.: 670538
Telex: 21501 PE

Military Geographic Institute
Av. Andres Aramburu 1198
Apartado 2038
Lima

Naval Bureau of Hydrography and Navigation
Calle Saenz Pena No. 590, La Punta
Callao, Lima
Tel.: 652995

Philippines
National Mapping and Research Institute Administration
NAMRIA Building
1201 Fort Bonifacio
Makati, Metro Manila
Tel.: 810-5468

Coast and Geodetic Survey Department
421 Barraca Street
Binondo 1006
Tel.: 47-5645/48-4679

Poland
Instytut Geologiczny
ul. Rakowiecka 4
00-975 Warszawa

Coordinative and Administrative Body
ul. Jasna 2/4
00-013 Warszawa

Instytut Geodezji i Kartografii
ul. Jasna 2/4
00-013 Warszawa

Instytut Geografii i
Przestrzennego
Zagospodarowania PAN
ul. Krakowskie Przedmiescie 30
00-325 Warszawa

Portugal
Instituto Geografico e Cadastral
Praca da Estrela
1200 Lisboa
Barata Pinto
Tel.: 66 61 12/60 98 20/60 99 25
Telex: 62638 IGC P

Qatar
Industrial Development Technical Center
P.O. Box 2599
Doha
Bin Jabor Al-Thani
Tel.: 832121
Telex: 4323 IDTC DH

Department of Petroleum Affairs
Ministry of Finance and Petroleum
P.O. Box 83
Doha
Tel.: 461444
Telex: 4233 QATFIN DH

Romania
Institutul de Geografie si Geologie
Str. Dimitrie Racovita no. 12
Bucharest
Tel.: 16 68 80

Rwanda
Ministry of Industry and Handicrafts
B.P. 73
Kigali
Tel.: 75417

Ministry of Public Works
B.P. 24
Kigali
Tel.: 86649

Saudi Arabia
Aerial Survey Department
P.O. Box 247
Riyadh
Tel.: (966-1) 478-1661
Telex: 201490/201615/201058

Senegal
Bureau for Cartography and Geology
7, rue Mermoz
B.P. 12.098 Colobane-Dakar
Dakar
Tel.: 21 62 26
Telex: 51274 SG

Senegal Geographic Service
14, rue Victor Hugo
B.P. 740
Dakar
Tel.: 21 65 67

Sierra Leone
Surveys and Land Division
Ministry of Lands
Youyi Bldg.
Freetown
Tel.: 40689

Singapore
Housing Development Board
Building and Development Division
3451 Jalan Bukit Merah, HDB Centre
Singapore 0315
Tel.: 2739090
Telex: RS 22020 SINHDB

Jurong Town Corporation
Technical Division
301 Jurong Town Hall Road
Jurong Town Hall
Singapore 2260
Tel.: 5600056
Telex: RS 35733 JTC

Ministry of Environment
Environmental Engineering Division
40 Scotts Road
Environment Building
Singapore 0922
Tel.: 7327733
Telex: RS 34365 ENV

Survey Department
Ministry of Law
8 Shenton Way
Dover Crescent
#28-01, Treasury Building
Singapore 0106
Tel.: 2259911

Urban Redevelopment Authority
45 Maxwell Road
Urban Redevelopment Authority Building
Singapore 0106
Tel.: 2216666
Telex: RS 20703 SINURA

Solomon Islands
Ministry of Agriculture and Lands
P.O. Box G 13
Honiara
Tel.: 23567

Somalia
Survey and Mapping Department
Ministry of Defense
P.O. Box 24
Mogadishu

South Africa
Directorate, Surveys and Mapping
Private Bag
Mowbray
7700
Tel.: (021) 689-9721/685-4070
Telex: 521-418

Soviet Union
Main Administration for Geodesy and Cartography
Applied Geodesy SRI, Novosibirk
NII Prikladnoy Geodezii

Spain
Universidad de Oviedo
Laboratorio de Geologia
Escuela Superior de Minas
Independencia 13
33004 Oviedo
Tel.: (985) 240-358

Direccion General del Instituto Geografico Nacional
General Ibanez Ibero, 3
28003 Madrid
Tel.: (91) 233-3800

Real Sociedad Geografica
Valverde 22
Madrid 13

Sri Lanka
Surveyor-General's Office
Kirula Rd.
Colombo 5
Tel.: 585569

Sudan
Ministry of Internal Affairs
Survey Dept.
P.O. Box 306
Khartoum

Suriname
Centraal Bureau Luchtkaartering
P.O. Box 971
Dr. Sophie Redmondstraat 131
Paramaribo
Tel.: 74421

Swaziland
Department of Surveys
Ministry of Works and Communications
P.O. Box 58
Mbabane
Tel.: 42321

Sweden
The Geological Survey of Sweden
Box 670
S-751 28 UPPSALA
Tel.: 46 18 179000

Switzerland
Federal Office for Topographical Survey
Seftigenstrasse 264
CH-3084 Wabern

Syria
National Remote Sensing Center
Ministry of Electricity
Sultan Salim St.
Damascus

Taiwan
Insitute of Earth Sciences
Academia Sinica
P.O. Box 23-59
Taipei

Central Geological Survey
P.O. Box 968
Taipei

Agriculture and Forestry Aerial Survey Team
Taiwan Provincial Forestry Bureau
101-10 Ping West Rd., Section 2
Taipei

Tanzania
Survey and Mapping Division
Ministry of Lands, Natural Resources and Tourism
P.O. Box 9201
Dar es Salaam
Tel.: (501) 21241

Thailand
Royal Thai Survey Department
Ministry of Defense
Rachinee Rd.
Bangkok 10200

Togo
Service Topographique
Ministere des Mines, de l'Energie, de Ressources
    Hydrauliques et des Travaux Publiques
B.P. 500
Lome

Tonga, Kingdom of
Ministry of Lands, Survey and Natural Resources
P.O. Box 5
Nuku'alofa

Trinidad and Tobago
Ministry of Planning and Mobilisation
Lands and Surveys Div.
Red House, St. Vincent St.
Port-of-Spain

Seismic Research Unit
c/o University of the West Indies
St. Augustine
Tel.: (809) 662-4659

Uganda
Department of Lands and Surveys
P.O. Box 7061
Kampala

Union of Myanmar (see Burma)

United Kingdom
Ordnance Survey
Ramsey Rd., Maybush
Southampton SO9 4DH
Tel.: (703) 79200

Directorate of Military Survey
Ministry of Defence
Elmwood Ave.
Feltham, Middlesex TW13 7AE
Tel.: (1) 890 3622

Uruguay
Servicio Geografico Militar
Avenida 8 de Octubre 3255
Montevideo
Tel.: 80 71 11

Direccion de Topografia
Ministerio de Transporte y Obras Publicas
Rincon 561, Piso 2
Montevideo
Tel.: 95 94 34

Servicio de Oceanografiaidrografia y Meteorologia de
    la Armada
Capurpo 980
Montevideo
Tel.: 39 92 20

Venezuela
Direccion de Cartografia Nacional
Edificio Camejo, Piso 2, Oficina 230
Centro Simon Bolivar
Caracas
Tel.: 408 1710

Western Samoa
Lands and Survey Department
Main Beach Rd.
P.O. Box 63
Apia
Tel.: 22481

Yemen, People's Democratic Republic of
Geological and Mineral Exploration Directorate
P.O. Box 5252
Aden

Yugoslavia
Insitute for Cartography
39, Bulevar Vojvode Misica
11000 Belgrade
Tel.: 011 651 255

Zaire
Zairian Geographic Institute
106, Blvd. du 30 Juin
B.P. 3086
Kinshasa
Tel.: 31 854

Zambia
Survey Department
Ministry of Lands and Natural Resources
P.O. Box 50397
Lusaka

Zimbabwe
Department of the Surveyor General
Ministry of Lands, Resettlement, and Rural
    Development
P.O. Box 8099, Causeway
Harare

APPENDIX D
Selected Map Stores

The list of map stores is constantly growing. The **International Map Dealers Association**, a trade association for retail map dealers, distributors, and manufacturers of maps and related products, can provide names and addresses of map dealers in your area. IMDA is located at P.O. Box 1789, Kankakee, IL 60901; 815-939-4627.

Alabama
Carto-Craft Maps Inc.
738 Shades Mountain Plaza
Birmingham, AL 35226
205-822-2103

Nautical Publications
2625 Highland Ave. S.
Birmingham, AL 35205
205-930-0171

Re-Print Corp.
2025 First Ave. N.
Birmingham, AL 35203
205-251-9171

Alaska
The Maps Place
700 E. Benson
Anchorage, AK 99503
907-274-7335; 907-284-6277

Arizona
Arizona Hiking Shack
11645 N. Cave Creek Rd.
Phoenix, AZ 85020
602-943-2722

Desert Mountain Sports
4506 N. 16th St.
Phoenix, AZ 85016
602-265-4401

Desert Mountain Sports
2824 E. Indian School Rd.
Phoenix, AZ 85016
602-955-2875

Earth Tracks Recreational Maps
3644 E. McDowell Rd.
Phoenix, AZ 85008
602-224-9578

Tucson Map & Flag Center
2590 N. First Ave.
Tucson, AZ 85719
602-623-1104

Wide World of Maps
2626 W. Glenrosa Ave.
Phoenix, AZ 85017
602-279-2323

Wide World of Maps
1526 N. Scottsdale Rd.
Tempe, AZ 85281
602-949-1012

Wide World of Maps
1440 S. Country Club Dr.
Mesa, AZ 85202
602-844-1134

Arkansas
AAA Map Company
6917 Geyer Spring Rd., #7N
Little Rock, AR 72209
501-562-6219

The Map Shop
402-B S. Thompson, P.O. Box 1571
Springdale, AR 72765
501-751-5863

Shepherd's Inc.
603 W. Markham St.
Little Rock, AR 72205
501-375-6937

California
Allied Services
966 N. Main St.
Orange, CA 92667
714-532-4300

A.L.S. Maps
610 N. Azusa Ave.
West Covina, CA 91791
818-915-5165

Bookends Bookstore
1014 Coombs St.
Napa, CA 94559
707-224-7455

Bucksport Sporting Goods
3650 Broadway
Eureka, CA 95501
707-442-1832

Cal-Gold
2569 E. Colorado Blvd.
Pasadena, CA 91107
818-792-6161

California Map Center
3211 Pico Blvd.
Santa Monica, CA 90405
213-829-7902

California Survey & Drafting Supply
4733 Auburn Blvd.
Sacramento, CA 95841
916-344-0232

Champion Map Corp.
9550-F Micron Ave.
Sacramento, CA 95827
916-366-6622

Compass Maps
1172 Kansas Ave.
Modesto, CA 95352
209-529-5017

Dustbooks and Fulton's Bookstore
P.O. Box 100
Paradise, CA 95967
916-877-6110

Easy Going Travel
1400 Shattuck Ave.
Berkeley, CA 94709
415-843-3533; 800-233-3533

Geographia
4000 Riverside Dr.
Burbank, CA 91505
818-848-1414

Geographic Maps & Travel Books
4000 Riverside Dr.
Burbank, CA 91505
818-848-1414

Global Graphics
2819 Greentop St.
Lakewood, CA 90712
213-429-8880

Global Map Store
35 N. Fulton
Fresno, CA 93728
209-266-9831

The Map Center
2440 Bancroft Way
Berkeley, CA 94704
415-843-8080

Map Centre
2611 University Ave.
San Diego, CA 92104
619-291-3830

Map Link
529 State St.
Santa Barbara, CA 93101
805-965-4402

Map Shop
12112 W. Washington Blvd.
Los Angeles, CA 90066
213-391-1848

MAPS etc.
21919 Sherman Way
Canoga Park, CA 91303
818-347-9160

Maps to Anywhere Travel Bookstore
1514 N. Hillhurst Ave.
Hollywood, CA 90029
213-660-2101

Map Shop
12112 W. Washington Blvd.
Los Angeles, CA 90066
213-391-1848

Mountain Sports
217 Main St.
Chico, CA 95926
916-345-5011

Pacific Coast Map Service
12021 Long Beach Blvd.
Lynwood, CA 90262
213-636-6657

Pasadena Map Co.
985 E. Colorado Blvd.
Pasadena, CA 91106
818-795-3626

Rand McNally Retail Store
595 Market St.
San Francisco, CA 94105
415-777-3131

Sonoma Bookends
201 W. Napa St., #18
Sonoma, CA 95476
201-938-5926

Thomas Bros. Map Store
603 W. 7th St.
Los Angeles, CA 90017
213-627-4018

Thomas Bros. Map Store
550 Jackson St.
San Francisco, CA 94104
415-981-7520

The Travel Center
3636 Atlantic
Long Beach, CA 90807
213-487-2835

Travel Market
130 Pacific Avenue Mall
San Francisco, CA 94111
415-421-4080

USA Maps
2974 First St., Ste. 1
La Verne, CA 91750
714-593-3601

Word Journeys
971-C Lomas Santa Fe Dr.
Solana Beach, CA 92075
619-481-4158

Colorado
Boulder Map Gallery
1708 13th St.
Boulder, CO 80302
303-444-1406

Chinook Bookshop Inc.
210 N. Tejon St.
Colorado Springs, CO 80903
719-635-1195

Crossroads Map Co.
2717 E. Louisiana Ave.
Denver, CO 80201
303-733-2131

Hotchkiss Inc.
4825 Oakland
Denver, CO 80239
303-371-3600

International Map Service
85 S. Union Blvd., Unit D-2
Lakewood, CO 80228
303-987-2747

Macvan Productions Inc.
809 N. Cascade Ave.
Colorado Springs, CO 80903
719-633-5757

Maps Unlimited
899 Broadway
Denver, CO 80203
303-623-4299

Mountain Maps
147 W. 3rd St.
Salida, CO 81201
719-539-4334

Pierson Graphics
899 Broadway
Denver, CO 80203
303-623-4299

Wilderness Society
4260 E. Evans Ave.
Denver, CO 80220
303-839-1175

Connecticut
Back Roads Ltd.
99 Nova Scotia Hill Rd.
Watertown, CT 06795
203-274-4806

Huntington's Book Stores
65 Asylum
Hartford, CT 06105
203-527-1835

Map House
1520 Rhey Ave.
Wallingford, CT 06492
203-269-0685

The Melko Corp.
620 Villa Ave.
Fairfield, CT 06430
203-367-8327

Whitlock's Inc.
17 Broadway
New Haven, CT 06511
203-562-9841

Delaware
First State Map Co.
12 Mary Ella Dr.
Wilmington, DE 19805
302-998-6009

Maps by Mail

Here are six one-stop-shopping sources for both foreign and domestic maps of all types. Each of the companies offers mail-order services. Contact each to obtain a current catalog and ordering information.

Access Maps & Gear (321 S. Guadalupe, Santa Fe, NM 87501; 505-988-2442) is a mail-order service offering a diverse but select catalog of "unique maps" and related products from around the world. Products include antique maps, astronomical charts, map display and storage items, tools for map use, map clothing, puzzles, and games. Access also offers map framing, and sells map-frame kits.

Forsyth Travel Library (9154 W. 57th St., P.O. Box 2975, Shawnee Mission, KS 66201; 913-384-3440) is a specialist in travel books and maps, offering mail-order service as well as a warehouse open to the public. Forsyth Travel Library carries a wide range of guidebooks and maps for travel destinations all over the world.

Geoscience Resources (2990 Anthony Rd., P.O. Box 2096, Burlington, NC 27216; 919-227-8300; 800-742-2677) distributes thousands of maps from more than 80 foreign map companies and more than 75 foreign government surveys. Well-known maps, such as Kummerly & Frey and Recta Foldex, are represented by Geoscience Resources, as well as maps from smaller publishers, such as the Automovil Club of Argentina and P.T. Pembina of Indonesia. City maps, road maps, topographic maps, and recreation maps are among the offerings.

ITMB Publishing Ltd. (736A Granville St., Vancouver, BC, Canada V6Z 1G3; 604-687-3320) is the distributor for more than 50 cartographic publishers worldwide, from the U.S.'s Rand McNally and Prentice Hall to Chile's Motouiti to Germany's Falk Verlag to Iran's Sahab Publishing. ITMB carries a wide range of tourist maps and road atlases for countries, regions, cities and towns in virtually every area of the world. ITMB also distributes guidebooks from nearly 40 international publishers, including many small publishing houses, such as Corax Press (Canada), Inca Press (U.S.), Progress Press (U.S.S.R.), and TT Publications (India). ITMB offers a selection of more than 200 guidebooks, from general guides to more specialized hiking guides, nature guides, ethnology guides, architecture guides, and boating guides, for travel destinations in Asia, Africa, Europe, the Pacific and the Americas.

Map Link (529 State St., Santa Barbara, CA 93101; 805-965-4402) is a map wholesaler and retailer, stocking more than 35,000 titles representing every region of the world. It distributes topographic, geologic, city, regional, and wall maps. Map Link is a good resource for otherwise hard-to-find maps. A catalog is available upon request. Map Link has also compiled "The World Map Directory," a guide to its extensive map inventory ($29.95). Map Link can provide a list of local map stores that distribute their maps and can also help in finding appropriate map collections.

McCarthy Map Company (1003 Main St., Boonton, NJ 07005; 201-316-5494) specializes in business-related products, although its product line includes a few less-than-serious items, too. Among products features in McCarthy's catalog are maps for sales analysis and market planning, business travel maps and road atlases, and zip code maps.

Newark Newsstand
70 E. Main St.
Newark, DE 19711
302-368-8770

Ninth Street Book Shop
110 W. 9th St.
Wilmington, DE 19801
302-652-3315

District of Columbia
The Map Store
1636 Eye St. NW
Washington, DC 20006
202-628-2608

National Geographic Bookstore
17th & M Sts. NW
Washington, DC 20036
202-857-7000

National Map Gallery & Travel Center
Union Station
50 Massachusetts Ave. NE
Washington, DC 20002
202-789-0100

Florida
Central Florida Map Company
2216 Vincent Rd.
Orlando, FL 32817
407-277-4408

Champion Map Corp.
200 Fentress Blvd.
Daytona Beach, FL 32114
904-258-1270; 800-874-7010; 800-342-1072 (in Florida)

Champion Map Corp.
332 N. Orlando Ave.
Maitland, FL 32751
305-629-5867

Map and Globe Store
1120 E. Colonial Dr.
Orlando, FL 32807
407-425-0185

The Map Shop
1718 College Pkwy.
Fort Myers, FL 33907
813-278-1949

Surf & Sand Map Co.
1978 Radcliff Dr. N
Clearwater, FL 34623
813-443-7146

Georgia
Borders Book Shop
3655 Roswell Rd. NE
Atlanta, GA 30342
404-237-0707

Bradford Map Co.
1873 Lawrenceville Hwy.
Decatur, GA 30033
404-633-7562

Latitudes
3393 Peachtree Rd.
Lenox Square Mall
Atlanta, GA 30326
404-237-6144

Latitudes
2246 Perimeter Mall, P.O. Box 467518
Atlanta, GA 30346
404-394-2772

Oxford Book Store
2345 Peachtree Rd. NE
Atlanta, GA 30305
404-262-3333

Travel Source
3815 Mill Creek Ct.
Atlanta, GA 30341
404-434-0739

Hawaii
Basically Books
169 Keawe St.
Hilo, HI 96720
808-961-0144

Pacific Map Center
647 Auahi St.
Honolulu, HI 96813
808-531-3800

Idaho
Hunt Enterprises
6208 Cassia St.
Boise, ID 83709
208-375-4200

Illinois
Genealogy Unlimited, Inc.
789 S. Buffalo Grove Rd.
Buffalo Grove, IL 60089
312-541-3175

Rand McNally Commercial Showroom (Business
    Maps)
1020 E. Higgins
Elk Grove, IL 60007
312-364-6866

Rand McNally Retail Store
23 E. Madison St.
Chicago, IL 60602
312-332-4628

The Savvy Traveller
50 E. Washington
Chicago, IL 60602
312-263-2100

Special Collections
1029 Greenleaf
Wilmette, IL 60091
312-251-4666

Suburban Map Store
910 Riverside, Unit #2
Elmhurst, IL 60126
312-941-7978

Indiana
Cram Co. Inc.
301 S. LaSalle
P.O. Box 426
Indianapolis, IN 46201
317-635-5564

Map World
645 Eastern Blvd.
Clarksville, IN 47130
812-283-6277

Odyssey Map Store
148 N. Delaware St.
Indianapolis, IN 46204
317-635-3837

Print Graphics
2505 E. 52nd
Indianapolis, IN 46205
317-577-6385

Riegel's
624 S. Calhoun St.
Ft. Wayne, IN 46802
219-424-1429

Iowa
Haunted Bookshop
227 S. Johnson St.
Iowa City, IA 52240
319-337-2996

Oak Ridge Sports Inc.
117 W. 11th St.
Dubuque, IA 52001
319-556-0861

Roberts' Maps
1100 Locust St.
Des Moines, IA 50309
515-243-5259

Travel Genie
3714 Lincolnway
Ames, IA 50010
515-292-1070

Kansas
Forsyth Travel Library Inc.
9154 W. 57th St., P.O. Box 2975
Shawnee Mission, KS 66201
913-384-3440

McLeod's
1818 W. 2nd
Wichita, KS 67203
316-263-3500

Rector's Bookstore
206 E. Douglas Ave.
Wichita, KS 67202
316-265-0611

Sportsmen's Maps Headquarters
333 E. English St.
Wichita, KS 67202
316-262-2192

Superior School Supply Center
241 N. Hydraulic
Wichita, KS 67214
316-265-7683

Kentucky
La Belle Gallery
741 E. Chestnut
Louisville, KY 40202
606-589-0621

Owl and Pussycat
314 S. Ashland Ave.
Lexington, KY 40502
606-266-7121

Louisiana
Beaucoup Books
5418 Magazine
New Orleans, LA 70130
504-895-2663

Globe Map Company
206 Milam St.
Shreveport, LA 71102
318-222-7453

McCurnin Nautical Charts
2318 Woodlawn Ave.
Metairie, LA 70001
504-888-4500

New Orleans Map Co.
3130 Paris Ave.
New Orleans, LA 70119
504-943-0878

Vidrine Office Supply
7730 W. Main, P.O. Box 7
Galliano, LA 70354
504-632-2163

Maine
Books-N-Things
Oxford Plaza
Oxford, ME 04270
207-743-7197

DeLorme Map Store
Lower Main St.
Freeport, ME 04032
207-865-4171

Kennebec Books
82 Western Ave.
Augusta, ME 04330
207-622-7843

Maryland
Bookstall
9927-B Falls Rd.
Potomac, MD 20854
301-469-7800

Exploration USA
14703 E. Baltimore Ave, Ste. 237
Laurel, MD 20707
301-490-2236

First Frame Graphics
P.O. Box 2179
Easton, MD 21601
301-820-4468
(Mail-order)

Greetings and Readings
809 Taylor Ave.
Towson, MD 21204
301-825-4225

Travel Books Unlimited
4931 Cordell Ave.
Bethesda, MD 20814
301-951-8533

Massachusetts
A2Z Science and Nature Store
150 Main St.
Northampton, MA 01060
413-586-1611

Arts & Cards Inc.
374 Boylston St.
Brookline, MA 02146
617-566-4984

Champion Map Corp.
186 Cedar Hills St.
Marlborough, MA 01752
508-481-8252; 800-922-9380

Eastern Mountain Sports
189 Linden St.
Wellesley, MA 02181
617-254-4250

Globe Corner Bookstore
1 School St.
Boston, MA 02108
617-523-6658; 800-358-6013

Grey Lady of the Sea
Old South West
Nantucket, MA 02554
508-228-9552

The Harvard Square Map Store
40 Brattle St.
Cambridge, MA 02138
617-497-6277

Michigan
Delta Maps
5800 12 Mile Rd.
Warren, MI 48092
313-573-9273

Geography Ltd.
912 Fountain St.
Ann Arbor, MI 48103
313-769-5152

Universal Map Enterprises
1606 E. Michigan Ave.
Lansing, MI 48912
517-484-1978

Minnesota
Books Abroad
25 University Ave. SE
Minneapolis, MN 55414
612-378-0961

Hudson Map Company
2510 Nicollet Ave.
Minneapolis, MN 55404
612-872-8818

Latitudes
3801 Grand Ave. S.
Minneapolis, MN 55409
612-823-3742

Latitudes
5101 Vernon Ave.
Edina, MN 55436
612-920-1848

The Map Store
348 N. Robert St.
St. Paul, MN 55101
612-227-6277

The Map Store
120 S. 6th St.
211 Skyway
Minneapolis, MN 55402
612-339-4117

Mississippi
George's Map Service
564 Dryden Ave.
Jackson, MS 39209
601-371-3875

Montana
Trail Head
501 S. Higgins
Missoula, MT 59801
406-543-6966

Nebraska
Stephenson School Supply
1112 O St.
Lincoln, NE 68508
402-476-7663

Nevada
Front Boy Service Co.
3340 Sirius Ave.
Las Vegas, NV 89102
702-876-7822

Good Directions
2633 Lenna St.
Las Vegas, NV 89102
702-871-3679

New Hampshire
Globe Corner Bookstore
Settlers Green, Route 16
P.O. Box 1756
North Conway, NH 03860
603-356-7063

Goodman's
383 Chestnut St.
Manchester, NH 03101
603-622-2153

New Jersey
Ardic Book Distributors Inc.
174 Route 206 S
Somerville, NJ 08876
609-924-2532

Geo Graphics Inc.
208 Glenridge Ave., P.O. Box 183
Montclair, NJ 07042
201-744-7873

Geographia Map Center
231 Hackensack Plank Rd.
Weehawken, NJ 07087
201-867-4706; 212-695-6585

Geostat Map & Travel Center
Caldor Shopping Center
Rte. 10 and 202
Morris Plains, NJ 07950
201-538-7707

Geostat Map & Travel Center
Wick Shopping Plaza
Rte. 1 and Plainfield Ave.
Edison, NJ 08817
201-985-1555

Geostat Map & Travel Center
Montgomery Shopping Center
Rte. 206 and 518
Rocky Hill, NJ 08558
609-924-2121

Geostat Map & Travel Center
910 N. Rte. 73
Marlton, NJ 08053
609-983-3600

Geostat Map & Travel Center
174 Rte. 206 S.
Somerville, NJ 08876
201-359-2828

International Map Company
547 Shaler Blvd.
Ridgefield, NJ 07657
201-943-5550

McCarthy Map Co. Inc.
1003 Main St.
Boonton, NJ 07005
201-316-5494

Universal Success Corp.
550 Cookman Ave.
Asbury Park, NJ 07712
201-774-2020

New Mexico
Access Maps & Gear
321 S. Guadalupe
Santa Fe, NM 87501
505-988-2920

Base Camp
121 W. San Francisco St.
Santa Fe, NM 87501
505-982-9707

Burnt Horses Book Store
307 Johnson St.
Santa Fe, NM 87501
505-982-4799

Holman's Inc.
401 Wyoming Blvd. NE
Albuquerque, NM 87123
505-265-7981

Page One Newsstand & Bookstore
11200 Montgomery Blvd. NE
Albuquerque, NM 87111
505-294-3054

New York
Book House of Stuyvesant Plaza Inc.
Stuyvesant Plaza
Albany, NY 12203
518-489-4761

Compete Traveler Bookstore
199 Madison Ave.
New York, NY 10016
212-679-4339

Hagstrom Map & Travel Center
57 W. 43rd St.
New York, NY 10036
212-398-1222

Jimapco Mapcenter
Route 9
Clifton Park, NY 12065
518-899-5091

Map Man
120 Bethpage Rd.
Hicksville, NY 11801
516-931-8404

Marshall Penn-York
538 Eric Blvd. W.
Syracuse, NY 13204
315-422-2162

New York Map & Travel Center
150 E. 52nd St.
New York, NY 10022
212-758-7488

New York Nautical
140 W. Broadway
New York, NY 10013
212-962-4522

Rand McNally Map Store
666 W. Third Ave.
New York, NY 10017
212-758-7488

Sanborn Map Co.
629 5th St.
Pelham, NY 10801
914-738-1649

Timesavers
One N. Transit, P.O. Box 229
Lockport, NY 14094
716-434-1234

North Carolina
Carolina Maps Inc.
210 W. Fourth St., P.O. Box 8026
Greenville, NC 27834
919-757-0279

Champion Map Corp.
4237 Raleigh St.
Charlotte, NC 28213
704-596-7165; 800-438-7406; 800-532-6675 (in North
    Carolina)

Geoscience Resources
2990 Anthony Rd., P.O. Box 2096
Burlington, NC 27216
919-227-8300

Treasure Hutch
5800 Yadkinville Hwy.
Pfafftown, NC 27040
919-945-3831

North Dakota
Book Fair
212 DeMers Ave.
Grand Forks, ND 58201
701-775-6491

Tel-E-Key
1014 18th St. NW
East Grand Fork, ND 56721
701-775-8266

Ohio
Duttenhofer's Map Store
210 W. McMillan St.
Cincinnati, OH 45219
513-381-0007

Leo's Book Shop
330 N. Superior
Toledo, OH 43604
419-255-5506

The Map Store
5821 Karric Square Dr.
Dublin, OH 43017
614-792-6277

Ohio Canoe Adventures Inc.
Backpackers Shop
5128 Colorado Ave.
Sheffield Lake, OH 44054
216-934-5345

Wilderness Trace
1299 Bethel Rd.
Columbus, OH 43212
614-457-8496

Oklahoma
Maps, Plaques & Laminating
3315 S. Harvard Ave.
Tulsa, OK 74135
918-744-4494

Mosher-Adams Maps
400 SW 25th St.
Oklahoma City, OK 73109
405-632-3321

Topographic Mapping Co.
6709 N. Classen
Oklahoma City, OK
405-843-4847

Traveler's Pack, Ltd.
9417 N. May
Oklahoma City, OK 73120
405-755-2924

Oregon
Captain's Nautical Supplies
138 NW 10th
Portland, OR 97209
503-227-1648

Libra Books, Inc.
856 Olive St.
Eugene, OR 97401
503-484-0512

Pittmon Map Co.
732 SE Hawthorne Blvd.
Portland, OR 97214
503-232-1161

Powell's Travel Store
Pioneer Courthouse Square
701 SW 6th Ave.
Portland, OR 97204
503-228-1108

Pennsylvania
Book Swap
316 Horsham Rd.
Horsham, PA 19044
215-674-3919

Franklin Maps
333 S. Henderson Rd.
King of Prussia, PA 19406
215-265-6277

Geostat Map & Travel Center
125 S. 18th St.
Corner of Sansom St.
Philadelphia, PA 19103
215-564-4700

Alfred B Patton Inc.
Swamp Rd. & Center St., P.O. Box 857
Doylestown, PA 18901
215-345-0700

Pilothouse
Pier 3 North
Philadelphia, PA 19106
215-351-4008

J. R. Wedlin Co.
415 Wood St.
Pittsburgh, PA 15222
412-281-0123

Rhode Island
Armchair Sailor Bookstore
Lee's Wharf
Newport, RI 02840
401-847-4252

The Map Center Inc.
204 Broad St.
Providence, RI 02903
401-421-2184

Outdoorsman
1000 Bald Hill Rd.
Warwick, RI 02893
401-823-3158

South Carolina
Capitol Map Supplies
619 12th St.
W. Columbia, SC 29169
803-796-3399

Luden Marine Supplies
Concord and Charlotte Sts.
Charleston, SC 29403
803-723-7829

The Map Shop
5-B E. Coffee St.
Greenville, SC 29602
803-271-6277

Texas
Allstate Map Co.
1201 Henderson St.
Ft. Worth, TX 76102
817-332-1111

Apache Trading Post
P.O. Drawer 929
Alpine, TX 79830
915-837-5149

Ferguson Map & Travel
8131 I-10 West, Ste. 219
San Antonio, TX 78230
512-341-6277

Key Maps Inc.
1411 W. Alabama St.
Houston, TX 77006
713-522-7949

Mapsco
5308 Maple Ave.
Dallas, TX 75235
214-521-2131

Mapsco
13536 Preston Rd.
Dallas, TX 75240
214-960-1414

One Map Place
11351 Harry Hines Blvd.
Dallas, TX 75229
214-241-2680

Southwest Map
2406 South Juniper Rd.
Garland, TX 75041
214-494-4443

Venture Map & Globe Co., Inc.
2130 Highland Mall
Austin, TX 78752
512-452-2326

Zdansky Map Store
5230 Kostoryz #16
Corpus Christi, TX 78415
512-855-9226

Utah
Map World
6526 S. State St.
Murray, UT 84107
801-262-1814

Vermont
Lost Mountain Bookshop
6 Main St.
Randolph, VT 05060
802-728-5655

Virginia
Globe and Map Technik, Inc.
11634 Busy St., Ste. A
Richmond, VA 23236
804-320-0719

Hudson Trail Outfitters
9683 Lee Hwy.
Fairfax, VA 22030
703-591-2950

Hudson Trail Outfitters
11750 Fair Oaks Mall
Fairfax, VA 22030
703-385-3907

Mapcom Systems Inc.
6947 Hull St.
Richmond, VA 23224
804-276-1502

Washington
Arnold Map Service
119 W. 24th St.
Vancouver, WA 98660
206-695-7897

Marysville Map & Flag Shoppe
4200 84th St. NE
Marysville, WA 98270
206-659-4827

Metsker Maps of Seattle
702 First Ave.
Seattle, WA 98104
206-623-8747

Metsker Maps of Tacoma
4020 S. Steele St., Ste. 107
Tacoma, WA 98409
206-474-6277

Northwest Map Service
W. 713 Spokane Falls Blvd.
Spokane, WA 99201
509-455-6981

Pacific Northwest National Parks Assn.
Mount Rainer Branch
Longmier, WA 98397
206-569-2211

Pioneer Maps
14125 NE 20th St.
Bellevue, WA 98007
206-746-3200

Wide World of Books
401 NE 45th St.
Seattle, WA 98105
206-279-2323

West Virginia
H. T. Hall Co.
3622 MacCorkle Ave. SE
Charleston, WV 25304
304-925-1117

Highway Maps Inc.
4838 MacCorkle Ave. SW
S. Charleston, WV 25303
304-768-0441

Wisconsin
A Global Affair
2768 Marshall Pkwy.
Madison, WI 53713

Milwaukee Map Service
4519 W. North Ave.
Milwaukee, WI 53208
414-445-7361

Northland Map Co.
711 S. Fisk
Green Bay, WI 54303
414-494-4904

Wyoming
Mountain Sports
543 S. Center St.
Casper, WY 82601
307-266-1136

Rio Colorado Trading Co.
P.O. Box 1121
Jackson Hole, WY 83001
307-733-2132

Teton Mountaineering
86 E. Broadway
Jackson, WY 83001
307-733-3595

CANADA
Alberta
Carter Mapping Ltd.
#430, 736-8 Ave. SW
Calgary, Alberta T2P 1H4
403-264-1230

Map World Services Inc.
#204-321 6th Ave. SW
Calgary, Alberta T2P 3H3
403-294-0393

Maps & Tourism International Inc.
12535 102 Ave.
Edmonton, Alberta T5N 0M4
403-453-6277

Map Town Ltd..
640 6th Ave. SW
Calgary, Alberta T2P 0S4
403-266-2241

Quillan Travel Store
#4B-112 11th Ave. SE
Calgary, Alberta T2G 0X5
403-246-2557

British Columbia
Travel Bug
1095 W. Broadway
Vancouver, BC V6K 2G2
604-737-1122

World Wide Books and Maps
736A Granville St.
Vancouver, BC V6Z 1G3
604-687-3320

Manitoba
Global Village Map & Travel Store
736 Osborne St.
Winnipeg, Manitoba R3L 2C2
204-453-7081

Ontario
A-1 Maps
1605 Bloor St. W.
Toronto, Ontario M6P 1A6
416-531-4108

Allmaps Canada Ltd.
390 Steelcase Rd. E.
Markham, Ontario L3R 1G2
416-477-8480

Canada Map Co.
211 Yonge St.
Toronto, Ontario M5B 1M4
416-362-9297

Gulliver's Travel Bookstore
609 Bloor St. W.
Toronto, Ontario M6G 1K5
416-537-7700

H M Dignam Corp. Ltd.
370 Dunlop St W., Unit 807
Barrie, Ontario L4N 5R7
705-721-1515

Open Air Books and Maps
25 Toronto St.
Toronto, Ontario M5C 2R1
416-363-0719

Oxford Books & Stationery
740 Richmond St.
London, Ontario N6A 1L6
519-438-8336

Perly's Maps
1050 Eglinton Ave. W
Toronto, Ontario M5V 1R5
416-593-6277

Place Bell Bookstore
175 Metcalfe St.
Ottawa, Ontario K2P 2E9
613-233-3821

Technicom Consultants
115 Randall Dr., Unit 5
Waterloo, Ontario N2V 1C5
519-747-1779

Worldwide Maps and Guides
202-316 rue Dalhousie
Ottawa, Ontario K1N F3F
613-230-4888

Quebec
Aux Quatre Points Cardinaux (AQPC)
551 Ontario E.
Montreal, Quebec H2L 1N8
514-843-8116

Conexfor Inc.
CP 697 Succ Desjardins
Montreal, Quebec H5B 1B8
514-849-5741

Dougherty Maps
762 Millington Ave.
Greenfield Park, Quebec J4V 1R7
514-672-5348

Distribution Ulysse
4176 rue St. Denis
Montreal, Quebec H2W 2M5
514-843-9447

ENGLAND
Bellows & Bown
7 Commercial Rd.
Gloucester GL1 1NW
011-44-452-21206

AUSTRALIA
Australian Mineral Foundation Inc.
63 Conyngham St.
Glenside 5065
South Australia
08-379-0444

Cartotech Services Pty Ltd.
51 John St., 1st Fl.
Salisbury, S. Australia 5108
618-221-2461

Hema Maps
8 Paxton St.
P.O. Box 724
Springwood, Brisbane 4127
07-29-00-322

Rex Map Centres
413 Pacific Hwy.
Artarmon, NSW 2064
61-2-428-3566

MEXICO
Guia Roji
Jose Moran #31
Miguel Hidalgo, Mexico DF 11850
905-515-0384; 905-515-7963

Sistemas de Informacion Geographica SA
San Francisco #1375
Mexico DF 03100
525-559-4644

JAPAN
Map House Co. Ltd.
5th Fl., Taiyodo Bldg.
1-10 Jimbocho, Kanda
Chiyoda-ku, Tokyo 101
03-295-1555

Teikoku-Shoin Co. Ltd.
29, Jimbocho 3-chome, Kanda
Chiyoda-ku, Tokyo 101
03-262-5039

Selected Map Libraries

GENERAL LIBRARIES

Alabama
Auburn University
Ralph B. Draughon Library
Special Collections Department
Auburn, AL 36841
205-844-1700

University of Alabama
Map Library, Dept. of Geography
Box 1982, University of Alabama
University, AL 35486
205-348-5095

Arizona
Arizona State University
Noble Science & Engineering Library
Map Collection
Tempe, AZ 85287
602-965-3582

University of Arizona
University Library, Map Collection
Tucson, AZ 85721
602-626-2596

Arkansas
University of Central Arkansas
Dept. of Geography, Map Library
Old Main
Conway, AR 72032
501-450-3164

University of Arkansas
University Libraries
Reference & Government Documents
Map Library
Fayetteville, AR 72701
501-575-4101

California
University of California, Berkeley
Earth Sciences Library
230 Earth Sciences Bldg.
Dept. of Geology & Geophysics
Berkeley, CA 94720
415-642-2997

University of California, Berkeley
General Library, Map Rm.
Berkeley, CA 94720
415-642-4940

California State University, Chico
Meriam Library—Maps
Chico, CA 95929
916-895-6803

University of California, Davis
Shields Library, Map Section
Davis, CA 95616
916-752-1624

California State University, Fresno
Henry Madden Library, Map Library
Fresno, CA 93740
209-294-2174

California State University, Los Angeles
The Geography and Map Library
Geography & Urban Studies Dept.
5151 State University Dr.
Los Angeles, CA 90032
213-343-2225

Los Angeles Public Library
Mary Helen Peterson Map Rm.
630 W. 5th St.
Los Angeles, CA 90071
213-626-7461

University of California, Los Angeles
Map Library
Los Angeles, CA 90024
213-825-3526

University of California, Los Angeles
Wm. C. Putnam Map Room
4697 Geology Bldg.
Los Angeles, CA 90024
213-825-1055

Oakland Public Library
125 14th St.
Oakland, CA 94612
415-273-3136

San Diego State University
University Library, Map Collection
San Diego, CA 92182
619-265-5832

University of California, San Diego
Map Section C-075P, University Library
San Diego, CA
619-452-3338

San Jose State University
Dept. of Geology, Map Room
San Jose, CA 95192
408-277-2387

University of California, Santa Cruz
Map Collection, University Library
Santa Cruz, CA 95064
408-429-2364

Colorado
University of Colorado
University Libraries, Map Library
Campus Box 184
Boulder, CO 80309
303-492-7578

Denver Public Library
Map Collection, Government Publ. Dept.
1357 Broadway
Denver, CO 80203
303-571-2000

Colorado School of Mines
Arthur Lakes Library, Map Rm.
Golden, CO 80402
303-273-3697

Connecticut
Wesleyan University
Science Library
Middletown, CT 06457
203-347-9411

Yale University Library
Map Collection
Box 1603A, Yale Station
New Haven, CT 06520
203-436-8638

Yale University
Geology Library
210 Whitney Ave., P.O. Box 6666
New Haven, CT 06511
203-436-2480

University of Connecticut
University Library
Map Library U-5M
19 Fairfield Rd.
Stoors, CT 06268
203-486-4589

District of Columbia
Defense Mapping Agency
Hydrographic/Topographic Center
Scientific Data Dept., Support Div.
6500 Brookes Lane
Washington, DC 20315
202-227-2109

Library of Congress
Geography & Map Div.
Washington, DC 20540
202-287-8530

National Geographic Society
Map Library
1146 16th St. NW
Washington, DC 20036
202-857-7000

Florida
Florida State University
R. M. Strozier Library
Maps Dept.
Documents-Maps-Micromaterials Dept.
Tallahassee, FL 32306
904-644-6061

Georgia
Georgia Institute of Technology
Price Gilbert Memorial Library
Dept. of Government Documents & Maps
225 North Ave.
Atlanta, GA 30332
404-894-4538

Hawaii
University of Hawaii, Manoa
Library, Map Collection
2550 The Mall
Honolulu, HI 96822
808-948-8539

Idaho
Boise State University, Library
Map Dept.
1910 University Dr.
Boise, ID 83725
208-385-3958

University of Idaho Library
Map Section
Moscow, ID 83843
208-885-6344

Illinois
Southern Illinois University
Map Library
Morris Library, Science Div.
Carbondale, IL 62901
618-453-2700

Rand McNally & Co.
Map Library
P.O. Box 7600
Chicago, IL 60680
312-673-9100

University of Chicago Library
Map Collection
1100 E. 57th St.
Chicago, IL 60637
312-962-8761

University of Illinois, Chicago
The University Library, Map Section
P.O. Box 8198, 801 S. Morgan St.
Chicago, IL 60680
312-996-5277

Northern Illinois University
Map Library
Davis Hall 222
DeKalb, IL 60115
815-753-1367

Southern Illinois University, Edwardsville
Lovejoy Library, Map Library
Box 63
Edwardsville, IL 62062
618-692-2422

Northwestern University
University Library, Map Collection
Evanston, IL 60201
312-492-7603

Western Illinois University
University Map Library
Geography Dept.
Macomb, IL 61455
309-298-1171

Illinois State University
Map Rm., Milner Library
Normal, IL 61761
309-438-3486

Illinois State Library
Centennial Bldg.
Springfield, IL 62756
217-782-5430

Indiana
Indiana University
Geography & Map Library
Kirkwood Hall, Rm. 301
Bloomington, IN 47401
812-335-1108

Indiana University
Geology Library
Geology Bldg., Rm. 601
Bloomington, IN 47405
812-335-7170

DePaul University
Roy O. West Library
Box 137
Greencastle, IN 46135
317-658-4514

Indiana State Library
Indiana Div. & Federal Documents Collection
140 N. Senate Ave.
Indianapolis, IN 46204
317-232-3686

Ball State University
Dept. of Library Science
Map Collection
Muncie, IN 47306
317-289-1241

University of Notre Dame
Memorial Library
Microtext Reading Room
Notre Dame, IN 46556
219-239-6450

Indiana State University
Dept. of Geography & Geology, Map Library
Terre Haute, IN 47809
812-232-6311

Valparaiso University
Moellering Memorial Library, Map Library
Valparaiso, IN 46383
219-464-5364

Purdue University Libraries
Map Collection
Stewart Center, Rm. 279
West Lafayette, IN 47907
317-494-2906

Iowa
Iowa State University
Library, Map Room
Ames, IA 50011
515-294-3956

Kansas
University of Kansas
Spencer Research Library
KU Map Library
Lawrence, KS 66045
913-864-4420

Kansas State University Library
Map & Atlas Unit
Manhattan, KS 66506
913-532-6515

Louisiana
Louisiana State University
School of Geoscience
Baton Rouge, LA 70803
504-388-6247

Maryland
Enoch Pratt Free Library
General Information Dept.
400 Cathedral St.
Baltimore, MD 21201
301-396-5472

Johns Hopkins University
Milton S. Eisenhower Library
Government Publications/Maps/Law Dept.
Baltimore, MD 21218
301-338-8360

University of Maryland
McKeldin Library
Documents/Maps Rm.
College Park, MD 20742
301-454-3034

Massachusetts
University of Massachusetts
University Library, Map Collection
Amherst, MA 01003
413-545-2397

Harvard College Library
Harvard Map Collection
Cambridge, MA 02138
617-495-2417

Massachusetts Institute of Technology
Stein Club Map Rm.
14S-200
Cambridge, MA 02139
617-253-5651

Tufts University
Wessell Library
Gov. Pub., Microforms & Maps Dept.
Medford, MA 02155
617-628-5000

Smith College Map Library
Dept. of Geology
Burton Hall
Northampton, MA 01063
413-584-2700

Clark University
Guy H. Burnham Map & Aerial Photograph Library
Worcester, MA 01610
617-793-7322

Michigan
University of Michigan
Hatcher Graduate Library, Map Rm.
Ann Arbor, MI 48109
313-764-0407

Detroit Public Library
History & Travel Dept., Map Rm.
5201 Woodward Ave.
Detroit, MI 48202
313-833-1445

Michigan State University
Map Library
East Lansing, MI 48824
517-353-4593

Michigan Technological University
Map Library
Government Documents Dept.
Houghton, MI 49931
906-487-2599

Western Michigan University
Map Library, Waldo Library
Kalamazoo, MI 49008
616-383-5952

Minnesota
Mankato State University
Memorial Library, Map Library
Mankato, MN 56001
507-389-6201

Carleton College
Geology Map Library
Northfield, MN 55057
507-663-4401

St. Cloud State University
Learning Resources
St. Cloud, MN 56301
612-255-2022

Missouri
University of Missouri
Geology Library
201 Geology Bldg.
Columbia, MO 65211
314-882-4860

Southwest Missouri State University
Duane G. Meyer Library, Map Collection
Box 175
Springfield, MO 65804
417-836-5104

Saint Louis University
Pius XII Memorial Library
3655 W. Pine Blvd.
St. Louis, MO 63108
314-658-3105

St. Louis Public Library
1301 Olive St.
St. Louis, MO 63103
314-241-2288

Washington University
Earth & Planetary Sciences Library
St. Louis, MO 63130
314-889-5406

Montana
Montana Tech Library
Documents Div.
Butte, MT 59701
406-496-4286

University of Montana
Map Collection, Documents Div.
Maureen & Mike Mansfield Library
Missoula, MT 59812
406-243-4564

Nebraska
University of Nebraska, Lincoln
Geology Library
303 Morrill Hall
Lincoln, NE 68588
402-472-3628

New Hampshire
Dartmouth College
Library Map Room, Baker Library
Hanover, NH 03755
603-646-2579

New Jersey
Hammond Inc., Editorial Dept. Library
515 Valley St.
Maplewood, NJ 07040
201-763-6000

Rutgers University
Library of Science & Medicine
Piscataway, NJ 08854
201-932-2895

Princeton University Library
The Richard Halliburton Map Collection
Princeton, NJ 08544
609-452-3214

Princeton University
Geology Library, Map Collection
Guyot Hall
Princeton, NJ 08544
609-351-2525

New Mexico
University of New Mexico
General Library Map Rm.
Albuquerque, NM 87131
505-277-7182

New York
New York State Library
Manuscripts & Special Collections
Cultural Education Center
Albany, NY 12230
518-474-4461

Brooklyn Public Library
History Div., Map Collection
Grand Army Plaza
Brooklyn, NY 11238
212-780-7794

Buffalo & Erie County Public Library
Lafayette Square
Buffalo, NY 14203
716-856-7525

State University of New York, Buffalo
University Libraries
Science & Engineering Library Map Collection
Buffalo, NY 14260
716-636-2946

Cornell University
John M. Olin Library
Dept. of Maps, Microtexts, Newspapers
Ithaca, NY 14853
607-256-5258

Columbia University Libraries
Lehman Library, Map Room
420 W. 118th St.
New York, NY 10027
212-280-5002

New York Public Library
The Research Libraries, Map Div.
Fifth Ave. & 42nd St.
New York, NY 10018
212-930-0587

United Nations Map Collection
Dag Hammarskjold Library
New York, NY 10017
212-754-7425

State University of New York, Stony Brook
Melville Library, Map Collection
Stony Brook, NY 11794
516-246-5975

Syracuse University Libraries
Map Collection
E. S. Bird Library
Syracuse, NY 13210
315-423-2575

North Carolina
Appalachian State University Map Library
Rankin Hall
Boone, NC 28608
704-262-3000

University of North Carolina, Chapel Hill
Watson Library, Maps Collection
Chapel Hill, NC 27514
919-962-3028

University of North Carolina, Chapel Hill
Geology Library
Mitchell Hall 029A
Chapel Hill, NC 27514
919-962-2386

Duke University
Perkins Library
Public Documents & Maps Dept.
Durham, NC 27706
919-684-2380

North Dakota
North Dakota State University Library
Fargo, ND 58105
701-237-8886

University of North Dakota
Geology Library
326 Leonard Hall
Grand Forks, ND 58202
701-777-3221

Ohio
Ohio University
Map Collection
Athens, OH 45701
614-594-5240

Public Library of Cincinnati & Hamilton County
Map Collection, History Dept.
800 Vine St.
Cincinnati, OH 45202
513-369-6909

University of Cincinnati Library
103 Old Tech ML 13
Cincinnati, OH 45221
513-475-4332

Cleveland Public Library Map Collection
325 Superior Ave.
Cleveland, OH 44114
216-623-2880

Ohio State University
Map Library
1858 Neil Ave. Mall
Columbus, OH 43210
614-422-2393

Kent State University
Map Library
406 McGilvrey Hall
Kent, OH 44242
216-672-2017

University of Toledo
William S. Carlson Library Map Collection
2801 W. Bancroft St.
Toledo, OH 43606
419-537-2865

Oklahoma
University of Oklahoma Geology Library
830 Van Vleet Oval, Rm. 103
Norman, OK 73019
405-325-6451

Oklahoma State University
Edmon Lowe Library, Map Rm.
Box 12927, Capitol Station
Stillwater, OK 74078
405-624-6311

Oregon
Oregon State University
William Jasper Kerr Library, Map Rm.
Corvallis, OR 97331
503-754-2971

Library Association of Portland
Literature & History Dept.
801 SW 10th
Portland, OR 97205
503-223-7201

Pennsylvania
Bryn Mawr College
Dept. of Geology
New Gulph Rd.
Bryn Mawr, PA 19010
215-645-5111

Free Library of Philadelphia
Map Collection
Logan Square
Philadelphia, PA 19103
215-686-5397

Temple University
Samuel Paley Library, Map Unit
Philadelphia, PA 19122
215-787-8213

University of Pennsylvania
Geology Map Library
Hayden Hall
Philadelphia, PA 19104
215-898-5724

Carnegie Library of Pittsburgh
Science & Technology Dept.
4400 Forbes Ave.
Pittsburgh, PA 15213
412-621-7300

University of Pittsburgh
G-8 Hillman Library, Map Collection
Pittsburgh, PA 15260
412-624-4449

Pennsylvania State University
Pattee Library, Maps Section
University Park, PA 16802
814-863-0094

Rhode Island
Brown University
Map Collection, Sciences Library, Box 1
Providence, RI 02912
401-863-3333

South Carolina
South Carolina Dept. of Archives & History
1430 Senate St., P.O. Box 11,669
Columbia, SC 29211
803-758-5816

University of South Carolina
Map Library
Columbia, SC 29208
803-777-2802

South Dakota
South Dakota State Univ
H. M. Briggs Lib, Documents Dept.
Brookings, SD 57007
605-688-5106

Tennessee
University of Tennessee
Dept. of Geography, Map Library
Knoxville, TN 37996
615-974-2418

Vanderbilt University Library
Science Library—Map Rm.
419 21st Ave. S.
Nashville, TN 37240
615-322-2775

Texas
University of Texas, Austin
Perry-Castaneda Library
Map Collection, PCL 1306
Austin, TX 78712
512-471-5944

Texas A & M University
Sterling C. Evans Library, Map Dept.
College Station, TX 77843
409-845-1024

Southern Methodist University
Edwin J. Foscue Library
Dallas, TX 75275
214-692-2285

University of Texas, El Paso
Library
El Paso, TX 79968
915-747-5685

Utah
Brigham Young University
Geography Dept., Map Collection
690 SWKT
Provo, UT 84602
801-378-3851

Brigham Young University
Harold B. Lee Library, Map Collection
1354 HBLL
Provo, UT 84602
801-378-4482

University of Utah
Science & Engineering Library
Map Collection
158 Marriott Library
Salt Lake City, UT 84112
801-581-7533

Vermont
University of Vermont
Bailey/Howe Library, Map Rm.
Burlington, VT 05405
802-656-2020

Middlebury College
Dept. of Geography
Science Center 402
Middlebury, VT 05753
802-388-3711

Virginia
Virginia Tech
Newman Library, Map Collection
Blacksburg, VA 24061
703-961-6101

George Mason University
Fenwick Library, Audiovisual Library
4400 University Dr.
Fairfax, VA 22030
703-323-2605

U.S. Army Corps of Engineers
Norfolk District
803 Front St.
Norfolk, VA 23510
804-441-3562

U.S. Geological Survey Library, Reston
950 National Center
Reston, VA 22092
703-860-6671

Virginia State Library
Archives Branch Map Collection
12th & Capitol Sts.
Richmond, VA 23219
804-786-2306

Washington
Western Washington University
Dept. of Geography & Regional Planning
Map Library
Bellingham, WA 98247
206-676-3272

Seattle Public Library
History Dept., Map Collection
1000 4th Ave.
Seattle, WA 98104
206-625-4894

University of Washington Libraries
Map Section FM-25
Seattle, WA 98195
206-543-9392

West Virginia
West Virginia University Library
Map Collection
P.O. Box 6069
Morgantown, WV 26506
304-293-3640

Wisconsin
University of Wisconsin, Eau Claire
Simpson Geographic Research Center
Dept. of Geography
Eau Claire, WI 54701
715-836-3244

Milwaukee Public Library
814 W. Wisconsin Ave.
Milwaukee, WI 53233
414-278-3000

University of Wisconsin, Milwaukee
Map Library
Rm. 385, Sabin Hall
Milwaukee, WI 53201
414-963-4871

University of Wisconsin, Stevens Point
Map Library
Geography-Geology Dept.
Stevens Point, WI 54481
715-346-2629

Wyoming
University of Wyoming
Map Collection, Coe Library
University Station, Box 3334
Laramie, WY 82071
307-766-2174

AERIAL PHOTOGRAPHY COLLECTIONS

California
University of California, Berkeley
Dept. of Geography Library
501 Earth Science Bldg.
Berkeley, CA 94720
415-642-3903

U.S. Geological Survey Library, Menlo Park
345 Middlefield Rd., MS 55
Menlo Park, CA 94025
415-323-8111

University of California, Santa Barbara
Map & Imagery Laboratory
Santa Barbara, CA 93106
805-961-2779

Whittier College
Dept. of Geological Sciences
Whittier, CA 90608
213-693-0771

Florida
Map Library
University of Florida
Gainesville, FL 32611
904-392-0803

Georgia
University of Georgia Libraries
Science Library, Map Collection
Athens, GA 30602
404-542-4535

Hawaii
Bernice P. Bishop Museum Library
Geography & Map Div.
P.O. Box 19000-A
Honolulu, HI 96819
808-847-3511

Iowa
University of Iowa Libraries
Special Collections Dept.
Map Collection
Iowa City, IA 52242
319-353-4467

Kansas
Kansas Dept. of Transportation
Bureau of Transportation Planning
State Office Bldg.
Topeka, KS 66612
913-296-3841

Minnesota
Bemidji State University
Geography Dept. Map Library
Roy P. Meyer Memorial Map Library
Bemidji, MN 56601
218-755-2000

University of Minnesota
Wilson Library, Map Library
309 19th Ave. S.
Minneapolis, MN 55455
612-373-2825

Nebraska
Kearney State College
Geography Map Library
Kearney, NE 68849
308-234-8356

Nevada
Nevada Bureau of Mines & Geology
Open Files Section
Room 311 SEM, University of Nevada
Reno, NV 89557
702-784-6691

New York
New York State Dept. of Transportation
Map Information Unit
State Campus Bldg., Rm. 105
Albany, NY 12232
518-457-3555

Oregon
University of Oregon
Map Library
165 Condon Hall
Eugene, OR 97403
503-686-3051

South Dakota
EROS Data Center
Data Management Section
Sioux Falls, SD 57198
605-594-6594

Texas
Lunar & Planetary Institute
Planetary Image Center
3303 NASA Rd. 1
Houston, TX 77058
713-486-2172

Utah
USDA—ASCS Aerial Photography Field Office
2222 W. 2300 South
Salt Lake City, UT 84119
801-524-5846

Washington
Washington Dept. of Natural Resources, Photos,
Maps & Reports
QW-21
Olympia, WA 98504
206-753-5338

Wisconsin
University of Wisconsin, Madison
Arthur M. Robinson Map Library
310 Science Hall
550 N. Park St.
Madison, WI 53706
608-262-1471

University of Wisconsin
Milwaukee Library
American Geographical Society Map Collection
P.O. Box 399
Milwaukee, WI 53211
414-963-7775

SATELLITE IMAGERY COLLECTIONS

Arizona
University of Arizona
Space Imagery Center
Lunar & Planetary Laboratory
Tucson, AZ 85721
602-621-4861

California
University of California, Berkeley
Dept. of Geography Library
501 Earth Science Bldg.
Berkeley, CA 94720
415-642-3903

U.S. Geological Survey Library, Menlo Park
345 Middlefield Rd., MS 55
Menlo Park, CA 94025
415-323-8111

Regional Planetary Image Facility
Jet Propulsion Laboratory
4800 Oak Grove Dr., Mail Stop 264-115
Pasadena, CA 91104
818-354-4321

University of California, Santa Barbara
Map & Imagery Laboratory
Santa Barbara, CA 93106
805-961-2779

Colorado
Colorado State University Libraries
Documents Div., Map Collection
Fort Collins, CO 80523
303-491-5911

Florida
Map Library
University of Florida
Gainesville, FL 32611
904-392-0803

Georgia
University of Georgia Libraries
Science Library, Map Collection
Athens, GA 30602
404-542-4535

Hawaii
Bernice P. Bishop Museum Library
Geography & Map Div.
P.O. Box 19000-A
Honolulu, HI 96819
808-847-3511

Iowa
University of Iowa Libraries
Map Collection
Iowa City, IA 52242
319-353-4467

Minnesota
University of Minnesota
Wilson Library, Map Library
309 19th Ave. S.
Minneapolis, MN 55455
612-373-2825

Missouri
Washington University
Regional Planetary Image Facility
Campus Box 1169
St. Louis, MO 63130
314-889-5679

New York
Spacecraft Planetary Imaging Facility
Cornell University
317 Space Sciences Bldg.
Ithaca, NY 14853
607-256-3833

Oregon
University of Oregon Map Library
165 Condon Hall
Eugene, OR 97403
503-686-3051

South Dakota
EROS Data Center
Data Management Section
Sioux Falls, SD 57198
605-594-6594

Wisconsin
University of Wisconsin, Madison
Arthur M. Robinson Map Library
310 Science Hall, 550 N. Park St.
Madison, WI 53706
608-262-1471

University of Wisconsin—Milwaukee Library
American Geographical Society Map Collection
P.O. Box 399
Milwaukee, WI 53211
414-963-7775

SPECIAL COLLECTIONS

Arizona
Grand Canyon Study Collection
Map Library
P.O. Box 129
Grand Canyon National Park, AZ 86023
602-638-7769

California
Hoover Institution on War, Revolution & Peace
Stanford University
Stanford, CA 94305
415-497-2058

District of Columbia
World Bank Cartography Library
1818 H St. NW
Washington, DC 20433
202-676-0229

Association of American Railroads
Economics & Finance Dept. Library
1920 L St. NW, Rm. 523
Washington, DC 20036
202-835-9387

Massachusetts
Woods Hole Oceanographic Institution
Data Library, McLean Laboratory
Quissett Campus
Woods Hole, MA 02543
617-548-1400

New Jersey
Motor Bus Society Inc., Library
P.O. Box 7058
West Trenton, NJ 08628

Texas
Amoco Production Co.
Library Information Center
P.O. Box 4381
Houston, TX 77210
713-931-2781

Utah
Latter Day Saints Genealogical Dept. Library
Map Collection
50 E. North Temple
Salt Lake City, UT 84150
801-531-3416

APPENDIX F
Selected Map Terms

aeronautical chart—a map showing recognizable features as seen from the air, required for air navigation or planning around airports.

atlas—a bound collection of maps.

base map—a map used in the construction of other maps. The term formerly referred to what are now called "outline maps." Base maps also are known as "mother maps."

bathymetric map—a map delineating the shape of the bottom of a body of water by the use of depth contours, called "isobaths."

bench mark—a relatively permanent object, natural or artificial, bearing a marked point whose exact elevation is known. It may appear as an official government seal embedded in rock or soil.

boundary monument—any object placed on or near a boundary line to preserve and identify its location.

cadastral map—a map showing the boundaries of subdivisions of land, often including the bearings and lengths of the boundaries and the areas of individual tracts, for purposes of describing and recording ownership. It also may show culture, drainage, and other features relating to land use and value.

cartography—the art or science of making maps and charts. The term may comprise all the steps needed to produce a map: planning, aerial photography, field surveys, photogrammetry, editing, color separation, and printing. Map-makers, however, tend to limit use of the term to map-finishing operations, in which the master manuscript is edited and color separations prepared for printing.

central meridian—the vertical meridian of a map projection around which the map is centered.

chart—any map used for nautical or aeronautical purposes, although the term is sometimes applied to describe other special-purpose maps.

contour—an imaginary line on the ground, all points of which are at the same elevation. The difference between two adjacent contour lines is known as the "contour interval."

coordinates—two-dimensional linear or angular quantities that designate the position a point occupies on a map.

cultural features—any man-made objects that are under, on, or above the ground, including roads, trails, buildings, canals, sewer systems, and boundary lines. In a broader sense, the term also applies to all names, identifications, and legends on a map.

diazo process—a rapid method of copying documents in which the image is developed by exposure to ammonia, used in some types of map reproduction.

elevation—the vertical distance of a point above or below a reference surface, usually mean sea level.

feature separation—the process of preparing a separate drawing, engraving, or negative for selected types of data in the preparation of a map or chart.

flood plain—the belt of low, flat ground bordering a stream that is flooded when runoff exceeds the capacity of the stream channel.

geodesy—the science concerned with the measurement and mathematical description of the size and shape of the Earth and its gravitational field. The term also refers to large-scale, extended surveys used to determine positions and elevations of points, in which the size and shape of the Earth must be taken into account.

globe—a spherical map of the Earth or heavens.

gore—a section of a globe printed on paper, intended to be cut out and pasted to the surface of a sphere; usually shaped like a football.

graticule—a network of lines on a map that represent the meridians of longitude and parallels of latitude.

grid—a network of uniformly spaced parallel lines intersecting at right angles. When superimposed on a map, it usually carries the name of the projection used for the map—the "universal transverse Mercator grid," for example. The numbers and letters used to describe specific points on the grid are known as "grid coordinates."

ground survey—a survey made at ground level, as distinguished from an aerial survey taken from above ground.

hachure—a series of lines used on a map to indicate the general direction and steepness of slopes. The lines are short, heavy, and close together for steep slopes; longer, lighter, and more widely spaced for gentle slopes.

hemisphere—any half of the Earth's surface.

hydrology—the scientific study of the Earth's waters, especially in relation to the effects of precipitation and evaporation on the character of ground water.

imagery—visible representation of objects as detected by cameras or other sensing devices. Recording may be on photographic film or on magnetic tape for subsequent conversion and display on a computer screen.

isobath—the contour lines designating the bottom of a body of water on a bathymetric map.

latitude—the angular distance north or south from the equator, measured in degrees, minutes, and second. Latitudes are sometimes called "parallels."

legend—an explanation of symbols and other information shown on a map, usually appearing as a list or table in the map's margin.

longitude—the angular distance east or west of the Greenwich meridian, measured in degrees, minutes, and seconds.

map—a graphic representation, on a plane, of certain selected features of a part or the whole of the surface of the Earth or any other entity.

map projection—an orderly, mathematical system of parallels and meridians used to prepare a map. Several different projections are used in cartography, with the Mercator and Robinson being the most common.

map scale—the relationship that exists between a distance on a map and the corresponding distance on the Earth. It may be expressed as an equivalence, one inch equals 16 miles; as a fraction or ratio, 1:1,000,000; or as a bar graph subdivided to show the distance that each of its parts represents on the Earth.

mean sea level—the arithmetic mean of hourly water levels observed over a specific 19-year cycle. Shorter series are specified in the name—"monthly mean sea level," for example

meridian—a great circle passing through the geographical poles and any given point on the Earth's surface. All points on a given meridian have the same longitude.

mosaic—an assembly of aerial photographs whose edges usually have been matched to the imagery on adjoining photographs to form a continuous representation of a portion of the Earth's surface.

mother map—another name for a "base map."

multispectral scanner (MSS)—a device for sensing radiant energy of the electromagnetic spectrum used in satellite imagery.

neatline—the line that bounds the body of a map, separating it from the margin.

oceanography—the science of the oceans, their forms, physical features, and phenomena.

offshore—a comparatively flat zone of variable width that extends from the outer margin of a shoreline to the edge of the continental shelf.

orthophotograph—a photograph having the

properties of an orthographic projection, derived from a conventional photo through a means that removes image displacements caused by camera tilt and terrain. An orthophotographic map is a map produced by assembling orthophotographs at a uniform scale.

orthophotomap—an orthophotograph containing contours and other cartographic information.

orthophotoquad—an orthophotomap in a standard quandrangle format with no contours and little or no cartographic treatment.

outline map—a simple map showing only the political divisions of a continent, state, or smaller region. Also refers to what is now called a "base map."

parallels—small circles on the Earth's surface, or lines on a map, perpendicular to the axis of the Earth, marking latitude north or south of the equator.

photogrammetry—the science or art of obtaining reliable measurements or information from photographs or other sensing systems.

photomap—an aerial photograph or assembly of photographs to which minimal descriptive data have been added.

planimetric map—a map that represents only the horizontal positions for features represented, distinguished from a topographic map by the omission of relief in measurable form.

plat—a diagram drawn to scale showing all essential data pertaining to the boundaries and subdivisions of a tract of land as determined by a survey or protraction.

prime meridian—the meridian of zero degrees, used as the origin for measurements of longitude. The meridian of Greenwich, England, is the internationally accepted prime meridian on most maps and charts, although other local or national prime meridians are sometimes used.

projection—see *map projection*.

quadrangle—a four-sided area, bounded by

parallels of latitude and meridians of longitude, used as an area unit in mapping. The dimensions of a quadrangle (or "quad") need not be the same in both directions.

reconnaisance map—a map not based on rigid trigonometric surveys, but possessing detailed data, generally made from rapid surveys.

relief map—any map that is, or appears to be, three-dimensional. "Relief" refers to elevations and depressions of the land or sea bottom.

rhumb line—a line that makes equal angles with all the meridians it crosses.

satellite imagery—see *imagery*.

scale—the relationship between a distance on a map, chart, or photograph and the corresponding distance on the Earth.

section—a unit of subdivision of a township; normally a quadrangle one mile square.

series—a group of maps produced simultaneously and designed in accordance with the same general specifications.

standard parallel—a parallel of latitude that is used as a control line in the computation of a map projection, and is, therefore, true to scale.

survey—an orderly process of determining data relating to any physical or chemical characteristics of the Earth.

thematic map—any map designed to provide information on a single topic, such as population, rainfall, or geology.

topography—the configuration of the land surface or sea bottom.

tropic—a line on a map or globe, usually broken or dotted, marking the limit reached by the overhead or vertical sun in its apparent annual migration. The northern line is called the "Tropic of Cancer," and the southern line the "Tropic of Capricorn."

zenith—the point in the celestial sphere directly over a given point on the Earth.

Index